NEUROMETHODS ☐ 7

Lipids and Related Compounds

NEUROMETHODS

Program Editors: Alan A. Boulton and Glen B. Baker

NEUROMETHODS

Program Editors: *Alan A. Boulton and Glen B. Baker*

NEUROMETHODS □ 7

Lipids and Related Compounds

Edited by

Alan A. Boulton

University of Saskatchewan, Saskatoon, Canada

Glen B. Baker

University of Alberta, Edmonton, Canada

and

Lloyd A. Horrocks

The Ohio State University, Columbus, Ohio

Humana Press • Clifton, New Jersey

Library of Congress Cataloging in Publication Data

Lipids and related compounds.

   (Neuromethods ; 7)
   Includes bibliographies and index.
   1. Lipids—Analysis.   2. Neurochemistry—Methodology.
I. Boulton, A. A. (Alan A.)   II. Baker, Glen B.,
Date.   III. Horrocks, Lloyd A.   IV. Series.
QP751.L55559   1988   612'.8042   88-8909
ISBN 0-89603-124-1

© 1988 The Humana Press Inc.
Crescent Manor
PO Box 2148
Clifton, NJ 07015

Printed in the United States of America

Preface to the Series

When the President of Humana Press first suggested that a series on methods in the neurosciences might be useful, one of us (AAB) was quite skeptical; only after discussions with GBB and some searching both of memory and library shelves did it seem that perhaps the publisher was right. Although some excellent methods books have recently appeared, notably in neuroanatomy, it is a fact that there is a dearth in this particular field, a fact attested to by the alacrity and enthusiasm with which most of the contributors to this series accepted our invitations and suggested additional topics and areas. After a somewhat hesitant start, essentially in the neurochemistry section, the series has grown and will encompass neurochemistry, neuropsychiatry, neurology, neuropathology, neurogenetics, neuroethology, molecular neurobiology, animal models of nervous disease, and no doubt many more "neuros." Although we have tried to include adequate methodological detail and in many cases detailed protocols, we have also tried to include wherever possible a short introductory review of the methods and/or related substances, comparisons with other methods, and the relationship of the substances being analyzed to neurological and psychiatric disorders. Recognizing our own limitations, we have invited guest editors to join with us on most volumes in order to ensure complete coverage of the field and to add their specialized knowledge and competencies. We anticipate that this series will fill a gap; we can only hope that it will be filled appropriately and with the right amount of expertise with respect to each method, substance or group of substances, and area treated.

Alan A. Boulton

Glen B. Baker

Preface

Lipids in the nervous system are major components of the membranes. The presence of glycolipids in high concentrations is unique for the nervous system. Recent discoveries of the functional and pathological importance of lipids such as diacylglycerols, polyphosphoinositides, prostaglandins, leukotrienes, docosahexaenoic acid, platelet activating factor, and gangliosides have markedly increased the number of publications on nervous system lipids. Many new methods have been developed. Thus, there is a need for this volume that is dedicated to that methodology. This volume places all of the methods for lipids into perspective with recommendations concerning the selection of a method for a specific purpose.

Lloyd A. Horrocks

Contents

Lipid Extraction
Norman S. Radin

Preparation and Analysis of Acyl and Alkenyl Groups of
Glycerophospholipids from Brain Subcellular Membranes
Grace Y. Sun

Quantitative Analysis of Acyl Group Composition
of Brain Phospholipids, Neutral Lipids,
and Free Fatty Acids
Victor L. Marcheselli, Burton L. Scott, T. Sanjeeva Reddy,
and Nicolas G. Bazan

Steroids and Related Isoprenoids
Thomas J. Langan, Robert S. Rust, and Joseph J. Volpe

Phospholipids
Yasuhito Nakagawa and Keizo Waku

Determination of Phospholipases, Lipases, and
Lysophospholipases
Akhlaq A. Farooqui and Lloyd A. Horrocks

Isolation, Separation, and Analysis of Phosphoinositides from
Biological Sources
Amiya K. Hajra, Stephen K. Fisher, and Bernard W. Agranoff

Analysis of Prostaglandins, Leukotrienes, and Related
Compounds in Retina and Brain
Dale L. Birkle, Haydee E. P. Bazan, and Nicolas G. Bazan

HPLC Analysis of Neutral Glycosphingolipids and Sulfatides
M. David Ullman and Robert H. McCluer

Methods to Study the Biochemistry of Gangliosides
Allan J. Yates

Contributors

BERNARD W. AGRANOFF • *University of Michigan, Neuroscience Laboratory, Ann Arbor, Michigan*

GLEN B. BAKER • *Neurochemical Research Unit, Department of Psychiatry, University of Alberta, Edmonton, Alberta, Canada*

HAYDEE E. P. BAZAN • *Louisiana State University Medical School, LSU Eye Center, New Orleans, Louisiana*

NICOLAS G. BAZAN • *Louisiana State University Medical School, LSU Eye Center, New Orleans, Louisiana*

DALE L. BIRKLE • *Louisiana State University Medical School, LSU Eye Center, New Orleans, Louisiana*

ALAN A. BOULTON • *Neuropsychiatric Research Unit, University of Saskatchewan, Saskatoon, Saskatchewan, Canada*

AKHLAQ A. FAROOQUI • *Department of Physiological Chemistry, The Ohio State University, Columbus, Ohio*

STEPHEN K. FISHER • *University of Michigan, Neuroscience Laboratory, Ann Arbor, Michigan*

AMIYA K. HAJRA • *University of Michigan, Neuroscience Laboratory, Ann Arbor, Michigan*

LLOYD A. HORROCKS • *Department of Physiological Chemistry, The Ohio State University, Columbus, Ohio*

THOMAS J. LANGAN • *Department of Pediatrics, Washington University School of Medicine, St. Louis, Missouri and Department of Neurology, School of Medicine, Health Science Center, State University of New York at Stonybrook, Stonybrook, New York*

VICTOR L. MARCHESELLI • *Louisiana State University School of Medicine, LSU Eye Center, New Orleans, Louisiana*

ROBERT H. McCLUER • *Department of Biochemistry, E. K. Shriver Center for Mental Retardation, Waltham, Massachusetts*

YASHUHITO NAKAGAWA • *Faculty of Pharmaceutical Sciences, Teikyo University, Sagamiko, Kanagawa, Japan*

NORMAN S. RADIN • *Mental Health Research Institute, The University of Michigan, Ann Arbor, Michigan*

T. SANJEEVA REDDY • *Louisiana State University School of Medicine, LSU Eye Center, New Orleans, Louisiana*

ROBERT S. RUST • *Department of Pediatrics, Washington University School of Medicine, St. Louis, Missouri*

BURTON L. SCOTT • *Louisiana State University School of Medicine, LSU Eye Center, New Orleans, Louisiana*

GRACE Y. SUN • *Sinclair Comparative Medicine Research Farm and Biochemistry Department, University of Missouri, Columbia, Missouri*

M. DAVID ULLMAN • *Research Service/GRECC, Veterans Administration Hospital, Bedford, Massachusetts*

JOSEPH J. VOLPE • *Departments of Pediatrics, Neurology, and Biological Chemistry, Washington University School of Medicine, St. Louis, Missouri*

KEIZO WAKU • *Faculty of Pharmaceutical Sciences, Teikyo University, Sagamiko, Kanagawa, Japan*

ALLAN J. YATES • *Department of Pathology, The Ohio State University, Columbus, Ohio*

Lipid Extraction

Norman S. Radin

1. Introduction

Given the nature of science, the organization of this chapter into sections has to be unsatisfying, so Table 1 has been prepared to assist the reader in finding specific kinds of information.

The extraction of lipids from tissues seems like a mundane operation, yet surprisingly many methods for doing this have been described. One reason for this diversity (besides the normal reluctance of chemists to follow a published procedure) is the analytical difficulty of deciding whether a given method has extracted all of the desired lipid. Simple weighing of the lipid extract, after solvent removal, does not give the yield of total lipid since the extract generally contains some nonlipids or substances not ordinarily considered to be lipids. Removing the nonlipids without losing polar lipids can be difficult. Identifying and measuring the amount of lipid that remains in the protein–nucleic acid–oligosaccharide residue is not simple.

The problem of removing nonlipids from an extract has become less urgent now that researchers are less interested in determining "total lipid." The main objection to the presence of nonlipids arises in contemporary research in which they produce column clogging or overloading or contamination of a particular polar lipid. A problem can also arise when nonlipid contaminants interfere with a colorimetric assay that is to be applied to the lipid extract.

Lipids occur in tissues to some extent as complexes with proteins and ionic substances. The anionic lipids are probably bound to metallic cations (Ca^{2+} and Mg^{2+}) or to the basic groups of proteins. Calcium ions can cross-link two different anionic groups, one from a protein and one from a lipid. During the extraction of an ionically bound lipid, an ion exchange reaction with available cations must occur. Under some conditions the protein is pulled into the organic solvent together with its acidic lipid. In samples that contain a relatively large proportion of ionic material (spinal fluid, buffered tissue homogenates), the cationic composition

1

Table 1
Comments About Specific Lipids and Procedures

Topic	Section number
Acetone use	7.2.5, 9.5, 11.1, 16.4
Acid, in extraction solvents	6.3, 7.2.3, 11.5
Acidic lipids (phosphates)	7.2.1
Acyl coenzyme A	11.11
Azeotropic distillation	15.1
Benzene use	2.4, 7.2.5, 16.1
Bile salt removal	9.4
Butanol	7.2.4, 9.1, 9.2
Centrifugation of extracts	5.2, 8.1
Ceramide oligohexoside	11.9
Cerebroside sulfate	11.6
Charcoal use	9.3, 12.3
Chloroform:methanol (Bligh/Dyer)	7.2
Chloroform:methanol (Folch)	7.1
Chloroform:methanol (polar mixtures)	7.3
Detergent removal	9.4
Dihydroxyacetone phospholipid	6.6, 7.2.3
Dolichol	11.8
Drying during extraction	10, 15
Ethanol use	9.5, 11.1
Ethyl acetate use	16.4, 16.5
Extraction reviews	1
Filtration of extracts	5.1
Free fatty acid	9.3, 11.5
Ganglioside	6.2, 6.5, 6.6, 7.1, 11.7, 11.9
Hexane/2-propanol	8
Homogenization	4
Hot extraction	1, 11.1
Ion exchange	1, 11.10
Isopropyl ether	9.1, 11.3, 11.5
Liquid/liquid columns	6.7
Liquid/liquid partitioning	7.1, 7.2.2, 11.5
Labile lipids	13
Lipidex	6.8
Methanol use	11.4, 16.1
Methanolysis	3, 15.2

might affect the ion exchange reaction or solubility in the organic solvent.

The importance of ion exchange during extraction was shown in a study by Spence (1969), who found that the extraction of ganglioside was impaired if soluble cations were removed from the tissue by dialysis or washing. This could be restored by adding back inorganic salts or organic amines.

The nature of the extracting solvent may control the *rates* of the required ion exchange reactions, affecting the amount of time required for complete extraction. A water-free extraction system (using lyophilized or acetone-extracted tissue) may not allow sufficient ion exchange to occur. It is a common practice to use an extraction solvent, such as chloroform: methanol (C:M) (2:1), at ratios of solvent to (wet) sample that are somewhat higher than originally proposed. The assumption is that the lower concentration of water in the final solution is unimportant. This may not be so. Note: Ratios of solvents to solvents are always presented in the usual way (v/v), whereas the ratios of solvents to tissues are presented as vol/g wet tissue (v/w).

Interesting attempts to extract ionic lipids at subzero temperatures together with their *normally bound* cations were reported by Katzman and Wilson (1961).

Of secondary interest is the behavior of the ionic lipids in subsequent separation procedures, since these may depend on the kinds of cations in the extract.

As a sample comes into contact with an extracting solvent, the proteins tend to assume new configurations and become denatured. Oxidative cross-linking of thiol groups may take place as the proteins lose their protected configurations. These processes can conceivably trap lipid molecules and render them difficult to extract. Erythrocytes tend to produce rubbery materials that trap lipids. Lipids containing high-energy linkages, such as acyl CoAs, may react covalently with exposed amino groups of proteins or lipids [phosphatidylethanolamine (PE), phosphatidylserine (PS)].

If the primary goal is simply to measure the total lipid content of a sample gravimetrically, it may be preferable to use an incomplete extraction technique. The small amount of nonlipid contaminants in the extract makes up for the loss of lipid.

If the sample is high in water content, as in an incubation mixture, a primary goal is to keep the volume of extracting solvent low. The high water content of the solvent mixture acts to increase the extract's content of nonlipid contaminants.

Although nonlipid residues are usually removed by filtration, small samples or a large number of samples may be handled better by centrifugation. In this case one ought to use a solvent of low density to produce a firm pellet.

In some cases one wants to process the lipid extract by a liquid/liquid partition technique to fractionate the lipids prior to measuring radioactivity or to simplify a chromatographic purification step. This goal can be decisive for choosing an extraction solvent.

If one wishes to store the extract and use portions of it for quantitative work, it is desirable to use an extracting solvent that is relatively nonvolatile and stable. In addition, the extraction and subsequent handling techniques should not result in excessive evaporation.

Although lipid chemists used hot solvents in the early days, sensitivity to the danger of oxidation has led most chemists to work at room temperature, or even at 0°C. The use of a low temperature may result in incomplete extraction. For many purposes there is no harm in using hot solvents, which allow one to handle large samples faster and with less solvent. One advantage of a hot solvent is that it denatures the lipophilic proteins, which otherwise can enter the organic solvent together with the lipids. The denatured proteins thus remain with the other proteins. The solubility of lipids rises rapidly with temperature, so much less solvent is needed. There is a danger of trapping lipids in the aggregates of denatured protein, however, and some analysts use Soxhlet or other continuous extractors for a prolonged period of time.

Warning: Do not heat organic solvents on the open bench—use a fume hood! Leaking flammable vapors are carried away from the hot plate by the draft—if the hood is a good one.

A practical question: does your experiment require *complete* extraction of all the lipids? For many purposes a simple extraction with hot isopropyl alcohol (2-PrOH; IP) gives adequate extraction. For preparative isolations, completeness of extraction is not as important as economy of solvent, in terms of volume and price.

A reminder for those planning to compare alternative extraction schemes: do not start with two different brains and extract each one separately. They might be distinctly different in composition. Pool several brains by rapid mincing or homogenizing, and weigh out representative portions for extraction.

Reviews on the extraction of lipids have recently been written by Nelson (1975) and Stein and Smith (1982). A paper by Hanson and Lester (1980) compares a variety of extraction methods.

2. Properties of Solvents

Although researchers used to redistill or otherwise purify solvents, they now typically use "pesticide grade," "HPLC grade," "glass-distilled," or "UV-grade" solvents. Some of these solvents contain stabilizers, which may need removal. Some are purified primarily to remove UV-absorbing materials, not reactive or high-boiling components. All solvents, however much purified, contain distinct amounts of nonvolatile impurities, as shown on the bottle label. These impurities may or may not be removed by chromatography or other purification step, and it may pay to "shop around" by comparing the analytical values from different suppliers.

Table 2 lists solvents that have reasonably low boiling points and prices. The densities are listed because they are relevant to plans involving centrifugal removal of proteins. They are also relevant to price considerations, since some solvents are sold by weight rather than volume. Extraction techniques normally are based on solvent volumes (in relation to sample weight or volume). When one buys purified solvents, the price differences become less noticeable because the extra purification costs have become a major component of the total cost.

Note the warnings in Table 2 about solvents that appear to be particularly dangerous to the user. Although some of the testing techniques used to establish the dangers may seem to have been excessively strenuous (in terms of dosages), it is nevertheless clear that lipid operations should be run in a hood as much as possible. An informative book on solvents is available (Burdick & Jackson, 1984).

Note a problem of finding solvents in chemical catalogs: nomenclature is still inconsistent, especially with symmetrical compounds. One may have to look under *di*ethyl ether instead of ethyl ether.

2.1. Ethers

Ethyl ether, dioxan, isopropyl ether, and tetrahydrofuran form peroxides on storage in air, especially in the presence of sunlight. They should always be kept in dark bottles. Dioxan can be stored as a frozen solid in a cold room, and thus be somewhat protected against peroxidation.

The easiest way to remove peroxides is to distill a small batch at a time from KOH pellets. But DO NOT DISTILL TO DRYNESS! Washing with aqueous ferrous ammonium sulfate or chromatography through a column of alkaline alumina has been recommended,

Table 2
Solvents for Extraction of Lipids[a]

	bp, °C	Density
Acetone	56	0.791
Acetonitrile[b]	82	0.786
Acetic acid	118	1.049
1-Butanol	118	0.810
2-Butanol	98	0.808
t-Butyl methyl ether	56	0.758
Benzene[c]	80	0.874
1-Chlorobutane	78	0.886
Chloroform[c]	62	1.492
Cyclohexane	81	0.779
1,2-Dichloroethane[c]	83	1.256
Dichloromethane	40	1.325
Dioxan[c]	102	1.034
Ethyl acetate	78	0.902
Ethyl alcohol 100%	79	0.789
Ethyl ether	35	0.706
n-Heptane	98	0.684
Hexanes (isomers)	69	0.670
IsoBuOH (2-methyl-1-PrOH)	108	0.803
Isooctane	99	0.692
Isopropyl ether	69	0.725
Methanol	65	0.791
1-Propanol	97	0.804
2-Propanol	82	0.785
Pyridine	115	0.978
Tetrachloroethylene[c]	121	1.623
Tetrahydrofuran	67	0.886
Toluene	111	0.867
2,2,4-Trimethylpentane (see Isooctane)		

[a]Boiling points can be affected by the formation of azeotropes with other solvents.
[b]Neurotoxin.
[c]Cancer-causing suspect.

but appreciable losses occur. Note that ethers can dissolve a good deal of water and HCl (and vice versa). Most of the water can be removed by washing with saturated aqueous NaCl or by diluting the ether with a solvent of low polarity (hexane, chloroform). A solid drying agent ($MgSO_4$) can then be used for further drying.

Sodium borohydride can be used as a preservative and reducing agent for peroxides, but it must be added carefully if the solvent is old. The container must be very slightly open to vent hydrogen. It seems likely that KOH is the best protective agent. Ethers that have been freed of peroxides can be stabilized to some extent by adding 2% ethanol (EtOH), which should not interfere with extraction or partitioning procedures.

t-Butyl methyl ether has been recommended because of its resistance to peroxidation, but it is expensive. It has the special merit of permitting chromatography of the extract while monitoring in the low UV wavelengths. It is better than ethyl ether with respect to explosion risk. Diisopropyl ether (DIPE) has found some use and is probably moderately resistant to peroxide formation.

2.2. Alcohols

Alcohols are strong solvents for most lipids. Methanol (MeOH) and 2-PrOH are the most popular. Ethanol has often been used, but, because of oxidation by air, it is contaminated with acetaldehyde (which forms self-condensation products). Thus the lipids containing amine groups—PE, PS, psychosine, sphingosine—gradually form adducts with the aldehydes. The higher primary alcohols, such as 1-propanol and 1-butanol, probably suffer from the same problem, but to a lesser extent. One would expect MeOH to contain formaldehyde for the same reason, but this probably condenses with the remaining MeOH to form the acetal (dimethoxymethane, a good solvent), which is stable except in acid. Secondary and tertiary alcohols do not seem to oxidize in air. This makes 2-PrOH a particularly useful extractant.

A special grade of denatured EtOH that is offered by Fisher Scientific and Aldrich Chemical probably deserves investigation as an extractant and HPLC solvent. It is called "alcohol, reagent" and it contains 5% MeOH, 5% 2-PrOH, and 90% EtOH. Perhaps the MeOH blocks formation of acetaldehyde or converts it to the acetal as fast as it forms.

Methanol is immiscible with hexane and cyclohexane.

2.3. Halocarbons

Chloroform is the most popular in this class, particularly for nonpolar lipids and lipids of intermediate polarity. It is probably the equivalent of an ether in terms of solubilizing power. It is not completely stable, forming phosgene and HCl in air, and should be stored away from light. Phosgene probably reacts with the amine lipids, and the HCl can cleave the plasmalogens. Low levels of strong acid in an anhydrous extract have the unexpected ability to catalyze solvolysis of certain lipids, such as the sulfate esters. Chloroform can be purchased with or without 0.5–1% EtOH, which is added as a stabilizer. One hopes that very little of the EtOH gets oxidized to acetaldehyde.

Methylene chloride (dichloromethane), butyl chloride (1-chlorobutane), 1,2-dichloroethane (ethylene dichloride), and tetrachloroethylene have some of chloroform's solvent power, but have not been used as extractants in the lipid laboratory. 1,2-Dichloroethane has been used for lipid extraction on an industrial scale (*see* section 15.1). Ethylene dichloride should not be confused with 1,2-dichlorethylene. Fluorochlorocarbons have some merit, but should not be used because of their destructive effect on the earth's ozone layer.

2.4. Hydrocarbons

Hexane is a good solvent for lipids of low polarity, but poor for compounds containing hydroxyl groups. It is only slightly miscible with MeOH and acetonitrile, but entirely miscible with almost every other solvent. In former days hexane was contaminated with unsaturated and thio analogs that oxidized and polymerized, but contemporary hexane seems very good. The hexane normally used for lipid work is a mixture of isomers and is accordingly more accurately called "hexanes." Petroleum ether, ligroin, benzine, skellysolve, and petroleum naphtha are the names assigned to mixtures of various hydrocarbons in the C_5 to C_8 range.

Hexane is better than chloroform in that it is less volatile and therefore better for storage of lipids, but it represents an explosion hazard in a cold box. It is also cheaper and less toxic than chloroform. (Hexane is metabolized somewhat inefficiently to 2,5-hexanedione, a neurotoxin.) "Isooctane" (2,2,4-trimethylpentane) is not much more expensive than hexane, but is less volatile and may deserve trials as an extractant.

Benzene is a useful solvent, having higher polarity and lower volatility than hexane, but it is now considered so toxic that it should not be used. Toluene has much the same solvent power, yet is less toxic and less volatile. The rotoevaporation of extracts containing toluene can be time-consuming if large volumes are involved, however. Toluene should certainly be tested in solvent schemes when benzene was prescribed.

A solvent deserving consideration is cyclohexane, which has the same boiling point as benzene. Both of these solvents can be removed by lyophilization, since they freeze at about 6°C and readily remain frozen. Extracts of lipids that contain cyclohexane (initially or added later) can be evaporated to a small volume to remove nonfreezing solvents, then frozen and lyophilized to yield a solvent-free, fluffy product. This is of no value, however, if the lipid mixture is an oil, since the mixture melts during evaporation of the cyclohexane.

Another solvent of interest is heptane, which is less volatile than benzene, but readily removed from an extract by rotoevaporation. Like hexane, it is immiscible with MeOH.

The aliphatic hydrocarbons are immiscible with acetonitrile.

2.5. Miscellaneous Solvents

Acetone, because of its high volatility and miscibility with water, has often been used to convert a wet tissue to a dry powder that can then be extracted with a relatively small amount of a more polar solvent. Acetone extracts the least polar lipids, such as triglycerides and cholesterol. Significant amounts of other lipids are also removed, however, so its primary use is in preparative work, where a 100% yield is unnecessary.

Acetonitrile is a good solvent, but rather expensive. A particular merit is its transparency in the UV region, an advantage for lipid HPLC. It is a distinct neurotoxin, but its popularity in HPLC seems to lead to careless handling.

3. Storage of Lipid Extracts

It is customary for lipid biochemists to avoid the storage of solvent-free extracts. The concentration of oxygen in a vessel containing a volatile solvent is lower than in a solvent-free vessel (because of the partial pressure of the solvent), so oxidative degradation should be slower in solutions. Moreover one would expect that diluting a polyunsaturated lipid with a solvent would reduce the rate of destruction by oxygen-produced free radicals. This kind of protection must depend on the chemical nature of the solvent, however. Probably it is safe to store solvent-free lipids in an oxygen-free vessel, especially in the cold, provided the pH of the sample is close to 7.

For solutions, air and light, especially direct sunlight, should be avoided. A freezer, particularly a "super-freezer" ($-70°C$), is the best storage site. Because of the water condensation that forms when refrigerated containers are warmed up in the laboratory, labels should be protected by an outer layer of the very clear, wide polyester tape that recently became available (Bel-Art Products, Pequannock, NJ).

Plastic vials or bottles, if they are not made from Teflon, are undesirable because they typically leach contaminants into the solution even though the plastic itself may appear to be unharmed. Most plastics contain their polymerization catalyst, an antioxidant, a UV-absorbing agent to prevent cracking, and incompletely polymerized (short-chain) material. They also allow solvent to diffuse out of the container, and it seems likely that some of the lipid in the extract also enters the plastic. Most plastic containers are fabricated by a molding process and may be contaminated on the surface by a "mold release" substance, a coating applied to keep the plastic from sticking to the mold. Presumably this can be washed off since it is on the surface.

Users of polycarbonate containers and polystyrene screw caps (from acid reagent bottles) quickly learn to avoid organic solvents.

Teflon bottles and vials are available, albeit quite expensive, and are probably satisfactory for storage, although I have found some of them to leak vapors through the lid threads. Tightness is easy to test by weighing a filled container after some period of time. Nylon is now being recommended by manufacturers for use with organic solvents, but the reader should check his solvent by TLC (after concentrating the solution).

Stored polyunsaturated lipids may oxidize, although tissue extracts normally contain antioxidants (e.g., vitamin E). Pro-oxidants (iron and ascorbic acid) are probably present at relatively low levels in lipid extracts from neural tissue. In my experience, there is no value in adding an antioxidant if the sample is kept very cold and worked up in a few days. Precise data on the stability of polyunsaturated fatty acid lipids seem to be scarce, and it may be noted that heat and pure oxygen are used when one wishes to form oxidation products from pure polyunsaturated fatty acids. Some researchers add 50 µg/mL of an antioxidant, such as 2,6-di-*t*-butyl-*p*-cresol (BHT). This may be particularly valuable for the extraction of enzyme incubation mixtures containing polyunsaturated compounds as metabolites.

The storage in MeOH-containing solvents of ester-type lipids, which constitute the major portion of a typical extract, might be thought to result in methyl ester formation by methanolysis. This reaction certainly goes well in dry C:M containing a trace of NaOH: only a few seconds are needed to cleave triglycerides with 70 mM NaOH. The stability of ester lipids as a function of pH (above 7) does not seem to have been studied, and tissue samples containing some alkali probably should be neutralized before extraction. Flint (soft) glass is often used to make screw-cap bottles and disposable Pasteur pipets. It readily raises the pH of unbuffered samples and should be avoided. Presumably transesterification of this sort would be avoided by using 2-PrOH instead of MeOH. As pointed out in section 2.3, extracts that contain HCl (but no water) could cleave sulfate esters.

Lysolecithin is said to be unstable when stored in C:M:W (W = water), but not in C:M (Mock et al., 1984). The 2-acyl group apparently hydrolyzed off.

Acidic extracts containing plasmalogen should not be stored because of the instability of the alkenyl ether bond. There is also some danger of solvolytic cleavage of sulfatide and cyclization or cross-linking of certain hydroxylic lipids.

Free fatty acids in extracts that contain methanol may gradually be converted to methyl esters. If this is a problem, it could be minimized by using 2-PrOH instead of MeOH.

A useful device for preventing oxidation during storage is the Firestone valve (from Sigma or Aldrich Chemical). It is suitable for samples that are in a vessel whose neck can be closed without exposing the contents to air (i.e., a vessel with a ground joint neck and a stopcock sealed to a corresponding ground joint that is stored with the container). The vessel must not be too full and must be vacuum-resistant (round-bottom flask, flat-bottom boiling flask <500 mL, Erlenmeyer <250 mL, or heavy bottle). The Firestone valve is connected via its three outlets to (1) a vacuum source, (2) a nitrogen source, and (3) the stopcock of the sample container. The nitrogen enters the valve at low pressure and flow rate, controlled by a screw clamp. After the parts are connected, the air in the container and solvent are exhausted *partially* (to prevent evaporation and bumping), nitrogen is allowed to enter, and the cycle is repeated a few more times to displace all the oxygen in the container. It takes a little while to pull air out of solution. A bit of stopcock grease must be used to seal the joint between the stopcock and vessel, but solvent must not contact it. Some stopcock greases, such as Apiezon AP 100, are claimed to resist organic solvents (silicone grease does not).

Some care is required when stoppering containers that hold lipid extracts. Rubber stoppers cannot be used, since they absorb solvent vapors and swell. Sometimes they absorb enough vapor to exude a lipid extract of the stopper itself into the sample. They are effective, however, if carefully covered with aluminum foil. Neoprene rubber stoppers are much less prone to swelling. Aluminum foil, or several layers of Saran Wrap, held with rubber bands, are temporarily satisfactory. Parafilm, which works so well in covering aqueous and alcoholic solutions, is readily attacked by organic solvents. Test tubes and bottles covered with screw caps are partially effective, but one must check to make sure the cap liner is Teflon, not rubber or paper. Paper cap liners are coated with a plastic that is vulnerable, like rubber. Unfortunately, Teflon liners are inelastic, making the seal unreliable. The container threads, before the cap is attached, can be covered with a layer of Teflon pipe-sealing tape, which is soft enough to form a seal. In my experience, these also leak vapors. Another approach is to cover the entire cap with a wide length of polyethylene self-sticking tape.

Containers having ground glass joints leak solvent vapors and let in air unless they are greased lightly. Of course they must not be tipped over or grease may get into the sample, and some of the sample will be lost in the joint. Some workers regrind their glass joints with very fine carborundum or alumina suspended in water; this should be done very gently to avoid spoiling the fit.

The lipid novice should note the danger of using plastic tubing in his apparatus. The dangers of plastics are noted above. The familiar Tygon contains a great amount of plasticizers, and several reports in the literature of their "natural occurrence" in tissues may arise from this source. Well-washed Neoprene, polyethylene, and polypropylene tubing probably will not contaminate samples, but they may absorb some. Neoprene and Viton tubing swell badly in many solvents. Teflon tubing, with compression-type fittings, is the safest.

4. Homogenizers

Teflon/glass devices tend to jam when tough tissues are present (cartilage or dura mater). If the tissue is cut up with scissors or a knife first, the tough tissues tend to mat together, but are probably completely extracted, and one should resist the temptation to prolong the extraction step in an effort to break up the mat.

Waring blenders should be avoided because they contain a rubber gasket in the rotary seal, which is attacked by the solvent. The larger blenders also contain a rubbery plastic in the cover as part of the upper seal.

Blenders that utilize a motor at the top and a cutting device at the bottom end of the motor shaft (Brinkmann Polytron, VirTis Ultra Shear) are very useful as homogenizers. They can be used in many kinds of containers, such as centrifuge tubes or bottles. It is important, however, to make sure that the lower rotary bearing, which is immersed in the solvent, does not contain any kind of rubber (such as Viton). Although this elastomer is stable against alcohols, halogenated and other solvents produce swelling and mechanical disintegration. Not every unit that is claimed to be "solvent-compatible" is indeed free of Viton rings. Teflon should be the only plastic in the bearing. Although it too gradually degrades, it last much longer, and the particles formed are insoluble and presumably free of extractables.

The design of these blenders is such that a small amount of lipid extract and nonlipid residue remains trapped inside the lower bearing. Probably this is unimportant for serial processing of samples, but they must eventually be given a thorough rinse.

The VirTis Co. (Gardiner, NY) also makes a blender that uses an overhead motor and sharp knives at the bottom end of the motor shaft. This shaft does not have a supporting lower bearing, so trapping of extract and tissue is less of a problem. Special "fluted" containers have to be used, together with a Teflon splash trap at the top of the container.

Tough tissues can be homogenized in solvent with a mortar and pestle with the aid of sand. Another approach is lyophilization of the intact tissue, grinding of the spongy material with a mortar and pestle, then homogenization with organic solvent in the ordinary way. Probably some water should be added with the solvent to improve the release of ionic lipids. Folch et al. (1957) suggested freezing peripheral nerve (with its attached cartilage) and pulverizing it at dry ice temperatures with a mortar and pestle.

Kitchen "food processors," such as the Cuisinart, or screw-feed meat grinders are fine for mincing large tissue samples prior to homogenization with organic solvent in a solvent-resistant apparatus. Probably they should be used only with frozen tissues to avoid the activation of hydrolases. A Gifford Wood colloid mill has been used for cutting up very large amounts of tissue (Furbish et al., 1977).

Homogenization with solvents should be performed in a hood to protect the operator from solvent vapor. When a flammable extraction solvent is used, the hood prevents fires or explosions arising from motor sparks. A flexible motor shaft extension would move the motor sparks further away, and might make a safe substitute for a hood.

Solvents that are highly insoluble in water (hexane, chloroform) do not readily penetrate intact tissues or membranes and tend to form gummy mixtures. Thus, when used without the addition of a water-miscible solvent, they require a high input of mechanical energy during the homogenization process. The same requirement exists for extraction solvents, such as MeOH, that are too polar to dissolve the nonpolar lipids. Such extraction methods require multiple extractions to obtain a good yield.

Ultrasonication, especially with an immersed dipping probe, ought to be useful for small samples since the probe is easily rinsed off. This approach has not found appreciable use to date.

5. Separating the Extract from the Tissue Residue

5.1. Use of Filtration

Although most workers use filtration with a suction flask, this produces evaporation of the solvent during filtration, so one cannot take out aliquots containing a known fraction of the total. Moreover, the vacuum tends to precipitate lipids and nonlipid contaminants in the funnel, making rinsing for complete transfer more difficult. A pressure funnel is much better: This is a funnel fitted with a stopper of some sort so that air (or nitrogen) pressure can be applied to the surface of the homogenate. With such a device one can filter into any vessel, without loss caused by transfer from a special suction flask. Solvent evaporation is relatively small, and nothing precipitates in the funnel. On the other hand, if the next step is to concentrate the filtrate, vacuum filtration will save some time.

One can buy such funnels from Aldrich Chemical and Fluka Chemical. Mine are made by a glassblower from standard sintered glass funnels (coarse porosity) by sealing a 28/35 ball joint to the top. Pressure is then applied from a 28/35 socket, which is attached with a ball joint clamp. If high air pressure (10 psi) is used, ground joints leak noticeably and cause evaporation of the extract, but ball joint clamps can be tightened with a screw. One type of commercially available pressure funnel is made with an O-ring ball joint, which is quite air-tight. Pressure funnels made with standard taper joints tend to leak more and pop apart. The disadvantages of pressure funnels are that large sizes are expensive and the narrow opening makes it necessary to use a conical funnel when adding suspensions.

If one wishes to use ordinary sintered funnels with a vacuum, yet catch the filtrate in any evacuable container, a vacuum-filtration adapter can be used (Aldrich Chemical).

Some homogenates are difficult to filter because the insoluble residue is soft and clogs the filter cake or sintered disc. Celite Analytical Filter Aid should be added to the suspension in generous amounts. The filter disc should first be coated with a thin layer of Celite. Strangely enough, the Celite (heavily washed inorganic skeletons of diatoms) contains some material that dissolves in C:M, so it should be washed with solvent before use.

5.2. Centrifugation

Centrifugation is preferred for small samples, but the extraction solvent must not be too dense. Chloroform:methanol (2:1) is too dense to sediment proteins firmly, but the addition of 0.2 vol of MeOH solves the problem.

The main objection to centrifugation, for quantitative isolations, is that one must resuspend the pellet with fresh solvent and sediment it again. This increases the total volume of extract.

Centrifuging a flammable solvent is hazardous, particularly with a motor that uses brushes. The centrifuge tube must be capped and the tube must not break. To reduce the danger of cracking test tubes one is well advised to fill the metal support cup with as much water as possible. This lowers the pressure on the bottom of the tube.

6. Removing Nonlipids from Lipid Extracts

Small molecules that occur at low concentrations dissolve to an unexpected degree in organic solvents, especially if water is present from the extracted tissue. I once isolated a crystalline "lipid" from erythrocytes that turned out to be urea. Another type of contaminant is the lipophilic protein or polypeptide, which is carried into the organic phase by the weakly attached lipids or through its outward-facing lipophilic amino acid side chains. Inorganic and organic cations are carried into the lipid extract as salts formed with anionic lipids. The organic cations may not be desirable for the next step. At low pH, many proteins and peptides become soluble in organic solvents.

6.1. Removal of Nonlipids by Evaporation and Reextraction

These contaminants can be removed somewhat by evaporating off the tissue water and solvent and re-extracting the lipid residue with dry solvent. This is the easiest way to remove much nonlipid, although one might worry about the possibility of physically trapping lipids in the dry particles. Perhaps suspension in an ultrasonic bath for a few minutes would improve recoveries. Lipophilic proteins tend to denature during the evaporation step and thus may not dissolve completely in the second extraction.

Many authors have described this process by saying that they "dissolved the residue in the solvent," but the dissolving step was surely incomplete, and it would be more correct to state that the residue was *suspended* in the solvent. Unfortunately, chemists seem to have no single word that means "part of the sample was dissolved and part was suspended." Such "solutions" can readily clog a column or let fine particles penetrate into the effluent.

6.2. Prewashing the Tissue Sample

Some chemists have used a simple wash of the tissue *before* extraction, by homogenizing it in water and centrifuging the membranes. This technique requires a high-speed centrifuge and causes a small loss of lipid in the soluble lipoproteins. A lower centrifugal speed (16,000g) can be used by the addition of a small amount of $CaCl_2$, which causes aggregation of microsomal membranes (Schenkman and Cinti, 1978). Washing a tissue is particularly useful if it has been preserved with formalin.

Phillips and Privett (1979) utilized a low pH to aggregate the homogenate and simplify the washing operation. In their procedure, the tissue was extracted twice with 0.25% HOAc under nitrogen. The pelleted residue, which had a pH of 4.0–4.4, was extracted twice with C:M (1:1) by stirring with a small metal spatula. The authors stated that 97% of the brain lipid is in this extract, and that a third extraction will gain 0.5–1% more.

The pellet, which retained some gangliosides, was washed with 1N HCl, water, and 50 mM ammonium hydroxide. Finally, the acidic lipids were extracted by high-speed homogenization with C:M (1:2) containing a little HOAc and then with MeOH. The four extracts were pooled and rotoevaporated almost to dryness, then taken up in 5 mL of C:M (2:1) and filtered through a glass fiber filter disc.

This lengthy procedure, which involves much handling and solvent, was described as being simple and faster than conventional methods in which the lipid *extract* is processed to remove nonlipids. Probably its main merit is for studying the polar gangliosides or tissues that contain radioactive precursors that are difficult to remove from the lipid being investigated. The initial wash at room temperature at pH 4.0–4.4 could possibly allow some enzymatic degradations to occur.

6.3. Removal of Nonlipids by Precipitating the Lipids with the Proteins

Instead of using aqueous HOAc as a tissue precipitant, Van Slyke and Plazin (1965) coprecipitated the sample proteins and lipids (of plasma) with zinc hydroxide. An acidic solution of zinc sulfate was added to the sample, followed by NaOH solution. The nonlipids in the supernatant fluid were removed by centrifugation, and the pellet was washed with water. They extracted the lipids with either C:M (2:1) or E:EtOH (1:3) (E = ethyl ether) ("Bloor's solvent," a popular pre-C:M mixture). The former appeared to be better.

This approach is particularly suited to dilute samples, such as spinal fluid or enzyme incubations, but it has not been tested with modern techniques. The gelatinous zinc hydroxide could possibly trap some lipids, keeping them from dissolving.

Trichloroacetic acid has been used in the same way with enzyme incubations, but in my experience the desired lipid did not precipitate with the proteins.

6.4. Liquid/Liquid Partitioning

This is one of the oldest and most common methods for removing hydrophilic nonlipid impurities from lipoidal solutions. The method, which generally involves shaking water or an aqueous solution with an immiscible organic solvent, is described in many of the procedures in this review. One problem that can arise is the formation of emulsions, generally as the result of precipitation of lipid caused by movement into the aqueous phase of a polar solvent from a mixed solvent (e.g., transfer of MeOH from a C:M extract). Emulsion formation tends to be prevented by including a substantial amount of salt or polar organic solvent in the system. Over-enthusiastic shaking can produce microdroplets of one phase inside the other, which may not settle back into their home bulk phase on simple standing. This can produce a variable radioactive blank from enzyme incubations.

I have noticed that students tend to have difficulty in seeing the interface between two solvents, and thus discard part of the lipid layer. This can be minimized by adding cresol red or bromcresol green to color the aqueous layer or oil red O to color the lipoidal layer. These dyes are readily removed, if necessary, in a subsequent step.

A problem with rotoevaporation of C:M:W upper layers containing gangliosides is severe foaming, which starts suddenly at some critical point. This is usually treated by adding an alcohol, such as isobutanol or 1-octanol, or toluene. Perhaps the most convenient method of evaporating such solutions would be to rotoevaporate with a large amount of cyclohexane to remove MeOH, then lyophilize the remaining water and cyclohexane.

One factor to consider in choosing solvent systems is whether you plan to wash the lipoidal layer more than once. If so, it is preferable to use a system in which the water-rich layer floats above the lipoidal layer. It is then easy to aspirate (nearly all of) the upper layer, then wash the lower layer with fresh upper layer. The use of a separatory funnel for such washings makes it essential to do a second wash, since a thin film of the discarded layer adheres to the glass walls and contaminates the lower layer when it is drained out of the funnel.

6.5. Removal of Nonlipids by Dialysis

Lipid extracts can be freed of small nonlipid molecules by simple dialysis against water. Although it may seem odd to dialyze an organic solution, the procedure works, and one obtains, in the case of C:M, an aqueous layer and a chloroform layer. Typically there is some precipitated protein and sphingolipid at the interface. The upper phase contains some lipid, so the entire contents of the dialysis bag may have to be evaporated to dryness before applying further processing. Ultimately the chloroform in the bag diffuses out.

Backwashing C:M extracts with water brings gangliosides and other very polar lipids into the water-rich layer (Folch upper layer, or FUL), leaving most of the other lipids in the less-polar layer. Removal of the nonlipids from the ganglioside fraction is often done by dialysis. Gangliosides probably exist in the initial water-rich solution primarily as Ca^{2+} and Mg^{2+} salts in polymeric, high-molecular weight (micellar) form, which do not dialyze out. As the methanol in the solution dialyzes out, making the solvent more polar, the polymers probably aggregate even further. The metallic cations eventually dissociate from the lipid, however, and the monomolecular forms (mol. wt. roughly 2000) dialyze out. The inclusion of Ca^{2+} in the dialysis water prevents dissociation, and some chemists use *unsoftened* tap water as a cheap (but slightly risky) substitute. Thus it becomes practical to dialyze the initial water-rich layer against tap water at first, concentrate the retentate by lyophilization or rotoevaporation, and remove the exogenous contaminants by dialysis against pure water.

In the case of C:M extracts that are dilute and that are then partitioned against water containing NaCl instead of divalent metal salt, the concentration of the endogenous Ca^{2+} might be insufficient to promote the formation of ionically cross-linked aggregates. In such a system there might be significant losses during dialysis.

Kanfer and Spielvogel (1973) showed that a *concentrated* solution of gangliosides from a brain C:M extract could be dialyzed against plain water with relatively little loss, evidently because of the preponderance of large micelles over monomers in such solutions. Ghidoni et al. (1978) concluded, however, that the ganglioside loss during dialysis was caused primarily by the MeOH in the FUL, since the MeOH enhances disaggregation. Some of the MeOH in such upper layers can be removed by rotoevaporation before foaming becomes serious. Perhaps the foaming can be blocked by adding calcium acetate.

The FUL may contain not only gangliosides, but also oligosaccharide ceramides (e.g., pentahexosyl), polypeptides, and small amounts of the polar phospholipids. These too are probably sufficiently polymeric to be retained during dialysis.

6.6. Separation from Nonlipids by Changing a Partition Constant

A partitioning study by Quarles and Folch (1965) offers a substitute for dialysis. They found that gangliosides could be pulled out of a FUL into a chloroform-rich layer in the system, C:M:W (8:4:3), if enough Ca^{2+} is present. The calcium salts of carboxylic acids are typically hydrophobic. In the case of polysialo-gangliosides, large hydrophobic aggregates can be formed with the divalent cation as a bridge. Thus it should be possible to separate the low-molecular weight, polar contaminants from the gangliosides in the FUL by adding $CaCl_2$ and chloroform.

An interesting finding from the partitioning experiments is that a *high* concentration of calcium ions (>0.26M) brought the gangliosides back into the upper layer. This is probably the result of forming a mixed calcium salt composed of one carboxylate group and one chloride ion. Such salts are highly hydrophilic.

Gatt (1965) also studied the partitioning properties of gangliosides and noted effects of various salts and lipids.

Carter and Kanfer (1975) applied the above observations in a somewhat long procedure for removing nonlipids (particularly labeled nucleotide sugars) from gangliosides. They not only adjusted the Ca^{2+} concentrations, but also used a low pH.

These transfer operations are typically called ion-pairing. The phenomenon is readily seen in the ability of a strongly acidic lipid (cerebroside sulfate) to pull a hydrophilic amine salt (methylene blue) into an organic layer (Kean, 1968; Radin, 1984). Thus liquid/liquid partitioning sometimes does not purify a lipid extract as expected.

Changing the pH of a partition system can influence ionization of acids and thus make them more or less hydrophilic. This was done by Das and Hajra (1984) to separate acyldihydroxyacetone phosphate from other lipids, after first extracting the total lipids with C:M:phosphoric acid (20:20:1).

6.7. Liquid/Liquid Partitioning in a Column with an Immobilizing Solid

Instead of partitioning the nonlipids and gangliosides into a water-rich layer, several chemists have utilized a liquid/liquid partitioning column. With Sephadex as the hydrophilic immobilizing phase, the packing is saturated with the water-rich phase, and the lipid extract, dissolved in the chloroform-rich phase, is passed through the column. This should help eliminate the danger of emulsion formation while giving the benefits of counter-current fractionation. A partitioning column is the equivalent of multiple liquid/liquid extractions with very small portions of wash solvent. This approach is particularly suited to working with very small volumes of extract.

A weakness of liquid/liquid columns is their sensitivity to room temperature: a decrease in temperature causes clouding (precipitation of one phase inside another), whereas an increase in temperature can strip the stationary phase of liquid, lowering its capacity for the nonlipids. If the stationary phase precipitates inside the moving, lipid-carrying phase, some of the nonlipid droplets may not make contact with the stationary beads and will fail to be retarded. Of course this problem is controllable to some extent by using jacketed columns and reservoirs, with thermostatted circulating water, but the publications from lipid chemists indicate that they have not used this technique.

An additional problem arises when soft packings are used to immobilize the stationary phase: An increase in operating pressure can compress the beads and squeeze out the stationary phase (with its nonlipids). Still another problem arises when a substantial amount of solute partitions from the moving phase into the stationary phase, causing its volume to increase. This may produce stripping of the stationary phase from the surface of the beads.

Wuthier (1966) described a procedure in which two equilibrated phases were used. A mixture of C:M:W (8:4:3) was shaken, and the resultant upper layer (FUL) was used to swell the Sephadex G-25 overnight. This was packed to form a column 1 × 10 cm. The lipid sample, freed of its extractant solvent, was dissolved in small portions of the lower layer (Folch lower layer, or FLL) and filtered directly onto the column packing. After a rinse with more FLL, the lipids were eluted with FLL, using a total of 25–30 mL. The elution was run under slight nitrogen pressure to give a flow rate of 1 mL/min. To wash out the column, 50 mL of FUL were passed through, followed by displacement and reequilibration with 20 mL of FLL. It is advisable to use a column having an adjustable flow inlet tube so that the wash-out step can be run in the opposite direction. Wuthier analyzed the crude and treated lipids for individual components by paper chromagraphy and phosphorus analysis, and found good yields in the partitioning process. Tests with added labeled phosphate, glucose, and glycine showed that less than 1% leaked into the moving phase.

Instead of C:M:W in the two phases, Terner et al. (1970) have proposed benzene:EtOH:W (50:40:10). The Sephadex G-25 (Fine) is swollen with the lower, water-rich phase, packed into a column with the lower phase, and washed with upper phase to displace the lower phase from between the beads. The lipids are added in the upper phase and eluted in the same phase in very high yield. As much as 100 mg of lipids could be handled by a 1 × 10 cm column.

6.8. Removal of Nonlipids by Lipid Adsorbents

A variant of the Wuthier procedure by Williams and Merrilees (1970) is similar to a liquid/liquid partitioning system, but is probably an example of adsorptive uptake of nonlipid. A homogenate made from wet tissue with C:M (2:1) was filtered and mixed with dry Sephadex G-25. The weight of resin used was equal to the estimated weight of water in the sample. The mixture was rotoevaporated until it appeared almost dry, at which point the residual water was bound somewhat to the Sephadex beads. The beads were resuspended in chloroform and re-evaporated. (It should be noted that an indeterminate amount of the water is removed by the evaporation step.) The beads were now transferred to a narrow column with chloroform and eluted with this solvent until all the lipids had been extracted. Recoveries of added labeled lipids were found to be 97 ± 5.5%. Since chloroform is such a poor solvent for the sphingolipids, an additional elution step would probably be needed for better yields. This method strongly resembles the drying/reextracting method that is often used to insolublize nonlipids (*see* section 6.1) since the sample is dried somewhat within porous beads instead of on a glass container's surface.

A different type of Sephadex, LH-20, has also been used (Maxwell and Williams, 1967). This is a lipophilic derivative of Sephadex that may allow better penetration by organic solvents than ordinary Sephadex. The beads, after washing with water, were air-dried and swelled in C:M (3:1). The column was packed with the same solvent, using a disc of filter paper to hold down the top of the bed. (Sand would do the same thing). The lipid extract, in C:M 4:1, was dried with a 1:1 mixture of Na_2HPO_4 and Na_2SO_4.

(This step alone probably removed some nonlipid material.) The extract was next evaporated to dryness, dissolved in C:M (3:1), and applied to the column. Elution with the same solvent yielded the lipids, followed in later fractions by some nonlipids. This procedure was not tested for the fate of specific lipids, such as gangliosides.

7. Extraction with Chloroform:Methanol Mixtures

7.1. The Methods of Folch et al.

Folch et al. (1951, 1957) introduced the use of C:M (2:1), still one of the best extractant solvents. The tissue was homogenized for 3 min with about 17 vol of the solvent, the suspension was filtered, and the homogenizer and funnel were rinsed with another 2 vol of solvent. Thus 1 g of tissue produced almost 20 mL of extract.

Centrifugation of the homogenate is not practical because the solvent is so dense, but the addition of 0.2 vol. (4 mL) of MeOH makes it possible to convert the proteinaceous residue to a compact pellet.

The first paper by Folch et al. (1951) utilized an unusual liquid/liquid partitioning technique to remove nonlipids. This was an overnight diffusion from the extract into a large volume of water, a visually attractive phenomenon. A faster procedure was developed by Folch et al. (1957) in a paper that is one of the most frequently cited references in the biochemical literature. One-fifth volume of water (4 mL) was shaken with each 20 mL of filtrate, producing a final mixture having the composition C:M:W (8:4:3). This produced two layers in which the volume of the upper one was 40% of the total. The upper layer (FUL = theoretical upper layer) had the composition C:M:W (3:48:47), and the lower layer, 86:14:1. Unlike the previous Folch partitioning method, this one yielded only a small amount of insoluble material at the interface since the lower layer contained much more MeOH. The last bits of upper phase were removed by carefully rinsing the upper parts of the vessel with FUL three times.

It is just as effective to use M:W (1:1) or M:saline (1:1) as the equivalent of FUL in such washing. Not only is it easier to prepare, but temperature changes cannot precipitate out a second phase. It does absorb some chloroform, however, lowering the volume of the lower phase, so it probably should not be used for multiple partitionings.

Suctioning is a very efficient way of removing the upper layer. A water aspirator or portable pump supplies the suction, and a suction flask or test tube is used as the receiver. For small samples we use a clamped 1/16-in od stainless steel tube and gradually raise the container so that the tube is immersed in the upper phase. For large volumes, we use a 1/16- or 1/8-in od Teflon tube for the transfer. For convenience in holding, the tube is jammed into a Pasteur pipet. The vacuum has to be weakened with an air leak tee-tube to prevent losses.

The 8:4:3 mixture of liquids seems to be particularly free of a tendency to emulsify, and many lipid biochemists have used that ratio for other lipid partitioning applications. In systems having a different ratio, one simply adds the required volume of extra component. Chloroform and water alone tend to produce bad emulsions with lipids. Part of this problem is caused by the relatively poor solubility of many lipids in such a system, and air bubbles get trapped in the soft lipid particles. The addition of MeOH generally clears up the emulsion, but evacuation to remove the bubbles can also be helpful (Jaxxon, 1986).

Folch et al. tested their washing procedure with brain tissue and made the important discovery that the upper layer contained gangliosides and possibly 0.3% of the total lipids. A second wash of the lower layer with FUL yielded additional gangliosides and an unexpectedly large amount of other lipids (2% of the total). It was evident that some component(s) of the initial extract had been transferred to the first upper layer; the component(s) prevented transfer of lipids, mainly acidic ones, to the upper layer. Thus was born the understanding that inorganic ions are needed to control the ion-pairing and, hence, the partition constants of the acidic lipids. The authors settled on FUL containing either 0.05% $CaCl_2$, 0.04% $MgCl_2$, 0.73% NaCl, or 0.88% KCl.

Many users of this partitioning system have used the same salt concentrations, but there is no reason to believe that a moderate increase to a rounded number would make any difference.

An important point to remember is that salt is needed in the first partitioning step, and that it is normally furnished by the sample itself. When a tissue sample is extracted with an unusually large volume of C:M (more than 20 mL/g), however, one must include salts in the initial wash and, perhaps, in the initial extraction solvent. Even enzyme incubation samples probably have too low an ionic strength to avoid the problem. Failure to remember this has probably led to many reports that certain methods gave a better yield of lipid than the method of Folch et al.

Workers who use C:M to extract subcellular cell fractions prepared in sucrose should note that the sugar substantially increases the amount of protein that enters the extract (Lees, 1966).

7.2. Biphasic Chloroform:Methanol Methods

Bligh and Dyer (1959) introduced the idea of carrying out the extraction and partitioning steps almost simultaneously, retaining the precipitated protein between the two liquid phases. This was done by extracting the sample with a relatively small volume of C:M (1:2) instead of a large volume of C:M (2:1), then forming a second liquid phase by adding additional water and chloroform. Important features of the method are that the water concentration is, at all stages, higher than in the Folch method and that the proteinaceous residue is removed *after* partitioning, not before. (Although most users refer to this as a two-phase system, there is actually a third phase, consisting of the nonlipid residue.)

Their basic procedure for a sample containing 1 mL of water (about 1.28 g of brain) went as follows: the sample was homogenized for 2 min with 3.75 mL of C:M (1:2). This produced a single liquid phase. Additional chloroform (1.25 mL) was added and homogenization was continued for 30 s more. Water (1.25 mL) was added, and the sample was homogenized (30 s) and filtered to remove the insoluble part of the tissue. The authors specify that the sequence of solvent additions and homogenizations should not be changed. The final extract consisted of two liquid layers, made of C:M:W (20:20:18) (total composition). Samples that were low in water content were supplemented with water before extraction. This method could be called Bligh:Dyer I, and apparently its properties were not studied in detail.

For a higher yield of lipid, the homogenizer, filter, and nonlipid residue were rinsed with a total of 1.88 mL more chloroform, which was filtered into the initial filtrate. The pooled filtrates were mixed and allowed to separate before removing the upper layer. This procedure resulted in a final solvent ratio of C:M:W (48:27:25) and total volume, before removal of the upper layer, of 7.3 mL. The extra wash with chloroform was shown to yield an additional 6% lipid. It seems likely that neural tissue would need a C:M rinse in order to extract the sphingolipids. The more complete procedure, with extra chloroform, could be called Bligh/ Dyer II.

Unfortunately, many of the researchers who have said they followed (or, more usually, modified) "the" method of Bligh and Dyer have not stated whether they used the extra wash. The compositions of the two layers are obviously different in the two methods, which could significantly affect the partitioning of polar lipids and nonlipids into the upper layer.

The second procedure was tested with fish muscle, and no consideration was given to the problem of extracting lipids typical of neural tissue. One test involved evaporating the lower (lipoidal) layer to dryness and extracting the residue with chloroform. All of the residue dissolved. Evidently the extract contained relatively little glycolipid, which is somewhat insoluble in chloroform. In another test, the upper (nonlipid) layer was concentrated and extracted with ethyl ether. The ether contained solid matter amounting to 1% of the total lipid. Ether does not extract gangliosides, however, which presumably are in the upper layer. A better test would be a backwash with C:M:W having the same composition as the lower phase.

Strong acid hydrolysis of the nonlipid residue yielded an additional 7% of lipoidal material, presumably fatty acids from acidic lipids that were tightly bound to protein. If such lipids contain 1/3 of their weight in the form of fatty acids, this experiment means that about 21% of the total lipid was not extracted by the Bligh/Dyer method. This somewhat severe weakness in the method has been corrected in later versions.

Although Bligh and Dyer did not anticipate the use of their method for extracting enzyme incubation samples or dilute cell suspensions, the low volume has proved to be a good selling point despite the incompleteness of extraction. The extraction can be carried out in a small test tube, the same one that was used for the incubation. In this case, the homogenate can be centrifuged to separate the phases (bringing the protein residue to the interface), the lower layer and interface can be washed with upper phase if necessary, and the lower layer can be evaporated to dryness. The protein in the residue is assumed not to bind any lipid and can be removed, if necessary, by dissolving the lipids and filtering or centrifuging. Although the addition of C:M to an enzyme incubation can be assumed to stop the reaction, some workers boil the system before adding the organic solvent.

7.2.1. Palmer's Modification of the Bligh/Dyer Method

Palmer (1971) found that the acidic phospholipids (PS, PI, phosphatidic acid) were extracted by the first Bligh/Dyer step with C:M (1:2), but that they reentered the tissue proteins when a second liquid phase was formed by the addition of chloroform and water. This phenomenon renders the two extraction systems—homogeneous vs biphasic—quite different. Presumably the uptake resulted from the exposure of hidden basic groups during denaturation of proteins. Palmer showed that the binding phenomenon could be blocked by including salts of divalent ions or by simply removing the insoluble proteins before adding the chloroform and water.

A lesson applicable to other extraction techniques may be seen here: It may be dangerous to add organic solvents to a tissue mince or suspension *slowly* since two liquid phases may form in the initial stages.

7.2.2. Use of Salts During Two-Phase Extraction

Ostrander and Dugan (1962) introduced the idea of adding zinc acetate to the initial, single-phase Bligh/Dyer extract. Their procedure was modified slightly by Melton et al. (1979). The detailed steps, applied to a rather dry material (soy bean meal), follow: 50 g of material were homogenized with 130 mL of MeOH for 5 min. Chloroform (65 mL) was added, and this was followed by 5 min of mixing. Another 65 mL of chloroform was added, and mixing was carried out for just 20 s. This was followed by a brief mixing with 65 mL of water containing 1.5 g of zinc acetate. The mixture was filtered and the residue and equipment rinsed with 100 mL of chloroform. This wash was filtered, catching the filtrate in the first filtrate. Rinsing was conducted again, with 75 mL of chloroform, then with 25 mL of MeOH. The filtrate was mixed and the two phases allowed to separate in the cold.

Unfortunately this last version of the Bligh/Dyer method was not evaluated with respect to the acidic lipids or nerve tissue. It does seem reasonable to believe, however, that the zinc acetate ensures solubility of the acidic lipids. Presumably they are converted to zinc salts, rather than calcium or potassium salts, a point that might bear on their chromatographic behavior.

Michell et al. (1970) used a neutral extraction system for the polyphosphoinositides. The common lipids were first extracted with C:M, then the residue was suspended in $2M$ KCl (5 mL/g brain). To this was added C:M (1:2) (19 mL/g brain), and the suspension was homogenized gently. The mixture was kept for 15 min, then partitioned by adding chloroform (6.25 mL/g) and then water (same volume). The mixture was shaken well and the upper layer was backwashed with chloroform; the two chloroform layers were pooled.

7.2.3. Two-Phase Extraction with Salt and Acid

Hajra, in a study of the acidic lipids acyl and alkyl di-hydroxyacetone phosphate (1974), found that they were lost to the aqueous phase in two-phase C:M:W partitionings. He introduced a modification of Bligh/Dyer I method as follows: add 4.5 mL of C:M (1:2) to 1.2 mL of enzyme incubation mixture and mix. Centrifuge the tube to sediment the protein, transfer the extract, and add 1.5 mL of 2M KCl containing 0.2M phosphoric acid and 1.5 mL of chloroform. Vortex the mixture, centrifuge, and remove the upper layer. The upper layer has the approximate composition C:M:W (1:12:12) (Amiya K. Hajra, personal communication).

This method differs from method I of Bligh and Dyer in that (1) the protein is removed *before* forming the two phases and (2) acidic salt water, instead of plain water, is used for the wash. It could equally well be called a variation of the Folch/Lees/Sloane-Stanley method, which also involves removing the protein residue before forming two phases for the wash. The method was shown to owe its success to the ability of strong acid, which yields pH <3, to convert ionized lipid (initially bound to ionic nonlipid) into partial-ly undissociated lipid. The ionic charge is –2 at pH 7, but only about –0.5 at pH 2. The KCl is added to salt the lipid out of the aqueous phase; the data show it to be unnecessary at this low pH, however. Hajra mentions the usefulness of this method for extraction of lysophosphatidic acid and other acidic lipids and the possibility that zwitterionic lipids do not stay with the chloroform layer at low pH.

Butanol (Daae and Bremer, 1972) was found to be less efficient as an extractant for the keto phospholipid and other polar lipids, but with some merit if plasmalogen stability is required.

Mogelson et al. (1980) also used HCl and an inorganic salt in their version of the Bligh/Dyer procedure. One gram of minced tissue is homogenized in 10 mL of 0.9% NaCl or 2% HCl in 0.9% NaCl, then extracted twice with C:M "in the proportions specified by Bligh and Dyer." A total of 81 mL of chloroform was used. It is not clear how much MeOH was used or if additional water was added. The nonacidic method was tested by adding labeled PC,

PE, lysoPC, and lysoPE; the recoveries ranged from 90 to 108%. Since the added lipids were not protein-bound, as in tissues, the test evaluated only the partitioning step and the possibility of binding to denatured protein.

In the case of the acidic extraction method, plasmalogen was found to be completely converted to lyso lipid. The matter of PI and PS extraction was not studied.

A less acidic mixture has been used for leukotriene B extraction from enzyme incubations (Prescott, 1984): 2.5. vol of C:M: formic acid (12:12:1) were added to each volume of sample. This produced two liquid phases, and the upper phase was backwashed with a little chloroform, which was pooled with the lower layer.

7.2.4. Two-Phase Extraction with Methanol, Butanol, and Chloroform

In this variant of the Bligh/Dyer procedure, 2 mL of incubation mixture were mixed with 0.3 mL of 0.5*M* EDTA, 2.8 mL of MeOH, 2.8 mL of 2-BuOH, 5.6 mL of chloroform, and 2.8 mL of water. The mixture was vortexed and centrifuged, and the upper layer was backwashed with 2 mL of chloroform, which was added to the first lower layer. The EDTA is needed to ensure extraction of PS, and the butanol is needed for the extraction of lyso phospholipids.

7.2.5. Two-Phase Extraction with Acetone, Benzene, Isopropanol, and Water

In a procedure by Allen (1972), a homogenate of tissue prepared in 0.9% NaCl was first shaken with 0.5 vol of acetone, then shaken with 1 vol of the upper phase prepared from a mixture of benzene:2-PrOH:W (2:2:1). The mixture was centrifuged and the upper layer, containing lipids, was removed. The lower layer plus interfacial material was extracted the same way three times, with the same volumes of the two solvents, and the four extracts were pooled.

This method was found to yield higher amounts of lipids than the "Folch standard method," which was applied to the intact tissue (nematodes), not to a homogenate. The explanation for the apparent superiority apparently lies in the mechanical difficulty of homogenizing the intact organisms, however. For neural tissue, the main potential merit lies in the method's applicability to aqueous homogenates or enzyme incubations.

7.3. Single-Phase Extraction with Other Chloroform: Methanol Mixtures

As researchers learned more about the nature of gangliosides and the existence of very acidic phospholipids, highly polar gangliosides, and oligosaccharide lipids, they reexamined the question of extraction by C:M (2:1). It is now generally agreed that a higher proportion of MeOH is helpful, sometimes with multiple extractions. Some workers include water, ammonium hydroxide, salts, or HCl in one of the extraction solvents.

LeBaron (1963) introduced the use of HCl with C:M to extract the polyphosphoinositides. Hauser and Eichberg (1973) recommended extracting nearly all of the tissue lipids first, with 15 vol of C:M (1:1) containing 60 μmol of $CaCl_2$ for each gram of tissue. The homogenate was filtered, and the nonlipid residue was washed twice with 3 vol of C:M (2:1). Finally the inositides were extracted with two portions of C:M:conc. HCl (266:133:1). The pooled filtrates were partitioned by adding 0.2 vol of 1M HCl, and the lower layer was washed with FUL in which the water was replaced by 1M HCl and (a second time) by 10 mM HCl. The Ca^{2+} had the effect of increasing the yield of PI-phosphate.

Byrne et al. (1985) recommended the use of 0.1M HCl in C:M:W (5:5:1). The water content was calculated on the assumed water content of the tissue (70% for white matter, 80% for gray), and additional water added to obtain the desired final ratio. After homogenizing the sample with C:M:W, they added conc. HCl to attain 0.1M HCl in the mixture. The mixture was vortexed, left 20 min, and centrifuged at low speed. Removal of nonlipids from the extract was performed by chromatography through a Sephadex LH-20 column, previously swelled and packed in C:M:W (5:5:1). During percolation of the sample into the packing, the tissue residue was reextracted with the same solvent, and the extract was added to the column. Further elution with this solvent brought out the gangliosides, partially overlapping the nonionic glycolipids.

Svennerholm and Fredman (1980) used C:M:W for extracting gangliosides from brain. The extract from this step was partitioned to separate the gangliosides from the other lipids by adding additional water. This method was found to give higher yields than other methods, especially for the polysialo lipids. Cerebral tissue (500 g) was homogenized in 1500 mL of water at 4°C, and the suspension was poured into 5400 mL of MeOH while stirring. To this was added 2700 mL of chloroform, and, after 30 min of stirring, the insoluble residue was removed by low-speed centrifugation. Additional nonlipid was removed by filtration through Celite 535. Additional ganglioside was extracted from the residue by homogenization in 1000 mL of water and 4000 mL of C:M (1:2). The filtrate from this was pooled with the first filtrate and partitioned (very gently) by adding 2600 mL of water. Separation of the phases took 6 h. The lower phase was clarified with 1500 mL of MeOH, then partitioned by adding 1000 mL of 10 mM KCl. The two upper phases were pooled.

It should be noted that the ratio of solvents in the first extraction is stated by the authors to be C:M:W (4:8:3), but inspection of the procedure (assuming that cerebrum contains 80% water) shows that the initial mixture has the composition 4:8:2.88. The liquid/liquid separation steps, which tend to form emulsions, could be speeded by centrifugation if run with smaller samples.

Gangliosides have also been extracted from brain with two steps: extraction with 19 vol of C:M (2:1), and then with 10 vol of C:M:W (33:66:5) (Suzuki, 1965). Enough chloroform (presumably 9.5 vol) was added to the pooled filtrates to bring the C:M ratio to 2:1 and two layers were formed by adding 0.2 vol of 0.1M KCl or NaCl. This brought much of the ganglioside into the upper layer. Further extraction was done by washing the lower layer with FUL containing salt and FUL without salt. This method partitions only part of sialosylgalactosylceramide (the least polar ganglioside) into the FUL.

7.3.1. Successive Extractions with Chloroform:Methanol

Researchers have used as many as four C:M successive extractions, some containing ammonium hydroxide or water. Dacremont et al. (1974) have extracted lyophilized samples with C:M (2:1), C:M (1:1), and C:M:W (47:47:5). The inclusion of aqueous salt solutions would probably be a good idea in view of findings relating ion exchange to extractability.

Successive extractions with C:M were used, some at 50–60°C, to be sure that all gangliosides were extracted from brain (Kawamura and Taketomi, 1977). As can be imagined, this type of extraction took out significant amounts of nonlipids, which were partially removed by the evaporation/reextraction technique (C:M 2:1) several times. The yield of gangliosides was stated to be 96% after chromatography with silica gel + diatomaceous earth containing $CaSO_4$.

8. Extraction with Hexane/2-Propanol

8.1. Initial Procedure

This system was introduced by Hara and Radin (1978) as an alternative to C:M in order to reduce the danger of toxicity and to offset the rapidly rising price of chloroform. In this procedure, brain tissue was homogenized with 18 vol of H:IP (3:2) for 30 s and filtered with a glass pressure funnel. The apparatus and tissue residue were then rinsed with 3 × 2 vol of the same solvent. Each rinse was allowed to soak into the residue and sintered disc for 2 min before applying pressure. Thus 1 g of tissue yielded about 24.5 mL of extract.

The resultant extract was shown to contain a negligible amount of protein, in contrast to C:M extracts. Comparison with an extract made with C:M (2:1) was done by treating the dried lipids with NaOH in C:M to methanolyze ester lipids, partitioning with water (to remove gangliosides, nonlipids, and methanolysis products), and chromatographing the products with a silica gel column. The H:IP method gave a slightly low yield of low-polarity lipids, mainly fatty acid methyl esters (2% low), but very similar yields of the more polar lipids. The discrepancy in methyl esters was caused in part by the failure of H:IP to extract the proteolipid proteins, which contain some covalently bound fatty acid. (Recent studies are disclosing an increasing number of proteins containing covalently linked fatty acid.) Tests with labeled ganglioside GMl showed that the H:IP method gave only partial extraction of gangliosides. The method was not investigated for its efficiency in extracting the strongly acidic phospholipids.

This method has several additional advantages over extraction with C:M. (1) The homogenate, because of the solvent's low density, can be centrifuged to remove the tissue residue. The residue is finely powdered and can be washed readily with more H:IP (but the solvent should include 2.5% water to improve the compaction of the pellet). (2) The low content of nonlipids makes it feasible to evaporate the extract to dryness, then apply an unfiltered solution of the lipids to a chromatographic column without clogging it, (3) H:IP extracts very little pigment, reducing the load on chromatographic packings. This is a particular advantage for extraction of liver and red cells. (4) The nonlipid residue is finely divided and probably easier to extract for other purposes. (5) H:IP is less volatile than C:M so lipid *concentrations* do not increase as readily during handling or storage.

Because of the flammability of hexane, homogenizer motors should be well ventilated.

For experiments in which nonlipid radioactive precursors or products must be washed out of the lipid extract, liquid/liquid partitioning can be accomplished by shaking the extract with a "washing solution," using 12 mL for each gram of tissue. This solution is made from 10 g of anhydrous Na_2SO_4 in 150 mL of water. Starting with 1 g of brain, one obtains the lipids in about 18 mL of *upper* layer. Since the lipids are present in the upper layer instead of the lower one, it is easy to aspirate the lipid extract, preferably with a vacuum transfer tube. Since the sodium sulfate solution pulls much of the 2-PrOH out of the extract, this is a convenient way to reduce the amount of solvent that is to be evaporated off. It also produces a solution that can be directly applied to a silica gel column for some isolation procedures, with monitoring in the far UV range. The lower layer from this washing step contains 2–3% of the polar lipids, which can be retrieved by a backwash with H:IP (7:2). An unusual situation occurs when the sodium sulfate concentration is too high or the mixture is cooled: three liquid layers are produced!

A later report (Radin, 1981) deals with the problem of extracting samples having a high water content, such as an incubation mixture. This calls for 33 mL of H:IP instead of 24 mL. A sign of an inadequate volume of extractant is the formation of a lumpy suspension instead of the usual fine suspension. This could result from samples with a high salt concentration and it might be necessary to increase the volume of 2-PrOH. In the case of plasma or blood, the sample should be added dropwise to a stirred or vortexed container of H:IP (Raymond J. Metz, personal communication).

No H:IP version of the Bligh/Dyer procedure has been described, nor has the possible use of HCl or salts during the extraction to increase the yield of acidic lipid been reported.

8.2. Modified Hexane/2-Propanol Procedure

In an independently derived approach, Yahara et al. (1982) described two modifications, one using a less polar mixture of H:IP, in the ratio 78:20, the other using a water-containing mixture. In the latter procedure (method A) the solvent was prepared by equilibrating H : IP : W (8 : 4 : 3) and using the upper phase, which was estimated to consist of H : IP : W (78 : 20 : 2). Fresh neural tissue was homogenized in an all-glass homogenizer with 20 vol of the solvent, left for 30 min, and centrifuged. The clear supernatant liquid was decanted from the mushy residue, and the residue was extracted twice more as above, but with 10 vol of solvent each time. For these steps, the mixture was centrifuged without waiting 30 min. The three pooled extracts were then evaporated to dryness.

In method B, the sample was homogenized with 4 vol of 2-PrOH and again after adding 15.6 vol of hexane. After centrifugal removal of the extract, the residue and apparatus were washed twice with 1/2 the total volume of the same solvent (H : P; 78 : 20). The pooled extracts were evaporated to dryness.

Method A was compared with C : M extraction, using samples of forebrain from two different rats (assumed to have very similar lipid contents). Both extracts were evaporated to dryness and weighed, without re-extraction to separate off denatured proteolipid protein. The H : IP method yielded 9% more residue, but identical amounts of cerebroside and sulfatide. Although the authors concluded that their method produced a nonproteinaceous extract, their data seem to point to a high degree of nonlipid contamination. It is possible that method A brings more of the gangliosides into the lipid extract. Method B was not evaluated in this report, but was stated to give essentially the same results.

A potential problem with method A—and perhaps with the other H : IP methods—is the possibility that some enzymes are not inactivated by the solvent. In method A, the water-saturated homogenate was left at room temperature for 30 min, during which time some complicating reactions may occur.

Kolarovic and Fournier (1986) compared five different extraction procedures starting with heart microsomes and concluded that H : IP was the easiest to use and produced the highest yields of lipids. The H : IP was used without liquid/liquid partitioning to remove nonlipids, whereas the other systems (C : M and BuOH) were used *with* partitioning. In the procedure used by the authors, a sample of microsomes containing 3.6 mg of protein was suspended in 0.4 mL of water and placed in a 50°C ultrasonic bath. To this were added 10 mL of H : IP (3 : 2), and sonication was continued for 1 min. The tube was then centrifuged 1 min, and the extract was evaporated to dryness in a centrifugal vacuum evaporator (Speed Vac, Savant Instruments). This method was equivalent to other methods for the major lipids and yielded the highest content of lysoPC, lysoPE, and the acidic lipids (PI, PS, cardiolipin, phosphatidic acid). As the authors point out, the need to remove nonlipids is not as serious as earlier workers thought, and the H:IP method dissolves less of them anyway.

The Hara/Radin method was tested with brain by Saunders and Horrocks (1984) with respect to prostaglandin and phospholipid yields. Comparison of the H:IP (3:2) extracts (without removal of nonlipids) vs C:M (2:1) extracts (with a washing step) showed that the former gave higher yields of prostaglandins and equal yields of various other lipids (PC, PE, PS, PI, plasmalogen, sphingomyelin, and phosphatidic acid).

9. Nondenaturing Extraction Methods

Lipid extraction solvents normally denature (or are assumed to denature) the proteins in the sample, but sometimes one wishes to study the proteins for their biological activity in the absence of lipids or in the presence of specified lipids only. In the early days of serum lipoprotein research, hexane was used to remove lipids. To prevent the interface denaturation of proteins that often occurs on shaking two liquid phases, the samples were rotated slowly, for a long time.

9.1. Use of Diisopropyl Ether and 1-Butanol

Cham and Knowles (1976) used a more polar solvent—isopropyl ether:butanol (60:40)—which makes closer contact with the protein-bound lipids. Serum or plasma (5 mL) containing 0.5 mg of EDTA was rotated end over end in closed test tubes with 10 mL of the solvent for 30 min. The tubes were then centrifuged briefly to allow removal of the upper layer. Residual dissolved BuOH (which interferes with protein assay by the Lowry method) was removed by brief extraction with the ether alone. Residual ether was then removed by agitation under water vacuum for 1 min at 37°C. Analysis of the serum or plasma showed that all of the triglyceride, cholesterol, free fatty acid, and phospholipids had been removed. The lipid extract was free of protein.

With regard to the question of application to nervous tissue, one must note that the ability of a solvent to extract lipids in a nondenaturing system must depend on the relative affinity of the proteins for the lipids.

A slight modification of this method was introduced in order to reduce the problem of interfacial denaturation (Wong and Ladisch, 1983). One volume of serum was mixed for 60 min with 2 vol of the solvent in screw-cap glass centrifuge tubes using a motor-driven rocker. After a 5-min settling period, the tubes were centrifuged at 150g for 5 min to complete the separation, and the lower (aqueous) phase was extracted with 2 vol of ethyl ether. This process was done by simple manual inversion five times, followed by centrifugation as before. The ether extraction was repeated.

This procedure was found to leave most of the gangliosides and about 14% of the lecithin in the lower layer. It may be useful for partial delipidation of lipophilic enzyme preparations and cytosolic extracts.

9.2. Use of 1-Butanol Alone

This solvent is more polar than the butanol/isopropyl ether mixture, so it may take out the polar lipids more readily. It has been used to remove lipids from tissue extracts in the isolation of an enzyme (Furbish et al., 1977).

Butanol was shown to remove much lipid from aqueous egg yolk homogenates containing 1M NaCl (Meslar and White, 1978). This produced an interfacial precipitate consisting of much protein, but 22% of the protein remained in the aqueous layer, and much of the biotin-binding activity remained intact. No lipid was detectable in the aqueous layer, but the interface was not analyzed, and the possibility remains that protein/lipid complexes are located there. The function of the added salt is to reduce emulsion formation. I think, however, that salt acts to denature proteins in two-phase extractions of this sort.

9.3. Removal of Lipids from Serum Albumin

Removal of lipids from serum albumin presents a case of the problem of nondenaturing lipid extraction. Serum albumin occurs in blood bound to free fatty acids, bilirubin, tryptophan (a somewhat lipoidal amino acid), and various trace lipids. There has been a demand for lipid-free albumin, and several methods have been proposed for its preparation.

Extraction of lyophilized human serum albumin has been accomplished by soaking the protein in isooctane : HOAc (95:5) for 6 h at 0°C, followed by a rinse with isooctane (Goodman, 1957). The acidic extraction was repeated as before, but followed this time with *three* washes with isooctane. The protein was then dried in vacuum, dissolved in water, and dialyzed against water. This preparation was reported to have physical properties very similar to those of the original protein.

An ion exchange resin and a lipophilic resin have been used (Scheider and Fuller, 1970). A 5% solution of albumin in acetate buffer, pH 3–4, with an acetate ion concentration of $0.1M$, was first dialyzed against the buffer for 24 h in the cold. After filtration or centrifugation to remove turbidity, when present, the solution was passed through a column packed with Amberlite CG-400 (strongly basic anion exchange resin) or Amberlite XAD-2 (porous, nonionic polystyrene). The columns were pre-equilibrated with the same buffer. A small (10%) decrease in sulfhydryl content was observed as the result of lipid removal. Perhaps one should do this operation in solutions containing dithiothreitol or under nitrogen.

It was not clear whether the anion exchanger acts by ion exchange or by lipophilic phase partitioning (section 12). Presumably some protein is lost on the surface of such beads and a low degree of porosity must be important for good recovery. For applications in which recovery of a rare protein is important, it might be useful to presaturate the protein-binding sites with serum albumin.

Chen (1967) introduced the use of charcoal at low pH as a fatty acid adsorbent. Although Chen's detailed comparison between native and defatted albumin disclosed no harmful effects on the protein, Scheider and Fuller (1970) reported that their ion exchange method produced a smaller decrease in sulfhydryl content.

In a recent version of the Chen method, Nakano et al. (1983) used a column containing charcoal embedded in agarose beads. Sepharose 6B was heated in water to 90°C, forming a sol. Activated charcoal was mixed in thoroughly, and the mixture was poured into a stirred hot solution of a nonionic surfactant in cyclohexane. The suspension was stirred while cooling, and the resultant beads were washed with ethyl ether and water. The packed column was loaded with the albumin solution, at pH 3 in $0.1M$ NaCl, and eluted with $0.1M$ NaCl/1 mM HCl. The NaCl was needed to prevent adsorption of protein by the agarose.

None of these procedures removed all of the fatty acid. Stearic acid is particularly difficult to extract.

9.4. Nondenaturing Extraction of Detergents and Bile Salts

Removal of surface-active lipids (nonionic detergents, bile salts) without accompanying protein denaturation is another important problem. These materials seem to bind fairly strongly to the lipophilic groups in many proteins and therefore find use in dissociating them from the lipids and membranes that normally bind to them. Once the protein/detergent complexes have been separated from residual membranes by centrifugation, and from the lipid-detergent micelles by other means, the need arises to extract the detergent from the solubilized protein. A summary of several methods is given here.

Butanol, as mentioned above in section 9.2., is relatively nondenaturing and a reasonably good solvent for lipids.

Solid phase partitioning is often used, with lipophilic column packings. All of these materials tend to bind protein, possibly as a detergent/protein complex. This is a serious problem with small protein molecules, which penetrate the pores. Amberlite XAD-2 has been mentioned most often; XAD-7 has also found use. Pierce Chemical Co. (Rockford, IL) sells a resin called Extracti-Gel D for this purpose. The resin has been used to absorb octyl glucoside, SDS, Lubrol, digitonin, and deoxycholate. Octadecyldimethylsilyl silica gel ("ODS") has also been used.

9.5. Extraction with Acetone and Ethanol

These solvents have a long history of use in extracting lipids without denaturing proteins, since many of the early enzyme isolation methods used a preliminary delipidation (and dehydration) step. The basic requirement is to do the extraction while homogenizing the tissue in a cold room, gradually adding acetone or EtOH that has been precooled to –20°C. The homogenate is then filtered in the cold, with very cold funnels, and dried in a desiccator in the cold room. Thus the material is kept below zero during the times it is in contact with solvent.

10. Extraction under Drying Conditions

A common method of lipid extraction at one time was the use of a mortar and pestle with anhydrous sodium sulfate and ethanol. The water in the tissue is extracted as the salt hydrates, the crystals act as a grinding aid (like sand), and the salt that dissolves at first in the tissue water acts to denature the proteins or dissociate the protein/lipid complexes. This approach has the merit of utilizing a relatively small volume of solvent, since the solvent's sole purpose is to dissolve the lipids, not the tissue water. It might be worth using a mixture of higher extraction power, such as H:IP.

Marmer and Maxwell (1981) described a modification of this approach. Wet tissue (5 g), previously minced thoroughly with a home-type food processor, was ground in a mortar with 20 g of sodium sulfate. Celite 545 (15 g) was then added and grinding was continued. The powder was added to a chromatography column (35 mm id) that had been previously packed with 10 g of a mixture of $CaHPO_4$ and Celite (1:9). The lipids were extracted by passing dichloromethane:MeOH (9:1) through the powder, collecting 150 mL. The nonlipid content of this extract, starting with muscle, was <0.1% of the 5-g sample (method of analysis not given).

Whether this relatively nonpolar solvent would give complete extraction of the acidic lipids and ceramide oligosaccharides remains to be seen. The purpose of the calcium phosphate at the bottom of the column is to prevent extraction of nonlipids.

In a variant of the method, Marmer and Maxwell used two different solvents to obtain partial separation of nonpolar and polar lipids. In this procedure the first solvent was dichloromethane alone, then the mixture with MeOH. The first eluent took out most of the low-polarity lipids, but none of the polar ones, so it may have merit for preparative isolations. It is possible that a solvent gradient would yield a sharper separation.

11. Selective Extractions

No method now available will extract a single family of lipids without some contamination with other types of lipids. Selective extraction can be very useful for reducing the load of contaminants to be removed in a subsequent step, however, or for large-scale preparative work, where moderate losses are not so important.

11.1. Use of Acetone

Selective extraction of relatively nonpolar lipids by acetone is easy, although the extract contains all the tissue water and significant amounts of other lipids. In a typical procedure (Stein and

Smith, 1982), the tissue was homogenized with 10 vol of cold acetone, filtered, and air-dried. The powder was then extracted with 5 vol of cold acetone and the two filtrates were pooled. Water can be removed by codistillation with toluene or hexane or by partitioning with hexane after much of the acetone has been removed by rotoevaporation. The polar lipids in the tissue residue can be extracted relatively effectively with only a small volume of polar solvent.

A version of this procedure was described by Ohashi and Yamakawa (1977), who extracted the dry residue with hot EtOH. This extract was used as a source of gangliosides. The tissue was first soaked in 3 vol of acetone for 3 d, then heated with acetone in a Soxhlet extractor. The residue was then extracted at 80°C for 30 min with 5 vol of 90% EtOH. A second extraction was done with 2 vol of 90% EtOH. Glycolipids precipitated on storage at –20°C.

11.2. Extraction of Dispersed, Dry Tissues with Sequential Solvents

Marmer and Maxwell (1981) have described a modification of their "dry column" extraction (*see* section 10), in which most of the low-polarity lipids are extracted before the polar solvents.

11.3. Extraction with Isopropyl Ether

Cham and Knowle (1976) noted that some serum or plasma nonpolar lipids could be specifically extracted with diisopropyl ether. After 24 h of gentle mixing, the plasma had lost all its free fatty acids and triglycerides and 94% of its cholesterol, but only 4% of its phospholipids. There is some danger of enzymatic changes during this leisurely extraction.

11.4. Selective Extraction with Methanol

Ramesh et al. (1979) have applied solvent specificity to the problem of extracting egg phospholipids without extracting the nonpolar lipids. They extracted 44 g of egg yolk with 150 mL of MeOH, which is a poor solvent for triglycerides and cholesterol esters. The insoluble residue was separated from the extract by decantation and filtration through a cotton-plugged funnel, then reextracted four more times with MeOH. The extracts were pooled and rotoevaporated to about 50 mL, yielding two layers. The lower one contained the phospholipids and cholesterol, and the upper one contained most of the water and nonlipids. Probably the method could be made more efficient by immobilizing the lipids in a large amount of Celite before extracting with MeOH.

Methanol was used at 63°C to extract gangliosides from brain (Folch-Pi and Lees, 1959). Nonlipids were removed from the cooled extract (at 4°C) by collecting the precipitated material, dissolving it in C:M, partitioning with water, and dialyzing the upper layer against tap water.

11.5. Selective Extraction of Fatty Acids

A popular method for extracting free fatty acids (and some other lipids) is the use of heptane:IP with plasma acidified with sulfuric acid (Dole and Meinertz, 1960). This is similar to the hexane:IP solvent described in section 8. The method was modified for use with tissue particles by Ko and Royer (1967).

A variant of the Dole method, particularly suited to extracting the short-chain free fatty acids, is the use of toluene:EtOH (2:1) (Cohen et al., 1969). Presumably one needs this more polar solvent for the more polar fatty acids.

Another variant is the use of heptane:chloroform (3:7), with the sample buffered to pH 6.2 (Elphick, 1975). At this pH, fatty acids are largely undissociated, and apparently it is not necessary to use strong (denaturing) acid to release them from protein. The method extracted only 92% of the free fatty acids from serum, however. The samples may not have been shaken with solvent long enough; for such an immiscible solvent, the need for mechanical dispersion is strong. The chloroform, in this case, has to be free of EtOH, apparently to reduce the extraction of more polar lipids.

Free fatty acid can be extracted from serum or plasma by gentle mixing with diisopropyl ether for 30 min (Cham and Knowles, 1976). The extract contained no phospholipid, but 16% of the cholesterol and 40% of the triglyceride came out of the sample. It is interesting that the free fatty acids must enter the organic phase as a salt (in part, at least), since the aqueous layer was not acidified. Since Na^+ salts are rather insoluble in ethers, it appears likely that the fatty acids enter as the salt of a divalent metal cation or an organic base.

Free fatty acids can be extracted from an initial H:IP extract into an aqueous phase by using a quaternary amine (Kashyap et al., 1980). The acids were first extracted from the sample in a H:IP:W (5:4:2.3) system containing sulfuric acid. The hexane layer was washed with water three times to remove 2-PrOH, then with 20 μL of trimethyl(trifluorotolyl)ammonium hydroxide solution. The amine extracted all the fatty acids and simultaneously concentrated them into a very small volume. This sequence of ionic manipulations produced very pure fatty acids. It is similar to the ion-pairing partitioning of sulfatide (*see* section 6.6).

Eicosanoids, which resemble fatty acids, can be extracted by similar techniques. Rouzer et al. (1982) diluted their samples with an equal volume of absolute EtOH, adjusted the pH to 3.0 with formic acid, and extracted the compounds with chloroform. As shown by Unger et al. (1971), the formic acid acts to keep proteins in solution in the EtOH:W mixture, thus preventing loss of prostaglandins.

11.6. Extraction by Specifically Binding Column Packings

Polyphosphoinositides are rather specifically bound by neomycin, an aminoglycoside antibiotic, which can be bound to activated porous glass beads (Schacht, 1978). Although the column packing is an anion exchanger, the conditions of use reduce nonspecific binding by other anionic lipids. Elution of PI-phosphate is accomplished with ammonium acetate in C:M:W. PI-bisphosphate comes out with ammonium hydroxide in C:M:W.

Other types of specific extraction ought to be feasible. One could bind a lipid-binding protein covalently to a porous solid support, then pass plasma or tissue cytosol or emulsified lipids through a column packed with the bound protein. Proteins of this sort include bovine serum albumin depleted of fatty acids, fatty acid-binding protein, lipid transfer proteins, some hydrolase activator proteins, and antibodies that bind to lipids (particularly glycolipids). Apovitellenin I, an egg yolk protein, might find use this way for taking up lecithin (Fretheim et al., 1986).

One would have to devise gentle elution solvents to recover the lipids if the columns are to be reused.

11.7. Extraction of Alkali-Stable Lipids after Hydrolysis of Ester Lipids

Serum gangliosides have been extracted by a procedure that takes advantage of their stability to alkali (Stathaki and Levis, 1981). The published method, slightly modified (Gabriel M. Levis, personal communication), goes as follows: 15 mL of 1M KOH are added to 10 mL of serum. The mixture is stirred 3 h to cleave esters, and 15 mL of acetonitrile are added with mixing. The pH of the mixture is adjusted to 5 with 2M HCl and centrifuged for 20 min at 5000g. The pellet is washed with 40 mL of acetonitrile:W (1:1) by centrifugation, and the pellet lipids are extracted with 2 × 5 mL of C:M (1:1), using vortexing and centrifuging. The pooled extracts are evaporated to dryness, the residue is taken up in 1 mL of C:M (1:1), and the insoluble material is removed by centrifugation; the solvent from the solution is then evaporated. This residue is now taken up in 5 mL of M:W (4:1) and extracted with 3 mL of hexane. The resultant M:W solution is now suitable for TLC.

The advantage of this method is that the volume of solvent is relatively small because the gangliosides are concentrated in a proteinaceous pellet. It may, however, be simpler to just work with larger volumes of solvent and use column chromatography to prepurify the gangliosides.

11.8. Extraction of Alkali-Stable Lipids after Tissue Saponification

This approach (Keller and Adair, 1977) was used to liberate tissue-bound dolichol. A 1-g sample is heated in boiling water for 60 min with 0.5 mL of 60% KOH and 1 mL of 0.25% pyrogallic acid in MeOH. The pyrogallic acid is added as an antioxidant. This treatment dissolves the tissue, and the dolichol can be extracted with ethyl ether. One might expect the potassium soaps in the sample to produce an emulsion when ether is added, but the MeOH prevents that. A revised saponification procedure for use with total dolichol and dolichol phosphate has recently been offered by Rip and Carroll (1987).

11.9. Extraction of Gangliosides and Nonionic Glycolipids with Buffered Tetrahydrofuran

Trams and Lauter (1962) described a sequence of steps starting with brain. The tissue (100 g) was homogenized with 200 mL of 10 mM K phosphate, pH 6.8, and then with 1200 mL of tetrahydrofuran. The pellet obtained from this was reextracted with 300 mL of tetrahydrofuran, and the pooled extracts were back-extracted with 450 mL of ethyl ether. The ether reduced the solubility of water in the extract, forming a water-rich phase that contained the gangliosides. The ether-THF layer was extracted with 100 mL of water to recover additional gangliosides.

Tettamanti et al. (1973) have described a modified version of this procedure. The pooled water-rich layers were found to be relatively free of phospholipids. The aqueous layers were also found to be a good source for isolating the oligohexosylceramides (Slomiany et al., 1976).

11.10. Extraction by Anion Exchange Resins

Ion exchange resins work readily in organic solvents if water is present, the first example being a method for extracting acidic brain lipids from an extract containing neutral lipids (Radin et al., 1956). *Strongly* basic resins in the hydroxyl form were found to react chemically with chloroform.

A weakly basic resin (DEAE-Sephadex) has been used this way in C:M:W (60:30:8) to take up anionic lipids (*see*, for example, Ledeen and Yu, 1982). This approach has the advantage over partitioning systems in that all gangliosides are bound to the resin.

The glass-based resin made from neomycin (Schacht, 1978) has been used to take up all the acidic lipids from a solution of total lipids (minus gangliosides) in which the acids were in the free acid, not salt, form (Palmer, 1981). The acidic lipids could be eluted differentially.

11.11. Extraction of Fatty Acyl Coenzyme A

This lipid family is lost to the aqueous phase when C:M extracts are partitioned with water. Mancha et al. (1975) used a sequence of precipitation and adsorptive extraction. Incubation samples were first acidified with acetic acid after adding bovine serum albumin. Free fatty acids were removed with petroleum ether that had been saturated with IP:W (1:1). The sample was then treated with ammonium sulfate and C:M (1:2), which precipitated the salt and protein. The acyl CoAs in solution were extracted with neutral alumina.

An alternative procedure (Polokoff and Bell, 1975) used coprecipitation with bovine serum albumin to pull out the acyl CoAs. An incubation sample (0.2 mL) was treated with 20 μg of BSA and 2 mL of 0.3M trichloroacetic acid, then the precipitate was taken out of the mixture by filtration through a Millipore filter. It would be interesting to see which other lipids can be pulled out this way.

12. Solid Phase Extraction

This is used for samples that are largely aqueous (serum, spinal fluid, tissue cytosol, enzyme incubation solutions), with the lipid component in micellar or monomeric form or bound to water-soluble protein. One approach uses a column containing a hydrophilic, inert powder having a high surface area, such as diatomaceous earth (kieselguhr). Another approach uses a column containing a lipophilic powder or beads to remove the lipids. A third approach uses an adsorbent, usually charcoal, which is simply stirred with the sample.

12.1. Uptake from Physically Absorbed Aqueous Samples

In the first technique, typically with E. Merck Extrelut QE columns, the sample is poured into the column and is entirely absorbed. The volume of sample is limited to the amount that can be absorbed easily (0.3–50 mL, depending on column size). The absorbing powder has such a large surface area that the sample is spread out into a thin film. When an organic, water-immiscible solvent is passed through the column, most of the lipids are readily extracted. The recommended volume of eluting solvent for a 10-mL capacity column is 24 mL. The eluting solvent should not contain more than 10% of a water-miscible solvent. Thus the method is limited to relatively nonpolar lipids. In effect, the powder furnishes the equivalent of the mechanical dispersing action of a homogenizer. The method closely resembles the methods in which a tissue is ground with a water-absorbing substance, although the water is not bound in a crytal hydrate (*see* section 10). The Merck columns include an uncharacterized "cleanup area" at the bottom of the column.

12.2. Reverse Phase Uptake of Lipids

In the second technique, the commercially available, pre-packed column is typically much smaller. Various packings are available: silica gel itself and silica gel linked to a dimethylsilyl group that is itself linked to an organic substituent-methyl, octyl, decyl, octadecyl ("ODS"), cyanopropyl, aminopropyl, phenyl, a diol, or ion exchange group. The manufacturers often do not disclose the full chemical names of their packings. The larger the substituent, the more strongly it binds low-polarity lipids. Other selectivity factors come into play too, such as the affinity of nitrile groups for double bonds. The amino-substituted packings are also ion exchangers, when water is present, binding acidic lipids. An unfortunate characteristic of the aminopropyl columns under nonaqueous conditions is their instability, perhaps through air oxidation. They seem to be useful only once. The packings are supplied with a porous polypropylene cover to keep the upper surface flat. The columns are small because the particles are expensive, so only small samples can be extracted. Supelco (Bellefonte, PA) provides "Supelclean" columns containing 1- or 3-mL packings.

The ODS columns have found use in extracting gangliosides (Kundu and Suzuki, 1981; Williams and McCluer, 1980). They have also been used by Figlewicz et al. (1985) to separate radioactive water-soluble precursors from enzymatically formed lipids. In the latter procedure, 1-mL columns were first washed with 5-mL portions of C:M (2:1), water, C:M (2:1), and FUL containing nonradioactive precursor. The incubation sample was mixed with 5 mL of FUL containing 0.1M KCl and passed through the column, which absorbed the lipids. Further elution with this solvent and water took out residual nonlipid; then the lipids were eluted with 7 mL of C:M (2:1).

In the case of gangliosides, the initial sample effluent was passed through the column twice more to ensure complete binding to the packing. After a 25-mL wash with water, the gangliosides were released with 3 mL of MeOH. The KCl is needed to salt the gangliosides out of solution, into the nonaqueous (solid) phase.

Ledeen and Yu (1982) have replaced the KCl by 0.8M NaOAc in M:W (1:1). The reliability of ODS columns for ganglioside extraction has been brought into question, however. I think that other lipids bind to the packing, affecting its affinity for gangliosides. This problem arises even when the ester-linked lipids have first been cleaved by alkaline methanolysis. It may be pointed out that Williams and McCluer (1980), in their description of the Sep-Pak method, did not apply any quantitative tests with natural samples.

A curious observation by Williams and McCluer (1980) is that the use of CaCl$_2$ instead of KCl lowered the amount of ganglioside that adhered to the packing. Perhaps the ability of Ca^{2+} ions to cross-link ganglioside molecules produced such large polymers that access to ODS binding sites in the silica gel pores was restricted.

It is interesting to consider whether the above method can be used for small intact tissue samples. Evidently the lipid that is formed enzymatically is loosely bound, readily extracted by FUL. Since the weight of lipid is small, FUL has sufficient solubilizing power to dissolve it. I think that most of the lipid in intact cells is bound somewhat more tightly, to different proteins, and would require a solvent containing more chloroform.

The custom is to operate these columns very quickly, with minimal elution volumes for the rinsing and elution solvents. Centrifuges, syringes, or special vacuum filtration manifolds are used for the absorption and elution steps. The columns are generally made of polypropylene, and one would therefore expect to find some contamination of the samples by small lipid molecules in the plastic. The duration of contact with solvent is so short, however, that the contamination may not be noticeable.

Although the manufacturers of the columns recommend very high flow rates, one ought to determine the appropriate rate for each particular application. For a lipid that is bound to a protein, allowance should be made for the time needed to effectuate dissociation. This kind of factor may explain the need to run the ganglioside solution through the column three times (see above).

The reverse phase columns are labeled "disposable," but there is no reason to believe that the packings cannot be used many times provided there is not too much trapped residue. Perhaps protein can be washed out with a mixture of dithiothreitol and detergent.

Amberlite XAD-7 has been used to take up leukotrienes from samples extracted with EtOH (Mathews et al., 1981). The extracts are evaporated to dryness and taken up in water. The Amberlite, first washed extensively with solvents to remove impurities, is packed into a column with water. The sample is added and the column is rinsed with 0.19 mM NH_4Cl; the leukotrienes are then eluted with EtOH:W (4:1).

Lipidex 1000, a porous packing resembling Sephadex LH-20, has been used similarly (Dahlberg et al., 1980). In this case, however, the beads were saturated with aqueous buffer. By this means, lipids could be taken up from cytosol and solutions of protein/lipid complexes. The material was used this way by Glatz et al. (1984). Evidently the resin has appreciable hydrophilicity too, since the delipidated proteins retained their original properties.

12.3. Uptake of Lipids by Adsorbents

The third type of solid phase extraction, using an adsorbent, has found use in the separation of free lipid from protein-bound lipid. This approach has been used in the determination of lipid-binding proteins, usually those that bind steroids, and in the immunoassay of lipids. Activated charcoal, pretreated with dextran solution, is used to absorb the free lipid molecules. Untreated charcoal will adsorb both free *and* bound lipid, a characteristic that might find use in extracting large volumes of sample while permitting elution with a small volume of ammonia-containing solvent.

13. Special Problem of Labile Lipids

Free fatty acids and diglycerides appear rapidly as soon as an animal is killed. Quick cooling is necessary to halt the enzymatic reactions. One approach is to decapitate the animal and freeze the intact head (or dissected tissue) in liquid nitrogen or in pentane cooled by liquid nitrogen. Direct immersion in liquid nitrogen is slower since the bubbles of nitrogen that form around the head act as insulators and slow the heat transfer. Removal of the brain from the skull is awkward with such cold material, but it can be done.

Slightly slower freezing can be obtained with pentane cooled by dry ice. Since dry ice may be contaminated by grease, it should be kept away from the pentane, in an outer beaker.

Another freezing method involves simple contact of exposed nerve tissue with two precooled blocks of metal. Ponten et al. (1973) recommended that intact animals be anesthetized, maintained on a respirator, and cooled by pouring liquid nitrogen directly onto the exposed skull. Abe and Kogure (1986) used the latter method for diglyceride determination after first anesthetizing gerbils with nitrous oxide: oxygen: halothane (70:30:1). Some workers have opened the anesthetized animal's skull before pouring on the liquid nitrogen. It would appear to be better to pour on pentane that has been precooled with the liquid nitrogen, as pointed out above. It would also seem advisable to use nonexciting anesthesia techniques.

With all these methods there is a problem in measuring the weight of the tissue prior to extraction, since cold materials pick up atmospheric water by condensation. This problem is probably best handled by using a rapid-weighing, electronic balance, but working inside a plastic glove box might be necessary.

Kramer and Hulan (1978) recommended pulverizing the frozen tissue in a precooled steel mortar, then dropping the powder into C:M (2:1) (held at room temperature) and dispersing it with a homogenizer. This yielded lower contents of free fatty acids and diglycerides than the use of unfrozen tissue (normal homogenization) or the use of frozen tissue that was cut into small pieces before homogenizing in cold C:M.

Several studies have found hydrolytic breakdown of some lipids even in tissues stored at –20°C, and it looks as though a –75°C freezer is a worthwhile investment. In the case of yeast cells, Hanson and Lester (1980) recommended washing them with 5% trichloroacetic acid at 0°C to inactivate the phospholipases. This may be a useful technique for neural tissue, but there is a danger of plasmalogen hydrolysis and excessive nonlipid contamination.

Several papers have described the use of special high-power microwave ovens for the rapid inactivation of destructive enzymes, but the technique has not been used for lipids to my knowledge.

14. Pigment Problems

Red cells and spleen yield much pigment when extracted with C:M systems. As mentioned in section 8.1, H:IP does not. Rose and Oklander (1965) have recommended C:IP (7:11). The red cells are hemolyzed, and 2-PrOH is added slowly, with mixing. After 1 h the chloroform is added and mixed. Although nerve tissue contains very little pigment, this solvent might have merit because it probably dissolves less proteolipid protein than C:M.

The presence of reddish pigment in a lipid extract suggests that an iron porphyrin complex is present, which predisposes to peroxidation by air.

15. Extraction with Simultaneous Dehydration

15.1. Azeotropic Distillation with 1,2-Dichloroethane

This method has been used on an industrial scale for many years to prepare wheatgerm oil and delipidated tissues, such as lung, wheatgerm, and whole fish (Levin and Finn, 1955). It consists of homogenizing tissues with dichloroethane and distilling off the tissue water and part of the solvent at atmospheric pressure. The water is efficiently removed as an azeotrope at a relatively low temperature. Since it is immiscible with the halocarbon, the solvent is readily recovered and recycled. The solvent remaining with the dry tissue, together with most of the lipid, is removed by filtration.

The extent of lipid extraction does not seem to have been studied with modern methods, but one may assume that the very polar lipids remain with the tissue. In the case of fish powder, which was washed with MeOH to remove odors, a lipid content of 0.5% was found by Soxhlet extraction with C:M.

This rather mild method may have merit for many extraction problems. The extract has a relatively small volume, since the water is removed by distillation instead of solubilization, and the solvent is rather easy to remove.

15.2. Extraction during Chemical Removal of Water

When the goal of the extraction is to analyze the fatty acids of the sample by gas-liquid chromatography, the following method could be used (Shimasaki et al., 1977): 10 mg of tissue was mixed with 1 mL of 2,2-dimethoxypropane containing 10 μL of conc. HCl. The sample tube was flushed with nitrogen and shaken gently for 20 min to allow the water to react with the ketal (forming MeOH + acetone). The solvents were removed by a stream of nitrogen, and the residue was heated in a closed tube, under nitrogen, with methanolic HCl to carry out the methanolysis.

It would be of interest to see if simple filtration, after the reaction with water, would yield a useful lipid extract. Dimethoxypropane plus MeOH and acetone is probably an efficient extractant. Evaporation of the filtrate—which is now water-free—would be very rapid. The small amount of HCl in the mixture (which catalyzes the reaction with water) should help to liberate the acidic lipids, possibly without catalyzing the cleavage of plasmalogens. It would probably be helpful to use less solvent for such a small amount of tissue, thereby raising the proportion of MeOH and acetone in the final mixture.

An even simpler version of this method was devised by Tserng et al. (1981), who used only dimethoxypropane and HCl. Plasma (50 μL), internal standard (pentadecanoic acid) in MeOH (25 μL), dimethoxypropane (1 mL), and conc. HCl (20 μL) were mixed and left 15 min. This converted the free fatty acids to methyl esters, which could be extracted into isooctane. Yet another use of dimethoxypropane in this way was described by Browse et al. (1986). The writer of this chapter is pleased to see that his method of esterifying fatty acids described nearly 30 years ago, has found use.

A variant of this procedure was recently described by Lepage and Roy (1986) for the combined lipid extraction and methylation of the total fatty acids in small samples. Plasma or bile (0.1 mL) or homogenized liver (0.1 g) was mixed with 2 mL of benzene : MeOH (1:4) containing internal standard (tridecanoic acid). While the mixture was being stirred, 0.2 mL of acetyl chloride was added during a 1-min period. (This forms methyl acetate, acetic acid, and dry HCl.) The tubes were closed and heated at 100°C for 60 min to methanolyze cholesterol esters and phospholipid esters. The lipid amides, checked only with sphingomyelin, were largely cleaved after 4 h of heating.

This method gave higher fatty acid values than the use of C:M extraction followed by methanolysis. It would be interesting to see the effect of the benzene : MeOH : acetyl chloride method *without* heating. It might be a very efficient extraction mixture (with toluene instead of benzene).

15.3. Dehydration by Silica Gel on a TLC Plate

This approach solves the problem, to a certain extent, of removing nonlipids without losing lipids. The method has been used for very small samples (Kaschnitz et al., 1968). A microhematocrit tube is drawn down at its closed end to a fine tip, and a 3–5 mg sample of tissue is inserted into the open end. About 70 μL of C:M (2:1) are added, and the mixture is homogenized with a loop of thin wire. The suspension is stored 30 min to complete the extraction, the tip at the bottom is broken, and the solvent is allowed to form a thin spot or streak on a TLC plate. This is then analyzed by the usual method. Since the hematocrit tube can be graduated, or compared to a ruler, one can apply approximately any desired volume of extract.

16. Miscellaneous Solvent Methods

16.1. Benzene + Methanol (1:1)

This mixture (Sobus and Holmlund, 1976) was recommended for yeast, an organism noted for its tough cell wall. Thus it may have little application to neural tissue. At any rate, the benzene should be replaced by toluene.

16.2. Phenol + Water + Chloroform + Petroleum Ether

This was used for the extraction of bacterial glycolipids, while avoiding the extraction of protein (Galanos, 1969). Phenol, which finds use in nucleic acid extractions, seems to have some kind of specificity, but produces serious blisters on skin contact.

16.3. Thirty-one Miscellaneous Solvents

Autilio and Norton (1963) tested a large assortment of solvents with lyophilized brain white matter. Some of the solvents were quite high-boiling and had to be removed by molecular distillation (high vacuum, short path evaporation). The paper is of interest for its descriptions of solvent purification methods and extraction yields for total lipids, nonlipid solutes, cholesterol, phospholipids (total phosphorus), and lipid galactose. Warm tetrahydrofuran was tentatively recommended for total lipid extraction, with exclusion of proteolipid protein.

16.4. Ethyl Acetate and Acetone

This mixture was proposed particularly for cultured cells (Slayback et al., 1977). In method II, the cells were detached from the culture dish with trypsin, and 2.5 mL of the suspension were transferred to a centrifuge tube containing 2.5 mL of acetone. The tube was capped, heated at 85–90°C until the contents came to a boil, then vortexed while hot for 45 s. Ethyl acetate (5 mL) was added, and the hot extraction was repeated. Now the mixture was heated for 20 min at 65°C, vortexed, and heated another 20 min. The tube was cooled and centrifuged to separate the phases. This procedure (and variants of it) were found to give somewhat higher yields than the use of hot C:M (2:1), followed by partitioning with water.

The authors noted that the partitioning of the C:M extract with water caused loss of lipid. This was probably because of the failure to add ionic material. In the ordinary use of C:M, the ratio of tissue to solvent is much higher, so that ample salts are present to prevent loss of lipid into the upper layer.

16.5. Ethyl Acetate

This solvent, which resembles chloroform in its properties, but is less dense than water, has been used to extract a dihydroxy fatty acid formed enzymatically from arachidonic acid (Wong, 1984). Proteins were precipitated by adding 3 vol of EtOH, and the filtrate was evaporated to dryness. The residue was taken up in water and adjusted with HCl to pH 3.5, then extracted with 10 vol of ethyl acetate.

16.6. A Low-Volatility Solvent

If one wants to analyze multiple portions of a lipid extract, it is desirable to use an efficient solvent with a relatively high boiling point. I would suggest trials with isooctane:chloroform:isoamyl alcohol:2-PrOH (3:1:1:1). This was used in a partitioning system (Radin, 1984), but not tested with tissues.

Acknowledgments

The author is holder of a Jacob Javits Neuroscience Investigator Award and has been supported by grants from the National Institutes of Health, NS-03192 and HD-07406.

References

Abe K. and Kogure R. (1986) Accurate evaluation of 1,2-diacylglycerol in gerbil forebrain using HPLC and in situ freezing technique. *J. Neurochem.* **47**, 577–582.

Allen P. A. (1972) New extraction method for nematode lipids. *Anal. Biochem.* **45**, 253–259.

Autilio L. A. and Norton W. T. (1963) Non-aqueous solvent extracts of lyophilized bovine brain white matter. *J. Neurochem.* **10**, 733–738.

Bligh E. G. and Dyer W. J. (1959) A rapid method of total lipid extraction and purification. *Can. J.Biochem. Physiol.* **37**, 911–917.

Browse J., McCourt P. J., and Somerville C. R. (1986) Fatty acid composition of leaf lipids determined after combined digestion and fatty acid methyl ester formation from fresh tissue. *Anal. Biochem.* **152**, 141–145.

Burdick & Jackson (1984) *High Purity Solvent Guide* American Scientific Products, McGaw Park, Illinois.

Byrne M. C., Sbaschnig-Agler M., Aquino D. A., Sclafani J. R., and Ledeen R. W. (1985) Procedure for isolation of gangliosides in high yield and purity: Simultaneous isolation of neutral glycosphingolipids. *Anal. Biochem.* **148**, 163–173.

Carter T. P. and Kanfer J. N. (1975) Removal of Water-Soluble Substances from Ganglioside Preparations, in *Methods in Enzymology* vol. 35, *Lipids* part B (Lowenstein J. M., ed) Academic, New York.

Cham B. E and Knowles B. R. (1976) A solvent system for delipidation of plasma or serum without protein precipitation. *J. Lipid Res.* **17**, 176–181.

Chen R. F. (1967) Removal of fatty acids from serum albumin by charcoal treatment. *J. Biol. Chem.* **242**, 173–181.

Cohen M., Morgan R. G. H., and Hofmann A. F. (1969) One-step quantitative extraction of medium-chain and long-chain fatty acids from aqueous samples. *J. Lipid Res.* **10**, 614–616.

Daae L. N. W. and Bremer J. (1970) The acylation of glycerophosphate in rat liver, a new assay procedure for glycerophosphate acylation, studies on its subcellular and submitochondrial localization and determination of the reaction products. *Biochim. Biophys. Acta* **210**, 92–104.

Dacremont G., Kint J. A., Carton D., and Cocquyt G. (1974) Glucosylceramide in plasma of patients with Niemann-Pick disease. *Clin. Chim. Acta* **52**, 365–367.

Dahlberg E., Snochowski M., and Gustafsson J. (1980) Removal of hydrophobic compounds from biological fluids by a simple method. *Anal. Biochem.* **106**, 380–388.

Das A. K. and Hajra A. K. (1984) Estimation of acyldihydroxyacetone phosphate and lysophosphatidate in animal tissues. *Biochim. Biophys. Acta* **796**, 178–189.

Dole V. P. and Meinertz H. (1960) Microdetermination of long chain fatty acids in plasma and tissues. *J. Biol. Chem.* **235**, 2595–2599.

Elphick M. C. (1975) Automated modification of Duncombe's method for the ultramicro determination of serum free fatty acids. *J. Lipid Res.* **16**, 402–406.

Figlewicz D. A., Nolan C. E., Singh I. N., and Jungalwala F. B. (1985) Pre-packed reverse phase columns for isolation of complex lipids from radioactive precursors. *J. Lipid Res.* **26**, 140–144.

Folch J., Ascoli I., Lees M., Meath J. A., and LeBaron F. N. (1951) Preparation of lipide extracts from brain tissue. *J. Biol. Chem.* **191**, 833–841.

Folch J., Lees M., and Sloane-Stanley G. H. (1957) A simple method for the isolation and purification of total lipides from animal tissues. *J. Biol. Chem.* **226**, 497–509.

Folch-Pi J. and Lees M. (1959) Studies on the brain ganglioside strandin in normal brain and in Tay-Sachs' disease. *AMA J. Dis. Children* **97**, 730–738.

Fretheim K., Sleigh R. W., and Burley R. W. (1986) Formation of complexes between lecithin and apovitellenin I, an avian egg-yolk apoprotein. *Lipids* **21**, 127–131.

Furbish F. S., Blair H. E., Shiloach J., Pentchev P. G., and Brady R. O. (1977) Enzyme replacement therapy in Gaucher's disease: Large-scale purification of glucocerebrosidase suitable for human administration. *Proc. Natl. Acad. Sci. USA* **74**, 3560–3563.

Galanos C., Luderitz, and Westphal O. (1969) A new method for the extraction of R lipopolysaccharides. *Eur. J. Biochem.* **9**, 245–249.

Gatt S. (1965) Partitioning of gangliosides into a chloroform-rich phase. *J. Neurochem.* **12**, 311–321.

Ghidoni R., Sonnino S., and Tettamanti G. (1978) Behavior of gangliosides on dialysis. *Lipids* **13**, 821–822.

Glatz J. F. C., Baerwaldt C. C. F., Veerkamp J. H., and Kempen H. J. M. (1984) Diurnal variation of cytosolic fatty acid-binding protein content and of palmitate oxidation in rat liver and heart. *J. Biol. Chem.* **259**, 4295–4300.

Goodman D. S. (1957) Preparation of human serum albumin free of long-chain fatty acids. *Science* **125**, 1296–1297.

Hajra A. K. (1974) On extraction of acyl and alkyl dihydroxyacetone phosphate from incubation mixtures. *Lipids* **9**, 502–505.

Hanson B. A. and Lester R. L. (1980) The extraction of inositol-containing phospholipids and phosphatidylcholine from *Saccharomyces cerevisiae* and *Neurospora crassa*. *J. Lipid Res.* **21**, 309–315.

Hara A. and Radin N. S. (1978) Lipid extraction of tissues with a low-toxicity solvent. *Anal. Biochem.* **90**, 420–426.

Hauser G. and Eichberg J. (1973) Improved conditions for the preservation and extraction of polyphosphoinositides. *Biochim. Biophys. Acta* **326**, 201–209.

Jaxxon N. E. D. (1986) Letter to editor. *Aldrichimica Acta* **19**, 2.

Kanfer J. N. and Spielvogel C. (1973) On the loss of gangliosides by dialysis. *J. Neurochem.* **20**, 1483–1485.

Kaschnitz R., Peterlik M., and Weiss H. (1969) A micro method for extracting tissue lipids. *Anal. Biochem.* **30**, 146–148.

Kashyap M. L., Mellies M. J., Brady D., Hynd B. A., and Robinson K. (1980) A micromethod using gas-liquid chromatography for measuring individual fatty acids liberated during interaction of triglyceride-rich lipoproteins and lipoprotein lipase. *Anal. Biochem.* **107**, 432–435.

Katzman R. and Wilson C. E. (1961) Extraction of lipid and lipid cation from frozen brain tissue. *J. Neurochem.* **7**, 113–127.

Kawamura N. and Taketomi T. (1977) A new procedure for the isolation of brain gangliosides, and determination of their long chain base compositions. *J. Biochem.* **81**, 1217–1225.

Kean E. L. (1968) Rapid, sensitive spectrophotometric method for quantitative determination of sulfatides. *J. Lipid Res.* **9**, 319–327.

Keller R. K. and Adair Jr. W. L. (1977) Microdetermination of dolichol in tissues. *Biochim. Biophys. Acta* **489**, 330–336.

Ko H. and Royer M. E. (1967) A submicromolar assay for nonpolar acids in plasma and depot fat. *Anal. Biochem.* **20**, 205–214.

Kolarovic L. and Fournier N. C. (1986) A comparison of extraction methods for the isolation of phospholipids from biological sources. *Anal. Biochem.* **156**, 244–250.

Kramer J. K. G. and Hulan H. W. (1978) A comparison of procedures to determine free fatty acids in rat heart. *J. Lipid Res.* **19**, 103–106.

Kundu S. K. and Suzuki A. (1981) Simple micro-method for the isolation of gangliosides by reversed-phase chromatography. *J. Chromatogr.* **224**, 249–256.

LeBaron F. N. (1963) The nature of the linkage between phosphoinositides and proteins in brain. *Biochim. Biophys. Acta* **70**, 658–669.

Ledeen R. W. and Yu R. K. (1982) Gangliosides: Structure, Isolation, and Analysis, in *Methods in Enzymology* vol. 83, *Complex Carbohydrates* part D (Ginsburg V., ed.) Academic, New York.

Lees M. B. (1966) Influence of sucrose on the extraction of proteolipids from brain and other tissues. *J. Neurochem.* **13**, 1407–1420.

LePage G. and Roy C. C. (1986) Direct transesterification of all classes of lipids in a one-step reaction. *J. Lipid Res.* **27**, 114–120.

Levin E. and Finn R. K. (1955) A process for dehydrating and defatting tissues at low temperature. *Chem. Eng. Progr.* **51**, 223–236.

Mancha M., Stokes G. B., and Stumpf P. K. (1975) The determination of acyl-acyl carrier protein and acyl coenzyme A in a complex lipid mixture. *Anal. Biochem.* **68**, 600–608.

Marmer W. N. and Maxwell R. J. (1981) Dry column method for the quantitative extraction and simultaneous class separation of lipids from muscle tissue. *Lipids* **16**, 365–371.

Mathews W. R., Rokach J., and Murphy R. C. (1981) Analysis of leukotrienes by high-pressure liquid chromatography. *Anal. Biochem.* **118**, 96–101.

Maxwell M. A. B. and Williams J. P. (1967) The purification of lipid extracts using Sephadex LH-20. *J. Chromatogr.* **31**, 62–68.

Melton S. L., Moyers R. E., and Playford C. G. (1979) Lipids extracted from soy products by different procedures. *J. Am. Oil Chem. Soc.* **56**, 489–493.

Meslar H. W. and White III H. B. (1978) Preparation of lipid-free protein extracts of egg yolk. *Anal. Biochem.* **91**, 75–81

Michell R. H., Hawthorne J. N., Coleman R., and Karnowsky M. L. (1970)

Extraction of polyphosphoinositides with neutral and acidified solvents. *Biochim. Biophys. Acta* 210, 86–91.

Mock T., Pelletier M. P. J., Man R. Y. K., and Choy P. C. (1984) Alterations in lysophosphatidylcholine levels of canine heart: Modes of extraction and storage. *Anal. Biochem.* **137**, 277–281.

Mogelson S., Wilson, Jr. G. E., and Sobel, B. E. (1980) Characterization of rabbit myocardial phospholipids with ^{31}P nuclear magnetic resonance. *Biochim. Biophys. Acta* **619**, 680–688.

Nakano N. I., Shimamori Y., and Nakano M. (1983) Activated carbon beads for the removal of highly albumin-bound species. *Anal. Biochem.*, **129**, 64–71.

Nelson G. J. (1975) Isolation and Purification of Lipids from Animal Tissues, in *Analysis of Lipids and Lipoproteins* (Perkins E. G., ed.) American Oil Chemists' Society, Champaign, Illinois.

Ohashi M. and Yamakawa T. (1977) Isolation and characterization of glycosphingolipids in pig adipose tissue. *J. Biochem.* **81**, 1675–1690.

Ostrander J. and Dugan Jr. L. R. (1962) Some differences in composition of recovered fat, intermuscular fat, and intramuscular fat of meat animals. *J. Am. Oil Chem. Soc.* **39**, 178–181.

Palmer F. B. S-.C. (1971) The extraction of acidic phospholipids in organic solvent mixtures containing water. *Biochim. Biophys. Acta* **231**, 134–144.

Palmer F. B. S-.C. (1981) Chromatography of acidic phospholipids on immobilized neomycin. *J. Lipid Res.* **22**, 1296–1300.

Phillips F. and Privett O. S. (1979) A simplified procedure for the quantitative extraction of lipids from brain tissue. *Lipids* **14**, 590–595.

Polokoff M. A. and Bell R. M. (1975) Millipore filter assay for long-chain fatty acid:CoASH ligase activity using ^{3}H-labeled coenzyme A. *J. Lipid Res.* **16**, 397–402.

Ponten U., Ratcheson R. A., Salford L. G., and Siesjö B. K. (1973) Optimal freezing conditions for cerebral metabolites in rats. *J. Neurochem.* **21**, 1127–1138.

Prescott S. M. (1984) The effect of eicosapentaenoic acid on leukotriene B production by human neutrophils. *J. Biol. Chem.* **259**, 7615–7621.

Quarles R. and Folch J. (1965) Some effects of physiological cations on the behavior of gangliosides in a chloroform-methanol-water biphasic system. *J. Neurochem.* **12**, 543–553.

Radin N. S. (1981) Extraction of Tissue Lipids with a Solvent of Low Toxicity, in *Methods in Enzymology* vol. 72, *Lipids* part D (Lowenstein J. M., ed.) Academic, New York.

Radin N. S. (1984) Improved version of the Kean partition assay for cerebroside sulfate. *J. Lipid Res.* **25**, 651–652.

Radin N. S., Brown J. R., and Lavin F. B. (1956) The preparative isolation of cerebrosides. *J. Biol. Chem.* **219**, 977–983.

Ramesh B., Adkar S. S., Prabhudesai A. V., and Viswanathan C. V. (1979) Selective extraction of phospholipids from egg yolk. *J. Am. Oil Chem. Soc.* **56**, 585–586.

Rip J. W. and Carroll K. K. (1987) Extraction and quantitation of dolichol and dolichyl phosphate in soybean embryo tissue. *Anal. Biochem.* **160**, 350–355.

Rose H. G. and Oklander M. (1965) Improved procedure for the extraction of lipids from human erythrocytes. *J. Lipid Res.* **6**, 428–431.

Rouzer C. A., Scott W. A., Hamill A. L., Liu F-T., Katz D. H., and Cohn Z. A. (1982) Secretion of leukotriene C and other arachidonic acid metabolites by macrophages challenged with immunoglobulin E immune complexes. *J. Exp. Med.* **156**, 1077–1086.

Saunders R. D. and Horrocks L. A. (1984) Simultaneous extraction and preparation for high-performance liquid chromatography of prostaglandins and phospholipids. *Anal. Biochem.* **143**, 71–75.

Schacht J. (1978) Purification of polyphosphoinositides by chromatography on immobilized neomycin. *J. Lipid Res.* **19**, 1063–1067.

Scheider W. and Fuller J. K. (1970) An effective method for defatting albumin using resin columns. *Biochim. Biophys. Acta* **221**, 376–378.

Schenkman J. B. and Cinti D. L. (1978) Preparation of Microsomes with Calcium, in *Methods in Enzymology* vol. 52, *Biomembranes* part C (Fleischer S. and Packer L., eds.) Academic, New York.

Shimasaki H., Phillips F. C., and Privett O. S. (1977) Direct transesterification of lipids in mammalian tissue for fatty acid analysis via dehydration with 2,2-dimethoxypropane. *J. Lipid Res.* **18**, 540–543.

Slayback J. R. B., Cheung L. W. Y., and Geyer R. P. (1977) Quantitative extraction of microgram amounts of lipid from cultured human cells. *Anal. Biochem.* **83**, 372–384.

Slomiany A., Slomiany B. L., and Horowitz M. I. (1976) Blood Group A-Active Glycolipid Variants from Hog Gastric Mucosa, in *Glycolipid Methodology* (Witting L. A., ed.) American Oil Chemists' Society, Champaign, Illinois.

Sobus M. T. and Holmlund C. E. (1976) Extraction of lipids from yeast. *Lipids* **11**, 341–348.

Spence M. W. (1969) Studies on the extractability of brain gangliosides. *Can. J. Biochem.* **47**, 735–742.

Stathaki S. and Levis G. M. (1981) A rapid procedure for the extraction of serum gangliosides. *J. Clin. Chem. Clin. Biochem.* **19**, 847.

Stein J. and Smith G. (1982) Extraction methods. *Techn. Lipid Membr. Biochem.* **B401**, 1–10.

Suzuki K. (1965) The pattern of mammalian brain gangliosides-II. Evaluation of the extraction procedures, post-mortem changes and the effect of formalin preservation. *J. Neurochem.* **12**, 629–638.

Svennerholm L. and Fredman P. (1980) A procedure for the quantitative isolation of brain gangliosides. *Biochim. Biophys. Acta* **617**, 97–109.

Terner C., Szabo E. I., and Smith N. L. (1970) Separation of gangliosides, corticosteroids, and water-soluble non-lipids from lipid extracts by Sephadex columns. *J. Chromatogr.* **47**, 15–19.

Tettamanti G., Bonali F., Marchesini S., and Zambotti V. (1973) A new procedure for the extraction, purification and fractionation of brain gangliosides. *Biochim. Biophys. Acta* **296**, 160–170.

Trams E. G. and Lauter C. J. (1962) On the isolation and characterization of gangliosides. *Biochim. Biophys. Acta* **60**, 350–358.

Tserng K-Y., Kliegman R. M., Miettinen E-L., and Kalhan S. C. (1981) A rapid, simple, and sensitive procedure for the determination of free fatty acids in plasma using glass capillary column gas-liquid chromatography. *J. Lipid Res.* **22**, 852–858.

Unger W. G., Stamford I. F., and Bennett A. (1971) Extraction of prostaglandins from human blood. *Nature* (Lond.) **233**, 336–337.

Van Slyke D. D. and Plazin J. (1965) The preparation of extracts of plasma lipids free from water-soluble contaminants. *Clin. Chim. Acta* **12**, 46–54.

Williams M. A. and McCluer R. H. (1980) The use of Sep-Pak C18 cartridges during the isolation of gangliosides. *J. Neurochem.* **35**, 266–269.

Williams J. P. and Merrilees P. A. (1970) The removal of water and nonlipid contaminants from lipid extracts. *Lipids* **5**, 367–370.

Wong C. G. and Ladisch S. (1983) Retention of gangliosides in serum delipidated by diisopropyl ether-1-butanol extraction. *J. Lipid Res.* **24**, 666–669.

Wong P. Y.-K., Westlund P., Hamberg M., Granstrom E., Chao P. H-W., and Samuelsson B. (1984) ω-Hydroxylation of 12-L-hydroxy-5,8,10,14-eicosatetraenoic acid in human polymorphonuclear leukocytes. *J. Biol. Chem.* **259**, 2683–2686.

Wuthier R. E. (1966) Purification of lipids from nonlipid contaminants on Sephadex bead columns. *J. Lipid Res.* **7**, 558–561.

Yahara S., Kawamura N., Kishimoto Y., Saida T., and Tourtellotte W. W. (1982) A change in the cerebrosides and sulfatides in a demyelinating nervous system. *J. Neurol. Sci.* **54**, 303–315.

Note Added in Proof

ODS silica gel (section 12.2.) has been used for extracting prostaglandins and related substances from aqueous samples. The best extraction efficiency was obtained when the sample, at neutral pH, contained 15% MeOH. Presumably the acidic lipid entered the organic phase as a mixture of Ca^{2+} and Mg^{2+} salts. An interesting feature of this paper was the use of the ODS in an HPLC *precolumn*, which was connected to a six-way valve. After the extraction/uptake step and a wash, the precolumn was connected to an analytical ODS column and eluted with a ternary solvent gradient for separation of individual acids (Powell, 1987).

The isopropyl ether/butanol method (section 9.1.) has been extended to samples in which the proportion of gangliosides in the lipids is very low. The solvent in a total lipid extract, prepared by a high-yield extraction method, was removed by evaporation, and the dried residue was suspended in isopropyl ether:1-BuOH (6:4). The addition of 50 mM NaCl yielded two layers, in which the gangliosides had partitioned primarily into the water-rich layer and the other lipids were primarily in the organic layer. Without the correct salt concentration, phospholipids entered the aqueous layer. This enrichment process greatly expedited analysis of the gangliosides (Ladisch and Gillard, 1985).

Two successive solvents were used to extract highly polar glycolipids (section 11.1.). First the sample was extracted with acetone twice to remove the less-polar lipids, and then the dry residue was extracted with H:IP:water (20:55:25) by stirring overnight. This method is an example of the versatility of H:IP solvent systems (section 8.) (Kannagi et al., 1982).

Kannagi R., Nudelman E., Levery S. B., and Hakomori, S. (1982) A series of human erythrocyte glycosphingolipids reacting to the monoclonal antibody directed to a developmentally regulated antigen, SSEA-1. *J. Biol. Chem.* **257**, 14865–14874.

Ladisch S. and Gillard B. (1985) A solvent partition method for microscale ganglioside purification. *Anal. Biochem.* **146**, 220–231.

Powell W. S. (1987) Precolumn extraction and reversed-phase high-pressure liquid chromatography of prostaglandins and leukotrienes. *Anal. Biochem.* **164**, 117–131.

Preparation and Analysis of Acyl and Alkenyl Groups of Glycerophospholipids from Brain Subcellular Membranes

Grace Y. Sun

1. Introduction

The brain is one of the most complex organs in the mammalian body, since the tissue is comprised of several cell types each with intricate subcellular organization. It is not surprising that the subcellular membranes isolated from brain may have their origin from several cellular sources. Most neurochemical studies, however, are confined to specific brain regions that are less complex in their cellular composition. For example, the cerebral cortex is largely occupied by neurons; consequently, this brain region is more suitable for isolation of nerve-ending particles. On the other hand, the brainstem is suitable for isolation of myelin. In spite of obvious limitations, unique brain subcellular membrane fractions have been isolated in highly purified form for various neurochemical studies. Since the procedures for isolation of brain subcellular fraction have been reviewed elsewhere (Whittaker, 1984; Henn and Henn, 1984), they will not be discussed in this chapter.

Brain membranes are enriched in complex lipids, including phospholipids, glycosphingolipids, and gangliosides, and these lipids are important integral membrane components. The phospholipids are asymmetrically distributed between the two lipid bilayers. Although a large portion of the phospholipids is needed to provide the membrane its structural integrity, there are specific lipid pools that are actively engaged in various membrane functions. Specific phospholipids are also known to play a role in receptor functions and in mediating ion transport activity. Phospholipids that are present in discrete pools are normally undergoing active metabolic turnover (Sun et al., 1979). Although the phospholipids and acyl group composition in brain and subcellular membranes have been well described (*see* review by Sun and

Foudin, 1985), most earlier studies have only provided information on the major phospholipids [such as phosphatidylcholine (PC) and phosphatidylethanolamines (PE)], and very few studies have separately analyzed the PE and ethanolamine plasmalogens (PEpl). Furthermore, because of problems regarding separation of phosphatidyserines (PS), phosphatidylinositols (PI), and phosphatidic acids (PA) on the TLC plate, information on the minor acidic phospholipids is generally lacking.

In view of the fact that distinct acyl group profiles are associated with individual phosphoglycerides and neutral glycerides, several methods are available for separation of these lipids and subsequent analysis of their acyl or alkenyl group composition. The aim of this chapter is to describe the methods that are currently in use by our laboratory. Special emphasis is placed on complete extraction and separation of the phospholipids (both major and minor) and analysis of their acyl and alkenyl group composition among different brain subcellular fractions. Because a review of the phospholipid and acyl group composition of mammalian brain during development and aging has appeared elsewhere (Sun and Foudin, 1985), the results presented in this chapter are confined primarily to those obtained from studies utilizing these procedures.

2. Methods

2.1. Extraction of Lipids from Brain Tissue Homogenates and Subcellular Membranes

Brain tissue is homogenized in 10 vol of 0.32M sucrose containing 50 mM Tris-HCl (pH 7.4) and 1 mM EDTA using a motor-driven homogenizer fitted with a Teflon pestle. The entire homogenization procedure is carried out in ice-cold conditions. For a complete analysis of the phospholipids and their acyl group composition, aliquots should contain at least 2 mg of protein (1 mg for myelin lipids). The membrane suspension is normally diluted to 2 mL with the sucrose-Tris buffer prior to addition of 4 vol of chloroform-methanol [2:1 (v/v)]. The tubes are thoroughly mixed, then centrifuged briefly to allow phase separation. The lower organic phase is removed and transferred to another test tube. The aqueous phase is subjected to a second extraction with 2 vol of chloroform:methanol:12N HCl (4:1:0.013, by vol). The tubes are mixed and centrifuged to allow phase separation. The acidic organic phase is transferred to another tube and neutralized with $16N$ NH$_4$OH (1 drop) before combining with the first organic extract. The combined solvent is filtered through anhydrous Na$_2$SO$_4$ and taken to dryness by evaporation in a rotary evaportor. Lipids are resuspended in chloroform:methanol [2:1 (v/v)] and stored refrigerated until use.

Using this procedure, a complete extraction of the neutral as well as acidic phospholipids may be obtained from tissue homogenates and membrane suspensions. The step-wise extraction scheme is necessary in order to protect the alkenylether group from degradation by the acidic solvent system (*see* the previous chapter for a complete discussion of extraction methods).

2.2. Separation of Nonpolar Lipids, Glycosphingolipids, and Phospholipids by Silicic Acid (Unisil) Column Chromatography

In the event that a large amount of the tissue or membranes is used for lipid extraction, it is desirable to carry out a column chromatography procedure to separate the major lipid classes prior to separation by TLC. Unisil silicic acid (200–325 mesh, Clarkson Chemical Co., Williamsport, PA) has been used widely for bulk separation of these lipids. For a typical separation, 2 g of the Unisil is suspended in chloroform, and the suspension is slurried into a column that is equipped with a solvent reservoir and a Teflon stopcock to adjust flow rate. Samples are dissolved in chloroform and applied to the column. In the first fraction, nonpolar lipids (cholesteryl esters, triacylglycerols, fatty acid methyl esters, cholesterol, free fatty acids, diacylglycerols, and monoacylglycerols) are eluted with 50 mL of chloroform. The glycosphingolipids (cerebrosides and sulfatides) are eluted with 50 mL of acetone, and the phospholipids are eluted with 100 mL of methanol (Sun and Horrocks, 1970). Individual fractions are collected in round-bottom flasks and taken to dryness with a rotary evaporator under reduced pressure.

2.3. Separation of Phospholipids and Neutral Glycerides by Two-Dimensional TLC

The two-dimensional TLC procedure for separation of phospholipids is essentially that described by Horrocks and Sun (1972),

except for minor modifications of the solvent systems used. The
TLC plates are either precoated silica gel G TLC plates, 250 μm coat
thickness (Analtech, Newark, DE), or those self-prepared with
silica gel 60 (Merck). Samples are applied to the left lower corner of
the TLC plate. The plate is first developed in a solvent system
containing chloroform:methanol:16N NH₄OH (135:60:10, by
vol). After development, solvent on the plate is removed by blowing with an air gun for 15 min. The TLC plate is then exposed to
HCl fumes for 3 min, after which the excess fumes are removed
by blowing with an air gun for 15 min. The plate is then turned
90° and placed in a second solvent system consisting of
chloroform:methanol:acetone:acetic acid:0.2M ammonium acetate (140:60:55:3.5:10, by vol). After solvent development, the
TLC plates are briefly dried. Two methods are available for
visualization of the lipid spots. If the lipid spots are to be recovered
for quantitation, such as for phosphorus determination and for
scintillation counting, the TLC plates are exposed to iodine vapor.
If the lipids are to be recovered for analysis of their acyl groups or
for further derivatization, the TLC plates are sprayed with 2',7'-
dichlorofluorescein (0.2% in methanol), and the lipid spots are
visualized under a UV lamp. The above described procedure has
been used successfully in our laboratory for separation of phospholipids in brain and other tissues. The procedure allows a clear
separation of PI from PS and plasmalogens from the respective
diacyl phospholipids. A diagram depicting the normal appearance
of the individual phospholipids is shown in Fig. 1. Perhaps the
only limitation with regard to this separation procedure is the
inability to separate the alkylacyl phospholipid species from the
diacyl species. More involved separations of these phospholipid
species have been described by El-Tamer et al. (1984).

The above described procedure can also be extended to include recovery of the neutral lipids for further separation. After
developing the TLC plate with the first-dimension solvent system,
the neutral lipids (with the exception of free fatty acids) usually
appear at the solvent front. This lipid spot can be removed by
scraping, then transferred to a test tube. The lipids are separated
from the silica gel by suspending the gel in 5 mL of chloroform:
methanol 5:1 (v/v), mixing and centrifuging briefly. An aliquot of
the organic phase can be taken for further analysis. Since the
procedure for separation of the neutral lipids has been described
elsewhere (see chapter by Marcheselli et al., this volume), it will not
be elaborated further.

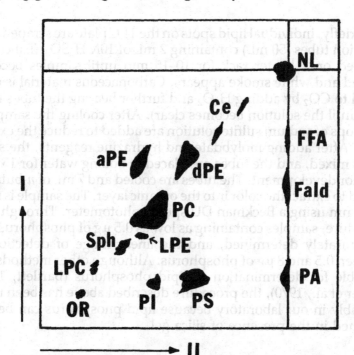

Fig. 1. A typical separation of the phospholipids in brain synaptosomes. Abbreviations: OR, origin with polyphosphoinositides; PA, phosphatidic acids; PS, diacyl-glycerophosphoserine; PI, diacyl-glycerophosphoinositol; Sph, sphingomyelin; LPC, acyl lyso-glycerophosphocholine; LPE, acyl lyso-glycerophosphoethanolamine; PC, diacyl-glycerophosphocholine; aPE, acyl lysoglycerophosphoethanolamine from alkenylacyl-GPE; dPE, diacyl-glycerophosphoethanolamine; Fald, fatty aldehydes from PEpl; FFA, free fatty acids; Ce, cerebrosides; NL, neutral lipids.

2.4. Determination of Phospholipids by Assaying the Phosphorus Content

The lipid phosphorus assay is carried out according to the procedure described by Gottfried (1967). In this procedure, the phospholipid together with the silica gel is digested by acid and converted to inorganic phosphorus. The inorganic phosphorus is allowed to complex with ammonium molybdate for color development. Sodium phosphate is used as standard. Since lipid phosphorus is determined in the presence of silica gel, the blank should also contain an appropriate amount of the silica gel.

Briefly, individual lipid spots on the TLC plate are scraped into digestion tubes (50 mL) containing 2 mL of $10N$ H_2SO_4. Tubes are digested on a heater rack for 10–15 min until samples become charred and white smoke appears. Carbonaceous material is converted to CO_2 by adding H_2O_2 and further heating the tubes for 2 min (until the solution becomes clear). After cooling the samples, 2–3 drops of sodium sulfite solution are added to reduce the excess H_2O_2. After adding molybdate and hydrazine reagents, the solution is mixed, and the tubes are placed in boiling water for 15 min for color development. The tubes are cooled and 7 mL of *n*-butanol is used to extract the color into the organic layer. The sample is read at 795 nm using a Beckman DU spectrophotometer. Through this procedure, samples containing as low as 0.5 µg of phosphorus can be accurately determined, and the linear range of detection is between 0.5 and 8 µg of phosphorus. Although other methods are available for determination of lipid phosphorus (Bartlett, 1959; Rouser et al., 1970), the procedure described above has been used favorably in our laboratory because lipid phosphorus can be determined in the presence of silica gel.

2.5. Analysis of the Glycerolipid and Neutral Glyceride Acyl Groups by Gas-Liquid Chromatography

The acyl groups linked to the glycerolipids can be converted to fatty acid methyl esters (FAME) by base-catalyzed methanolysis using sodium hydroxide in methanol (Sun and Horrocks, 1968). This procedure is specific for the glycerolipids and not for the acyl groups that are attached to sphingomyelin, cerebrosides, cerebroside sulfates, cholesteryl esters, and free fatty acids. For derivatization of acyl groups of these other compounds, it is necessary to use an acid-catalyzed methanolysis procedure.

For base methanolysis, silica gel spots corresponding to individual glycerolipids are scraped into test tubes, and a known amount of heptadecanoic acid methyl ester (C17:0 ME) is added to each tube as internal standard. The methanolysis is nonselective with respect to different types of glycerolipids, and the extent of methanolysis is essentially complete. Thus, the use of C17:0 methyl esters as internal standards for all glycerides has been found satisfactory. After adding the fatty acid methyl ester standard, 2 mL of the NaOH-methanol reagent ($0.5M$ NaOH in anhydrous methanol) is added. The tubes are mixed and allowed to stand at room temperature for 10 min, after which 4 mL of chloroform and 1.5 mL of water are added and mixed. At this step, the tubes can either be centrifuged briefly to allow a phase separation or left at room temperature overnight. The lower organic phase containing the fatty acid methyl esters is removed and filtered through anhydrous Na_2SO_4. The solvent is evaporated to dryness under nitrogen, and the residue is redissolved in hexane. Samples can be stored prior to evaporation, or they can be stored in hexane until the time of analysis by GLC.

For acid methanolysis, it is necessary to use heptadecanoic acid (C17:0) as internal standard. Methylating agents such as BF_3:methanol can be purchased commercially (Supelco, Bellefonte, PA), or 5 mL of conc. HCl in 95 mL of methanol can be freshly prepared in the laboratory just before use. We have successfully used the HCl:methanol as reagent for most acid methanolysis. To initiate the methanolysis, 3 mL of the HCl-methanol is added to the sample in a test tube fitted either with a ground glass stopper or a Teflon-lined screw cap. The mixture is heated in a water bath at 80°C for 1 h. Care is taken that the solvent does not evaporate to dryness during the heating step. After heating, the tubes are cooled in an ice bath, and two volumes of water are added. The FAMEs are extracted with hexane three times, and the pooled hexane phase is taken to dryness. Samples are redissolved in a small amount of hexane for GLC.

Separation of FAME by GLC is carried out using a Hewlett Packard gas chromatograph (model 5840A) equipped with dual flame ionization detectors and a plotter-integrator unit. Separations are carried out with two 6' × 1/8" stainless steel columns packed with 10% SP2330 on 100/120 chromosorb AW (Supelco, Bellefonte, PA). The oven temperature is programmed to increase from 185 to 240°C, with an increment rate of 3°C/min. The injection temperature is set at 225°C, and the flame ionization detector temperature, at 290°C. The carrier gas is prepurified nitrogen. Normally, a typical separation requires 18–25 min. A computer program is used to analyze the data. Under normal conditions, variance between repeated separations of the same sample is normally less than 5%.

Using this procedure for quantitation and separation of FAME from individual phosphoglycerides, a complete separation including analysis of PI and PA can be accomplished from 1 mg of protein. Nevertheless, the two poly-PIs, which remain at the origin of the TLC plate, are not separated. Separation of the poly-PIs can be accomplished by applying the acidic chloroform:methanol extract to a HPTLC plate (Silica gel 60, Merck) impregnated with 1% potassium oxalate and developing the plate in a solvent system containing chloroform:methanol:$4N$ NH_4OH (9:7:2, by vol) (*see* chapter by Hajra et al. for detailed methods for separation of phosphoinositides).

2.6. Analysis of the Alkenyl Group Composition of Ethanolamine Plasmalogen

When phosphoglycerides are in the plasmalogen form, acid methanolysis would result in simultaneous conversion of the acyl groups to FAME and the alkenyl groups to dimethylacetals (DMAs). If the FAMEs were not separated from the DMAs by TLC, analysis of the mixture by GLC would be complicated because of co-elution of some of the FAME with DMA. Although the FAME and DMA can be separated by TLC, the FAME from the acid methanolysis procedure would include acyl groups from both alkenylacyl and diacyl forms. On the other hand, if the alkenyl groups are first converted to the cyclic acetal (CA) derivatives, which is a relatively mild procedure, the acyl groups from the remaining alkenylacyl-GPE (as lyso-GPE) and diacyl-GPE can be separately analyzed.

The procedure for making the CA derivatives is based on that described by Rao et al. (1967) with modifications to accommodate smaller sample size (Sun and Horrocks, 1969). Briefly, phospholipids are placed in a Teflon-lined screw-cap culture tube containing 0.5–1.0 mg of p-toluene sulfonic acid. The solvent for dissolving the phospholipids is evaporated under nitrogen. After evaporation, 2 mL of methanol containing 50 µL of 1,3-propanediol is added to the tube. Tubes are capped and heated at 80°C for 2 h. After cooling the tubes, chloroform and H_2O are added to allow a phase separation. The upper phase containing the excess propanediol is discarded. The solvent in the lower phase is evaporated to dryness. Lipids are applied to TLC plates and developed in chloroform:methanol:$16N$ NH_4OH (65:25:4, by vol). Using this solvent system, diacyl-GPE (R_f 0.46) and 2-acyl-GPE (R_f 0.30) are separated from each other, and the CAs appear in the solvent front. The CAs are recovered from the plate, eluted, and further purified by TLC developed with toluene. The CAs are separated by GLC under the same conditions used for FAME. The 2-acyl-GPE and diacyl-GPE can be recovered from the TLC plates and used for further analysis of their acyl group composition.

3. Information on Phospholipid, Acyl, and Alkenyl Group Composition in Brain and Subcellular Membranes

3.1. Phospholipid Composition

Although the phospholipids of mouse brain are enriched in PC and PE with smaller amounts of PS, PI, cardiolipin (CL), and sphingomyelin, the distribution of these phospholipids varies depending on the brain region and the type of subcellular membranes used for analysis. In an earlier study in which mouse brain myelin, microsomes, and synaptosome-mitochondria fractions were analyzed for their phospholipid content, it became obvious that there was a high content of PEpl in the myelin, whereas the synaptosome-mitochondrial fraction showed an enrichment of CL (Sun and Horrocks, 1970). Although PS and PI were not separately analyzed in this study, together they comprised 20–24% of the total phospholipids in the various subcellular fractions. In a subsequent study in which subcellular fractions were isolated from human brain cortex, results indicated the predominance of PS in the PS–PI mixture (Sun, 1973). The data in Table 1 reveal the differences in phospholipid composition among myelin, microsomes, and synaptosome-mitochondiral fractions isolated from human brain autopsy samples (Sun, 1973). In this study, the high level of PEpl in myelin is again illustrated. The alkenylacyl-GPE to diacyl-GPE ratio in myelin is 4, whereas that in synaptosomes is less than 1. Unlike the mouse brain myelin, human brain myelin is also characterized by a high proportion of sphingomyelin. The data further show that phospholipids from microsomes obtained from the white matter resemble those of myelin, and those from the gray matter resemble those of the synaptosomes.

Because of the recent improvements in procedures for isolating brain subcellular membranes and for extraction of the membrane lipids, a more detailed analysis of the phospholipids in individual subcellular membranes was performed (Sun et al., 1987). In this study, the "classical" microsomal fraction (i.e., 100,000g pellet of the P_2 supernatant fraction) is now further separated by differential centrifugation to yield a somal plasma membrane (PM) fraction and a microsomal fraction. Following the procedure described by Booth and Clark (1978), it is also possible to separate the free mitochondria from the synaptosomes by Ficoll-sucrose gradient centrifugation. Data in Table 2 depict the phospholipid composition in these subcellular membranes after extrac-

Table 1

Phospholipid Compositions of Subcellular Fractions from Human Brain Cortex[a]

| | Myelin (5)[b] | Microsomes | | Synapto-some-rich fraction |
		White matter (10)	Gray matter (10)	
	Total lipid phosphorus ± SEM,%			
PS	15.6 ± 0.53	18.1 ± 0.73	11.3 ± 0.60	10.6 ± 0.52
PI	2.8 ± 0.09	4.4 ± 0.23	5.0 ± 0.15	4.2 ± 0.17
Sphingomyelin	12.3 ± 0.80	9.4 ± 0.65	5.6 ± 0.22	5.0 ± 0.25
Diacyl-GPC	26.4 ± 0.19	27.8 ± 0.42	37.4 ± 0.50	37.1 ± 0.52
Alkenylacyl-GPE	32.5 ± 1.00	27.9 ± 0.88	17.3 ± 0.25	17.9 ± 0.52
Diacyl-GPE	8.3 ± 0.48	10.3 ± 0.36	21.1 ± 0.40	21.9 ± 0.47
Others (PA + CL)	0.9 ± 0.22	2.4 ± 0.21	1.8 ± 0.16	4.1 ± 0.15

[a]Data taken from Sun (1973). Results for myelin were obtained from analyses of autopsy brain material from subjects ranging from 61 to 67 years old. Results for the other subcellular fractions were obtained from analyses of autopsy material from subjects ranging from 61 to 87 years old. Phospholipids were separated by TLC as described in the text. Phosphorus content of individual lipids was determined by the method of Gottfried et al. (1967).

[b]Numbers in parentheses represent the number of analyses.

tion by the stepwise extraction procedure. As shown previously, unique phospholipid profiles can be found in the myelin and mitochondrial fractions, but the profiles in the synaptic membranes and subsynaptic membranes are not greatly different. For the first time, we observed an enrichment of PA in myelin and mitochondria. The stepwise extraction procedure apparently resulted in a more complete extraction of the acidic phospholipids. The results also reveal differences in phospholipid profiles between mitochondrial membranes and synaptic membranes (SPM and SV). In this regard, mitochondrial phospholipids are characterized by a low level of PS and PEpl and the unique presence of cardiolipin (Table 2).

The phospholipid profiles of myelin isolated from different animal species and humans is shown in Table 3. Although there is a discernible pattern among the species, subtle differences can also

Table 2
Percent Distribution of Phospholipids
in Rat Brain Subcellular Membranes[a]

Phospholipids	Weight, %							
	BH	Mye	Mit	Mic	PM	Syn	SPM	SV
Origin[b]	1.5	0.8	1.0	0.7	0.6	0.6	0.4	0.3
PI	3.4	2.1	4.8	5.5	4.2	3.6	4.3	2.9
PS	10.8	16.2	5.8	11.5	13.1	14.7	12.2	14.4
PA	2.2	4.8	3.1	0.9	1.8	0.8	0.8	0.6
Sph	3.9	4.2	2.4	3.2	5.4	5.3	5.6	3.9
PC	43.8	21.6	36.8	41.7	37.1	36.8	39.3	38.0
PEpl	13.6	37.1	11.5	18.7	19.0	17.4	18.8	20.1
PE	18.4	12.6	27.5	17.6	17.5	20.1	18.3	19.5
CL	1.9	—	6.9	0.3	—	0.8	0.3	0.2

[a]Results are obtained from Sun et al. (1987). The procedure for subcellular fractionation is also described in Sun et al. (1987). Lipids were extracted by the two-step extraction procedure as described in section 2. Phospholipids from individual subcellular fractions were separated by reactional two-dimensional TLC as described by Horrocks and Sun (1972). Lipid spots were taken for assay of phosphorus according to the method of Gottfried (1967). Abbreviations: BH, brain homogenates; Mye, myelin; Mit, mitochondria; Mic, microsomes; PM, plasma membranes; Syn, synaptosomes; SPM, synaptic plasma membranes; SV, synaptic vesicles.
[b]Lipids in the origin are comprised largely of polyphosphoinositides.

be detected. This pertains especially to the proportion of sphingo-myelin, which is lower in the rodent myelin (4–6%) as compared to that in ox, monkey, and humans (12–17%). On the other hand, there is a higher proportion of PE in the rodent myelin than in that from other species.

The phospholipid profile of myelin isolated from rat spinal cord shows some differences from that isolated from brain (Sun et al., 1983). In this regard, phospholipids in the cord myelin are comprised of a lower proportion of PS and a higher proportion of sphingomyelin than those from brain. The phospholipids in both microsomes and myelin isolated from rat spinal cord indicated obvious compositional changes during development and matura-tion (Table 4). In particular, the PEpl in microsomes increased nearly threefold between 7 and 40 d, but this dramatic increase was not found in the cord myelin. In the cord myelin, the decrease in

Table 3
Comparison of the Phospholipid Composition of Myelin
Among Different Animal Species

Phospholipids	Weight %				
	Mouse[a]	Rat[b]	Ox[a]	Rhesus monkey[c]	Human[d]
PS	24	16.2	20	16.9	15.6
PI		2.1		16.2	2.8
Sph	6	4.2	17		12.3
PCpl	—	—	—	2.1	—
PC	25	21.6	21	25.0	26.4
PEpl	32	37.1	33	28.8	32.5
PE	14	12.6	9	9.5	8.3
PA	—	4.8	—	1.5	0.9

[a]Data taken from Sun and Horrocks (1970).
[b]Data taken from Sun et al. (1987). TLC analysis was carried out after extracting myelin with both neutral and acidified chloroform-methanol. The improved extraction procedure may account for the higher level of PA observed.
[c]Data taken from Sun and Samorajski (1973).
[d]Data taken from Sun (1973).

PC during development is accompanied largely by the increase in PEpl, whereas in the microsomal fraction, both PE and PEpl are increased on account of the decrease in PC.

3.2. Alkenyl Group Composition of Ethanolamine Plasmalogen in Brain

The alkenyl groups in brain are largely associated with the ethanolamine plasmalogens. Analysis of the alkenyl groups in brain as CA derivatives revealed the presence of three major types, namely, 16:0, 18:0, and 18:1 (Table 5) (Sun and Horrocks, 1969). The alkenyl group profiles between the ox and mouse brain are quite similar (Sun and Horrocks, 1969). The alkenyl group profile in myelin is different, however, from that in other subcellular fractions. Similar to the acyl groups, the alkenyl groups of myelin are enriched in 18:1, but those in synaptosomes and microsomes are high in 18:0 instead.

Table 4

Phospholipid Distribution in Rat Spinal Cord Microsomes and Myelin During Development and Maturation[a]

Phospholipids	Age, d							
	Microsomes				Myelin			
	7	13	30	40	7	13	30	40
Diacyl-GPS	9.7	10.1	9.7	12.1	7.0	6.4	8.4	7.8
Diacyl-GPI	—	—	2.2	—	3.0	2.5	1.6	2.2
Sphingomyelin	4.3	6.7	11.3	9.7	7.0	6.9	8.6	9.6
Diacyl-GPC	59.6	48.9	37.5	33.5	47.2	43.2	34.5	31.7
Alkenylacyl-GPE	6.9	13.0	11.3	17.4	19.1	23.2	30.2	28.1
Diacyl-GPE	13.7	16.5	25.4	25.8	14.3	16.3	14.3	16.1
Others	5.8	4.8	2.6	1.5	2.4	1.5	2.4	2.5

[a]Data taken from Sun et al. (1983).

75

Table 5
Alk-1-enyl Group Composition in Brain and Subcellular Fractions

Chain length and unsaturation	Weight, %					
	Mouse[a] brain	Ox[a] brain	Ox[b] myelin	Mouse[b] myelin	Mouse[b] synaptosomes-mitochondria	Mouse[b] micro-somes
16:0	19.3	25.4	24.3	16.6	22.2	22.1
16:1	<0.5	—	—	0.9	—	0.6
17:0	0.6	1.9	1.7	—	1.0	—
18:0	40.3	33.0	22.5	31.5	50.1	48.9
18:1	39.1	39.5	51.3	51.0	26.7	29.5

[a]Data taken from Sun and Horrocks (1969).
[b]Data taken from Sun and Horrocks (1970).

3.3. Acyl Group Composition of Phospholipids in Brain Subcellular Membranes

A number of studies have been carried out to examine the acyl group composition of brain phosphoglycerides at cellular and sub-cellular levels (*see* review by Sun and Foudin, 1985). Unlike the case with other body organs, unique acyl group profiles are found associated with each phosphoglyceride in brain (Sun and Horrocks, 1968). For example, the acyl groups of PC are enriched in 16:0 and 18:1, but contain only a small proportion of the polyunsaturated type. In contrast, the acyl groups of PE are enriched in the polyunsaturated type such as 20:4 and 22:6 (Sun and Horrocks, 1969). There is also some indication that the acyl groups found in the C-2 position of the alkenylacyl-GPE are different from those present in the C-2 position of the diacyl-GPE (Sun and Horrocks, 1969, 1970). In most instances, a higher proportion of polyunsaturated fatty acids is found in the plasmalogens as compared to the diacyl type, although the exact functional role of plasmalogens in brain membranes has not been clearly delineated.

With the exception of myelin, unique acyl group profiles of individual phospholipids in brain are found among different sub-cellular fractions. Data in Table 6 depict the acyl group composition of PC, PE, and PEpl in three major subcellular fractions of mouse brain, namely, myelin, microsomes, and synaptosome-mitochon-

Table 6

Acyl Group Composition of Diacyl-GPC, Diacyl-GPE, and Alkenylacyl-GPE in Mouse Brain Subcellular Membranes[a]

| | Weight, % | | | | | | | | |
| Acyl groups | Diacyl-GPC | | | Diacyl-GPE | | | Alkenylacyl-GPE | | |
	Myelin	Microsomes	Synaptosomes-mitochondria	Myelin	Microsomes	Synaptosomes-mitochondria	Myelin	Microsomes	Synaptosomes-mitochondria
16:0	30.0	42.7	40.5	8.4	9.4	6.2	2.4	1.1	2.0
18:0	18.5	13.7	15.5	25.6	32.9	36.0	0.9	1.3	4.5
18:1	42.8	30.7	26.5	35.3	16.0	13.7	42.8	21.7	10.1
18:2	0.8	0.7	0.9	0.9	—	—	—	—	—
20:0	1.2	0.4	—	2.8	—	—	2.6	0.6	—
20:1	3.8	1.9	1.4	7.2	1.6	0.5	20.6	9.0	2.6
20:3(n-6)	—	—	—	0.5	0.5	—	1.2	0.6	—
20:4(n-6)	2.5	4.7	8.4	9.7	11.7	13.7	11.6	14.2	18.4
22:4(n-6)	—	1.1	—	3.2	3.5	3.1	11.8	11.0	10.4
22:5(n-3)	—	—	—	3.5	—	—	—	—	—
22:6(n-3)	—	4.0	7.7	5.0	23.3	27.0	5.8	39.6	51.6

[a]Data taken from Sun and Horrocks (1970), with slight modifications and rearrangements.

77

drial fractions. It is obvious from this data that myelin phospholipids are particularly enriched in 20:1, and this acyl group is especially linked to the alkenylacyl-GPE. Similarly, the proportion of 18:1 is also greater in myelin as compared to these other subcellular fractions. Interestingly, both diacyl-GPE and alkenylacyl-GPE in microsomes and synaptosomes-mitochondria are enriched in 22:6(n-3), but the phosphoglycerides in these fractions contain a very low proportion of 20:1.

When acyl groups of PS and PI are separately analyzed, data show that these two phospholipids have distinctly different acyl group profiles. The acyl groups of PS are enriched in 18:0, 18:1, and 22:6, whereas those in PI are high in 18:0 and 20:4(n-6) (Table 7). Although there is a higher proportion of 18:1 in PS from myelin, very little 20:1 is detected in this phospholipid. Surprisingly, the acyl group profile of PI in myelin is not greatly different from that in other subcellular fractions. The acyl groups of phosphatidic acids (PA) are comprised largely of 16:0, 18:0, 18:1 and 20:4. Distinct differences in PA acyl groups are found between myelin and microsomes (Table 7), however. The PA in myelin is enriched in 18:1 and contains very little 22:6(n-3), whereas a considerable amount of 22:6(n-3) is found in the PA in microsomes. Although not separately analyzed, the acyl group profiles of polyphosphoinositides closely resemble that of PI. Similar to the PIs, only small variations in the acyl group profiles of poly-phosphoinositides are found among different subcellular fractions.

An interesting correlation is found by comparing the acyl group composition of myelin alkenylacyl-GPE isolated from different brain species (Table 8). In this regard, the proportion of 22:4 is lowest in the rodent myelin and is apparently higher in the myelin isolated from Rhesus monkey and humans. The increase in 22:4 seems to be compensated for by a decrease in 20:1 (Table 8). The physiological meaning of these species differences is presently not known.

4. Conclusions

Using the combined TLC and GLC procedures described in the methods section, it is possible to analyze the major and minor phospholipids and their acyl and alkenyl group composition in brain tissue and subcellular membranes. Results of these various studies have indicated that the brain membranes are unique in

Table 7
Acyl Group Composition of Phosphatidylserine and Phosphatidylinositol from Rat Brain Subcellular Fractions[a]

	Diacyl-GPS			Diacyl-GPI			Poly-PI			Phosphatidic acid		
	My	Mic	Syn-mito	My	Mic	Syn-mito	My	Mic	Syn-mito	My	Mic	Syn-mito
16:0	0.6	3.7	1.8	9.4	9.9	20.0	8.8	9.6	8.4	11.0	7.2	14.6
18:0	39.7	47.7	41.2	33.3	34.6	33.7	40.5	45.8	47.8	24.1	33.2	27.1
18:1	44.9	11.3	20.3	8.4	6.5	13.5	11.8	9.8	7.8	40.9	23.2	37.1
20:1	1.9	0.5	0.4	—	—	—	3.6	—	—	5.1	3.0	2.0
20:4	4.6	2.3	2.5	43.2	44.9	32.9	35.2	30.2	33.0	11.0	10.0	12.7
22:4	2.1	3.4	2.9	2.9	1.8	—	—	4.7	1.2	7.1	7.4	2.0
22:5	1.0	4.5	5.5	—	—	—	—	—	—	—	—	—
22:6	5.2	26.6	25.5	2.8	2.3	—	—	—	1.8	0.9	16.0	4.3

[a]Data taken from Sun et al. (1987).

79

Table 8
Acyl Group Composition of Alkenylacyl-GPE of Myelin
from Different Animal Species

	Alkenylacyl-GPE					
	Mouse[a]	Rat[b]	Ox[a]	Squirrel[c] monkey	Rhesus[d] monkey	Human[e]
16:0	2.4	4.2	2.0	3.2	3.2	2.7
18:0	0.9	1.0	1.0	0.7	0.5	—
18:1	42.8	52.9	48.3	39.3	34.1	50.0
18:2	2.6	—	—	—	—	—
20:1	20.6	14.1	16.4	17.3	11.6	11.2
20:2	—	—	0.9	—	2.3	1.2
20:3	1.2	—	0.9	—	2.1	0.7
20:4	11.6	10.7	8.3	12.0	10.9	7.5
20:5	—	—	2.4	2.0	1.7	2.3
22:4	11.8	10.7	15.6	17.7	22.3	19.5
22:5	—	—	1.0	0.6	1.1	—
22:6	5.8	4.3	2.9	5.7	2.5	1.9
24:4	—	—	—	—	6.6	0.9

[a]Data obtained from Sun and Horrocks (1970).
[b]Data taken from Sun et al. (1987).
[c]Data taken from Sun and Sun (1972).
[d]Data taken from Sun and Samorajski (1973). Myelin was isolated from corpus callosum of 6–10-yr-old Rhesus monkeys.
[e]Data taken from Sun (1973). Myelin was isolated from white matter of autopsy samples from 60–70-yr-old humans.

their phospholipid composition and the acyl groups that are linked to individual phospholipids. A high content of ethanolamine plasmalogens is particularly noted in brain, and this characteristic feature is especially demonstrated in myelin. In addition, there is a prevalence of 22:6(n-3) in the phosphoglycerides of neuronal membranes, and the phosphoglycerides of myelin are enriched in 18:1 and 20:1. The alkenyl groups that are linked to the plasmalogens are comprised mainly of the 16:0, 18:0, and 18:1 type. Unlike the case in other body organs, individual phosphoglycerides in brain are shown to have very specific acyl group profiles. In this regard, the near exclusive presence of 20:4(n-6) and 22:6(n-3) in the 2-position of PI and PS, respectively, is of special interest. The specificity of acyl groups and phosphoglycerides associated with different subcellular membranes suggests diverse roles for these phospholipids in neuronal membrane functions.

References

Bartlett G. R. (1959) Phosphorus assay in column chromatography. *J. Biol. Chem.* **234**, 466–468.

Booth R. F. G. and Clark J. B. (1978) A rapid method for the preparation of relatively pure metabolically active synaptosomes from rat brain. *Biochem. J.* **176**, 365–370.

El-Tamer A., Record M., Fauvel J., Chap H., and Douste-Blazy L. (1984) Studies on other phospholipids. I. A new method of determination using phospholipase A_1 from guinea pig pancreas. Application to Krebs-II ascites cells. *Biochim. Biophys. Acta* **793**, 213–220.

Gottfried E. L. (1967) Lipid of human leukocytes: relation to cell type. *J. Lipid Res.* **8**, 321–327.

Henn S. W. and Henn F. A. (1984) The Identification of Subcellular Fractions of the Central Nervous System, in *Handbook of Neurochemistry* vol. 2. 2nd Ed. (Lajtha A., ed.) Plenum, New York.

Horrocks L. A. and Sun G. Y. (1972) Ethanolamine Plasmalogens, in *Research Methods in Neurochemistry* vol. 1 (Marks N. and Rodnight R., eds.) Plenum, New York.

Rao P. V., Ramachandran S., and Cornwell D. G. (1967) Synthesis of fatty aldehydes and their cyclic acetals (new derivatives for the analysis of plasmalogens). *J. Lipid Res.* **8**, 280–390.

Rouser G., Fleischer S., and Yamamoto A. (1970) Two-dimensional thin layer chromatographic separation of polar lipids and determination of phospholipids by phosphorus analysis of spots. *Lipids* **5**, 494–496.

Sun G. Y. (1973) Phospholipids and acyl groups of cellular fractions from human cerebral cortex. *J. Lipid Res.* **14**, 656–662.

Sun G. Y. and Foudin L. (1985) Phospholipid Composition and Metabolism in the Developing and Aging Nervous System, in *Phospholipids in the Nervous Tissues* (Eichberg J., ed.) John Wiley, New York.

Sun G. Y. and Horrocks L. A. (1968) The fatty acid and aldehyde composition of the major phospholipids of mouse brain. *Lipids* **3**, 79–83.

Sun G. Y. and Horrocks L. A. (1969) Acyl and alk-1-enyl group compositions of the alk-1'-enyl acyl and the diacyl glycerophosphoryl ethanolamines of mouse and ox brain. *J. Lipid Res.* **10**, 153–157.

Sun G. Y. and Horrocks L. A. (1970) The acyl and alk-1-enyl groups of the major phosphoglycerides from ox brain myelin and mouse brain microsomal, mitochondrial and myelin fractions. *Lipids* **5**, 1006–1012.

Sun G. Y. and Samorajski T. (1973) Age differences in the acyl group composition of phosphoglycerides in myelin isolated from the brain of the Rhesus monkey. *Biochim. Biophys. Acta* **316,** 19–27.

Sun G. Y. and Sun A. Y. (1972) Phospholipids and acyl groups of synaptosomal and myelin membranes isolated from the cerebral cortex of squirrel monkey *(Saimiri sciureus). Biochim. Biophys. Acta* **280,** 306–315.

Sun G. Y., Su K. L., Der O. M., and Tang W. (1979) Enzymic regulation of arachidonate metabolism in brain membrane phosphoglycerides. *Lipids* **14,** 229–235.

Sun G. Y., DeSousa B. N., Danopoulos V., and Horrocks L. A. (1983) Phosphoglycerides and their acyl group composition in myelin and microsomes of rat spinal cord during development. *Int. J. Devl. Neurosci.* **1,** 59–64.

Sun G. Y., Huang H-M., Kelleher J. A., Stubbs E. B. Jr., and Sun A. Y. (1987) Marker enzymes, phospholipids; and acyl group composition of a non-synaptic plasma membrane fraction isolated from rat cerebral cortex: A comparison with microsomes and synaptic plasma membranes. *Neurochem. Int.* (Submitted).

Whittaker V. P. (1984) The Synaptosomes, in *Handbook of Neurochemistry* vol. 7, 2nd Ed. (Lajtha A., ed.) Plenum, New York.

Quantitative Analysis of Acyl Group Composition of Brain Phospholipids, Neutral Lipids, and Free Fatty Acids

Victor L. Marcheselli, Burton L. Scott,
T. Sanjeeva Reddy, and Nicolas G. Bazan

1. Introduction

The quantitative extraction, separation, and estimation of brain lipids should include a number of precautions to avoid or limit potential artifacts. It is particularly important to avoid these pitfalls when the objective is to assess the pool size and composition of minor, but metabolically very active, components such as free fatty acids (FFAs) and diacylglycerols (DGs) (Bazan, 1970, 1982, 1983; Bazan et al., 1983, 1985a). The quantitative estimation of these brain lipids poses several problems not found with other lipid classes. For example, if proper care is not taken during brain sampling, FFAs and DGs accumulate (Bazan, 1970; Aveldano and Bazan, 1975a). At the same time, polyphosphoinositides undergo rapid postmortem degradation (Eichberg and Hauser, 1967; Hauser et al., 1971). In addition, the unsaturated components of FFAs and DGs deserve special attention. Although their levels are lowest under resting conditions (Bazan, 1970; Cenedella et al., 1975; Bazan et al., 1983), they show by far the largest increase during brain ischemia or hypoxia (Bazan, 1970; Bazan et al., 1971; Aveldano and Bazan, 1975b; Cenedella et al., 1975) and are subsequently used as precursors for eicosanoids if sufficient oxygen is available. Thus, special care must be taken to avoid artifactual peroxidative changes of highly unsaturated fatty acids. Various methods for the isolation and estimation of these lipids have been described previously (Bazan and Joel, 1970; Bazan and Cellik, 1972; Bazan and Bazan, 1975; Rodriguez de Turco and Bazan, 1977). The main aim of this article is to describe in detail different procedures that are used in our laboratory for the quantitative extraction, separation, and estimation of brain lipid classes. Emphasis is placed on the quantitative extraction and estimation of minor lipids, such as FFAs, DGs,

83

phosphatidic acid, and polyphosphoinositides. These procedures include modifications and improvements of methods described previously, such as the use of quantitative capillary gas-liquid chromatography (GC), solvent systems, and ready-made thin-layer chromatography (TLC) plates from commercial sources to isolate and estimate various lipids from brain and retina.

2. Sampling of Brain

Rapid fixation of brain tissue is necessary for the estimation of basal levels of FFAs, DGs, and polyphosphoinositides within seconds after death (Bazan, 1970; Banschbach and Geison, 1974; Aveldano and Bazan, 1975b; Hauser et al., 1971). The effect of ischemia can be managed in different ways to obtain reproducible results. In all cases, it is important to keep constant the time between killing the animal and the actual time of arresting brain metabolic reactions (e.g., homogenization in chloroform: methanol); if this time interval is accurately known, reproducible results can be obtained (Bazan, 1970; Bazan et al., 1971). Rapid fixation of brain can be achieved by various means. One way is to fix the whole animal (if using small rats or mice) or the head alone (after decapitation) by plunging into liquid nitrogen, which has a temperature of $-196°C$. Faster freezing is achieved by vigorously swirling the tissue in the coolant. Further improvement in freezing rate is achieved by plunging the tissue into a bath of Freon (or isopentane) cooled by an outer bath of liquid nitrogen, because the heat transfer properties of liquid Freon are superior to those of liquid nitrogen. The use of Freon minimizes the gaseous interface between coolant and tissue, and the tissue is cooled faster. Freezing the whole animal results in slower fixation because of the larger tissue mass involved. The fixation method employing decapitation followed by plunging the head in a freezing medium would appear adequate; the effect of decapitation *per se* may alter chemical components of the brain, however. To counteract this problem, *in situ* freezing of brain with liquid nitrogen has been used by Ponten et al. (1973), whereby animals are anesthetized, cranial bones opened, and liquid N_2 poured onto the brain via a plastic funnel until the tissue is frozen. A modification of this method (Levy and Duffy, 1975) involves a scalp incision without craniotomy, followed by pouring liquid N_2 onto the calvarium.

Another procedure that eliminates the problem of decapitation effects is fixation by head-focused microwave irradiation (Cenedella et al., 1975). Galli and Spagnuolo (1976) have used this approach and confirmed the early onset and rapidity of the FFA release phenomenon in brain during ischemia (Bazan, 1970). Moreover, these studies support the hypothesis that accumulation of FFAs in brain is triggered by a receptor-linked phospholipase A_2 (Bazan, 1970, 1971). Recently, more powerful microwave-generating equipment has become available that can inactivate brain enzymes even more rapidly. Using this equipment, the rat brain can be fixed *in situ* within 2 s (6.0 kW at 2450 mHz, Cober Electronics, Inc., Stamford, CT) and brain FFAs, DGs, and polyphosphoinositides studied. Very low levels of FFAs and high levels of polyphosphoinositides were observed using this method, indicating that few, if any, reactions promoting accumulation of FFAs have been activated by ischemia, and that postmortem hydrolysis of polyphosphoinositides is avoided. Recently, Nishihara and Keenan (1985) reported higher values for polyphosphoinositides in brain fixed using a freeze-blowing method, as compared with microwave-fixed tissue. [Note: Because the head of the animal reaches high temperatures during microwaving, it should be immersed in ice-cold water containing crushed ice immediately after microwave fixation. Thus, depending upon the experimental model investigated (species of animal, body weight, and so on), a suitable method can be chosen to rapidly inactivate brain enzymes for the study of minor lipid pools.]

3. Extraction of Lipids from Brain

Lipid extraction can be performed on fresh tissue, frozen tissue, or tissue obtained from microwaved animals. Frozen tissue is more difficult and time consuming to dissect and work with. On the other hand, microwave-fixed neural tissue can be more precisely dissected, and specific neuroanatomical regions more easily isolated. In addition, the higher temperatures involved in microwave fixation speed up the subsequent lipid extraction step.

Brains are obtained from frozen crania after splitting the skull by a blow with a hammer and chisel. Often, the freezing process itself will crack the skull longitudinally at its base. After the skull is opened, the brain can be removed using a chisel and chip cutter. Care should be taken to keep the pieces frozen by reimmersing them periodically in liquid nitrogen. In practice, it is best to minimize fragmentation of the frozen brain and observe anatomical landmarks as a guide to identification of specific brain regions.

One gram of brain (fresh tissue or tissue from microwaved animals) is homogenized in 20 vol (20 mL) of chloroform:methanol (CM) mixture, 2:1 by vol, at 2000 rpm for 1–2 min using a Potter-Elvehjem homogenizer featuring a motor-driven Teflon pestle. Frozen tissue is pulverized with a chilled mortar prior to addition of CM. Lipids are extracted using the procedure of Folch et al. (1957). Homogenates are placed in 50-mL glass tubes, which are then filled with nitrogen gas, sealed with Teflon-coated screw caps, and stored in a refrigerator (4°C) overnight. Tubes are centrifuged at 2000 rpm for 10 min in a table-top centrifuge, and the supernatants transferred to fresh tubes. Ten milliliters of CM is added to the pelleted residues, the tubes are filled with nitrogen gas and capped, and the suspensions are mixed well by vortexing for about 20 s. Tubes are centrifuged again at 2000 rpm for 10 min. The first and second supernatants (total vol = 30 mL) are combined and mixed with 0.2 vol (6 mL) of 0.05% $CaCl_2$, vortexed, and centrifuged at 1000 rpm for 10 min. The presence of Ca^{2+} in this wash step increases the yield of acidic lipids (i.e., phosphatidic acid and phosphatidylinositol) in the lower, organic phase. The upper, aqueous phase (containing lipoproteins, gangliosides, sulfatides, and salts) is removed with a Pasteur pipet and the lower, lipid-containing phase is washed with 14.4 mL of Folch theoretical upper phase (chloroform:methanol:0.05% $CaCl_2$, 3:48:47 by vol). The washed lower phase contains the purified lipid extract, which is subsequently placed in a 30–35°C waterbath and dried under a stream of nitrogen gas. A small, known amount of CM is added to each dried extract, and the tubes are filled with nitrogen, capped with Teflon-lined screw caps, and stored at −20°C until separation of lipid classes or other analyses can be carried out.

When working with water-rich samples such as subcellular fractions, lipids are extracted initially by adding 20 vol of 1:1 (by vol) CM, vortexing for 1–2 min, and leaving the mixture at room temperature (23–25°C) for 2 h. The lower chloroform:methanol ratio employed in this step is required to prevent formation of a two-phase system, which would otherwise compromise the lipid extraction. A two-phase system can often be converted to a single homogenous phase by the addition of a small amount of pure methanol. The amount of methanol added must be noted so that the lipid extract can be adjusted to 2:1 CM (by vol) later (*see below*). Each tube is centrifuged, the supernatant transferred to a second tube, and the residue extracted with one-half the original volume (10 vol) of 2:1 CM. After centrifugation, the two supernatants are combined and diluted with sufficient pure chloroform to raise the final chloroform:methanol ratio to 2:1. Subsequent purification steps are as described above.

Polyphosphoinositides are extracted from delipidated brain residue using a CM mixture containing acid (0.5% concentrated HCl), according to the method of Hauser and Eichberg (1973).

4. Separation of Phospholipids

Separation of phospholipids such as phosphatidylethanol-amine (PE), phosphatidylcholine (PC), phosphatidylserine (PS), phosphatidylinositol (PI), and phosphatidic acid (PA), as well as sphingomyelin (SM), can be achieved by two-dimensional TLC [Fig. 1; method of Rouser et al. (1970), as modified in our laboratory (Bazan et al., 1984; Bazan and Bazan, 1984; Reddy et al., 1985)]. Redi-Coat-2D plates (Supelco, Inc., Houston, TX), coated with a 0.25-mm or 0.50-mm thick layer of silica gel H, are activated for 1 h at 110°C immediately prior to use. Each 20 × 20 cm plate is scored into four 10 × 10 cm squares, using a syringe needle and spotting guide. A single 20 × 20 cm plate can then accommodate four samples of lipid extract, applied about 2.5 cm diagonally inward from each corner of the plate. An alternative spotting method is useful for the study of a single, minor phospholipid such as phosphatidic acid. Plates are divided in half to accommodate two samples each, and larger amounts of lipid extract are applied as oblique streaks, instead of as single spots (Rodriguez de Turco and Bazan, 1977). In order to prevent deactivation of plates, sample application is carried out in a Plexiglass box under a constant stream of nitrogen gas (Cruess and Seguin, 1965). For estimation of phosphorus in individual phospholipids (method of Rouser et al., 1970), aliquots containing 20–30 μg of lipid phosphorus (1–1.5% of the total lipid extract obtained from 1 g brain) are spotted onto 0.25-mm thick silica gel plates, and two-dimensional TLC is carried out (*see below*). For fatty acid methyl ester (FAME) analysis of individual phospholipids, each sample typically contains 40–50 μg of lipid phosphorus (2–2.5% of the total lipid extract from 1 g of brain) and is spotted onto a 0.5-mm thick silica gel plate. The two

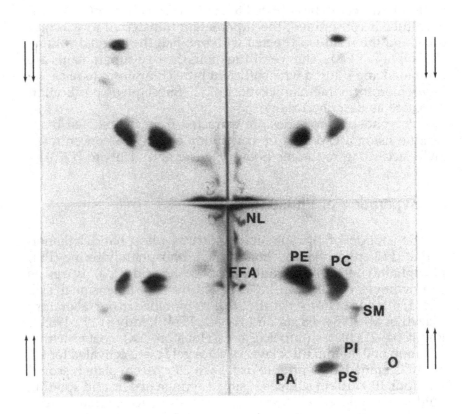

Fig. 1. Two-dimensional thin layer chromatographic separation of individual phospholipids. The Redi-Coat-2D plate (Supelco, Inc., Bellefonte, PA) was divided into four 10 × 10 cm squares and spotted with total lipid extract obtained from 10 mg of brain (lower left and upper right quadrants) and 20 mg of brain (upper left and lower right). Plates were chromatographed twice in the first dimension (marked with two arrows; solvent system consisting of chloroform:methanol:ammonia, 65:25:5) and once in the second dimension (one arrow; solvent system of chloroform:acetone:methanol:acetic acid:water, 3:4:1:1:0.5). Spots are visualized by exposure to iodine vapor. Labels in lower right quadrant: O, origin; PS, phosphatidylserine; PI, phosphatidylinositol; PA, phosphatidic acid; SM, sphingomyelin; PC, phosphatidylcholine; PE, phosphatidylethanolamine; FFA, free fatty acid; NL, neutral lipid.

solvent chambers (chloroform:methanol:ammonia, 65:25:5 by vol in the first chamber; chloroform:acetone:methanol:acetic acid:water, 3:4:1:1:0.5 by vol in the second) are allowed to saturate with solvent prior to introduction of the plates. Each chamber containing 100 mL of solvent can be used to run 16 samples on a total of four plates (two plates at a time), before the solvent system should be replenished. Chamber saturation is achieved by placing two 20 × 20 cm saturation pads in each tank (Alltech Associates, Inc., Deerfield, IL). The glass top of each chamber is sealed onto its base with Teflon stopcock grease (Dupont Inc.), and the chamber is allowed to saturate for about 30 min. Saturation is complete when the solvent system completely wets the saturation pads. One half of each plate (two 10 × 10 cm squares) is chromatographed once in the first dimension (first solvent system; chloroform:methanol:ammonia, 65:25:5 by vol), dried for 30 s with cold air from a hair dryer, chromatographed a second time in the first dimension (first solvent system), and dried as before. This process is repeated with the opposite half of the plate (the remaining two 10 × 10 cm squares). Less than 15 min is required for each run. After running plates twice in the first solvent system, plates are dried under nitrogen gas if fatty acid composition is to be determined. Alternatively, if phosphorus or radioactivity are to be measured in the separated lipids, the plates can be dried using cool air from a hair dryer.

In order to obtain successful separation of individual phospholipids by two-dimensional TLC, it is of critical importance to totally remove the ammonia present in the first solvent system before developing the plates in the acetic acid-containing, second solvent system. In addition, high laboratory humidity can partially deactivate the TLC plates during the drying step between solvent changes, resulting in poor separation of lipids. On humid days, we achieve good, reproducible separations by drying TLC plates in a desiccating chamber opened just enough to allow cool air from a hair dryer to contact the plate. The drying step usually takes 10–15 min. After drying, each plate is rotated 90° and placed in a second chamber which has been pre-saturated with 100 mL of the second solvent system: chloroform:acetone:methanol:acetic acid:water (3:4:1:1:0.5 by vol). After running the second solvent system through the lower two 10 × 10 cm squares, the plate is dried with a hair dryer for 30 s, rotated 180°, and placed back into the tank in order to develop the opposite two 10 × 10 cm squares. The plates are then dried until they are free of the odor of acetic acid (10–15 min).

For the estimation of either phosphorus or radioactivity in phospholipids, plates are exposed to iodine vapor after drying. Lipid spots visualized with iodine are outlined with a syringe needle, and the iodine is allowed to sublimate prior to scraping the spots into tubes for further processing.

5. Separation of Free Fatty Acids (FFAs), Diacylglycerols (DGs), and Triacylglycerols (TGs)

Neutral lipids (FFAs, DGs, and TGs) from brain lipid extracts are separated by a modified method employing mono-dimensional TLC (Bazan and Bazan, 1975; Roughan et al., 1979). Plates pre-coated with a 0.25-mm thick layer of silica gel GHL (Uni-plates, Analtech, Inc., Newark, DE) are divided into four equal lanes and activated for 1 h immediately prior to spotting. Samples consisting of 1/10 of the total lipid extract from 1 g of brain are spotted on three of the lanes, and reference standards (used for identification purposes) are spotted on the fourth. Spotting is carried out under a stream of nitrogen gas in a Plexiglass box. The plates are run in a chamber presaturated with 100 mL of petroleum ether:diethyl ether:acetic acid (60:40:2.3 by vol; Fig. 2). Hexane can be substituted for petroleum ether with minimal effect on R_f values. Alternative systems we use for separation of neutral lipid classes include (1) petroleum ether:diethyl ether:acetic acid (40:60:2.3), (2) petroleum ether:diethyl ether:acetic acid (40:60:1.5; Fig. 3), (3) petroleum ether:diethyl ether:acetic acid (70:30:1.3; Fig. 4), and (4) chloroform:acetone (96:4; Fig. 5). The solvent system selected depends on which lipids one wishes to separate. Free fatty acids and diacylglycerols are well resolved by the chloroform:acetone (96:4) solvent system (Fig. 5); with this system, however, free fatty acids overlap with monoacylglycerols and cholesterol esters overlap with triacylglycerols. With petroleum ether:diethyl ether or hexane:diethyl ether solvent systems, diacylglycerols and free fatty acids migrate more closely together. In many experimental designs, particularly when radioactively labeled fatty acids are used, it is critical to separate FFAs very well from diacylglycerols. Cross-contamination of these two lipid classes is generally difficult to prevent using petroleum ether- or hexane-based solvents; we find, however, that good separations are achieved using a solvent

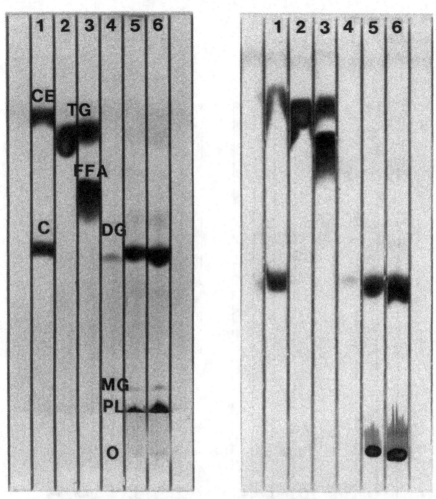

Fig. 2. Comparison of two types of chromatography plates. Separation of neutral lipids using a solvent system of hexane:diethyl ether:acetic acid, 60:40:2.3, by vol. (A) Silica gel LK5D plate. (Whatman Chemical Separation, Inc., Clifton, NJ). (B) Silica gel GHL Uni-plate. (Analtech, Inc., Newark, DE) Samples: lane 1, cholesterol (C) and cholesterol ester (CE); lane 2, triacylglycerol (TG); lane 3, free fatty acid (FFA) and triacylglycerol; lane 4, diacylglycerol (DG), lanes 5 and 6, total lipid extract from 10 and 20 mg of brain, respectively. Origin is labeled "O," and the solvent front is at the top of the plates. Spots were visualized by spraying plates with 3% cupric acetate in 8% orthophosphoric acid and charring at 110°C (Fewster et al., 1969). Diacylglycerols are not resolved from cholesterol using this system.

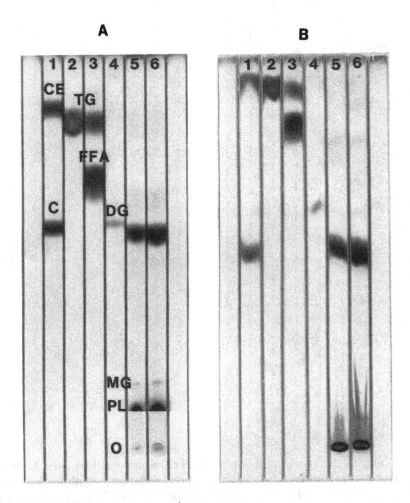

Fig. 3. Separation of neutral lipid classes by mono-dimensional thin layer chromatography using a solvent system of hexane:diethyl ether:acetic acid, 40:60:1.5. Comparison of two different types of chromatography plates. (A) Silica gel LK5D plate (Whatman Chemical Separation Inc., Clifton, NJ). (B) Silica gel GHL Uni-plate (Analtech, Inc., Newark, DE). Samples and labels are the same as in Fig. 2. Spots were visualized by charring, as described in Fig. 2. Monoacylglycerols (MGs) are better resolved in (A). Phospholipids are streaked, caused by overloading, in (B), however, diacylglycerols are better resolved. Neither system separates cholesterol esters from triacylglycerols.

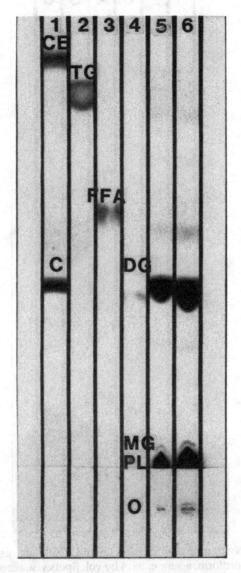

Fig. 4. Neutral lipid separation using a solvent system of petroleum ether:diethyl ether:acetic acid, 70:30:1.3 by vol. Silica gel LK5D plate. Samples and labels as in Fig. 3. Spots visualized by charring as described in Fig. 3. Triacylglycerols are well resolved from cholesterol esters.

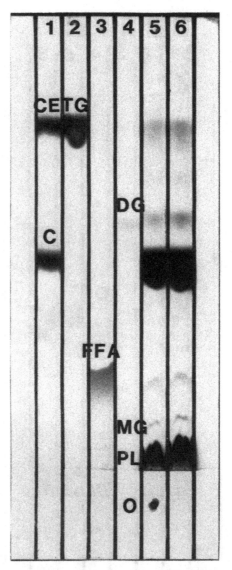

Fig. 5. Neutral lipid separation on an LK5D plate, using a solvent system of chloroform:acetone, 96:4 by vol. Spots visualized as in Fig. 3. Free fatty acids are found between spots for phospholipid and cholesterol. Diacylglycerols are well separated from cholesterol, but triacylglycerol and cholesterol esters are not well resolved. Samples and labels as in Fig. 3. MG, monoacylglycerol.

system of hexane: diethyl ether: acetic acid (40:60:1.5) in combination with Analtech Uniplates (Fig. 3b).

We obtain good separation of cholesterol esters from triacylglycerols using petroleum ether: diethyl ether: acetic acid (70:30:1.3; Fig. 4). In addition, gradient-thickness thin-layer chromatography (Bazan and Joel, 1970), which combines in a single run the advantages of preparative and analytical separations, can also be used to separate lipid classes. The solvent systems mentioned above can all be used with gradient thickness TLC plates.

Four plates (running two plates simultaneously per chamber) containing a total of 12 sample and four reference lanes can be developed using 100 mL of a solvent system. About 30 min is required to run a single plate. Spots are visualized by different methods, as described above, depending on whether fatty acyl group composition or radiolabeling is to be determined in the isolated lipids.

For estimation of lipid phosphorus or radioactivity in separated lipids, the dried plates are exposed to iodine vapor. Because iodine has a high affinity for double bonds, lipids containing unsaturated fatty acyl groups are visualized particularly well by this method. Iodine visualization cannot be used for lipids intended for analysis by GC because of interference by even trace amounts of halogen. Lipid spots are outlined with a needle and the iodine allowed to sublimate prior to scintillation counting so that quenching by the halogen is reduced. Radiolabeling of separated lipids is measured by scintillation counting after spots are scraped and transferred via a small plastic funnel to scintillation vials. Small spots (≤ 2 cm^2) are placed in 7-mL scintillation vials containing 0.4 mL of water—the water deactivates the silica gel—and 4.5 mL of Ready-Solv EP scintillation fluid (Beckman Instruments Inc., Fullerton, CA) added. Alternatively, 0.4 mL of 1% $Na_2S_2O_3$ (or 0.4 mL of 1% $NaHSO_3$) can be substituted for water, in order to reduce iodine to iodide ion, thereby decolorizing the mixture and minimizing quenching (modification of the method of Horrocks and Ansell, 1967). Large spots (≤ 8 cm^2) are scraped into 20-mL scintillation vials containing 1 mL of water, and 12 mL of Ready-Solv EP is added. Samples are vortexed and radioactivity measured with a Beckman LS 7500 liquid scintillation counter. Scintillation counts per min (cpm) are compensated for quenching and converted to disintegrations per min (dpm). Lipid phosphorus is estimated by spectrophotometric assay (method of Rouser et al., 1970) after scraping spots into disposable test tubes (13 × 100 mm).

When fatty acid composition is to be assessed, plates are sprayed (TLC sprayer, Analtech, Inc., Newark, DE) with 0.1% (w/v) 2′,7′-dichlorofluorescein (Sigma Inc., St. Louis, MO) in ethanol. Plates are dried for 60 s under a hair dryer and viewed with UV illumination (366 nm, Ultra-Violet Products Inc., San Gabriel, CA). Lipid classes appear as yellow-green fluorescent spots and are scraped into separate 16 × 125 mm Pyrex culture tubes (Corning Glass Works Inc., Corning, NY). The method of Morrison and Smith (1964) is used to convert lipid classes to fatty acid methyl esters (FAME): One milliliter of 14% boron trifluoride in methanol (Sigma Inc., St. Louis, MO) and 0.5 mL of benzene are added to each tube. Nitrogen gas is blown into the tubes before sealing them tightly with Teflon-lined screw caps. Tubes are vortexed and heated for 90 min at 100°C. A good seal between tube and cap is very important to prevent evaporation and oxidation of the sample during methanolysis. After heating, tubes are chilled briefly on ice and 3 drops of $3M$ HCl, 1.5 mL of water, and 3 mL of hexane are added. Tubes are flushed with nitrogen gas, vortexed vigorously, and centrifuged at 1000 g for 10 min. The upper, hexane-rich phase contains the FAMEs and is collected. The lower phase is re-extracted as above with 3 additional mL of hexane, and the combined hexane-soluble extracts subjected to GC (*see below*).

6. Separation and Quantification of Fatty Acid Methyl Esters by Gas-Liquid Chromatography

Fatty acid methyl esters (FAMEs) are routinely separated and quantitated in our laboratory using a Varian Vista Model 6000 gas-liquid chromatograph equipped with a CDS-401 integrated data module system, dual columns, and dual flame ionization detectors. Open glass columns measuring 1.8 m long × 2 mm inner diameter and packed with cyanosilicone stationary phase (10% SP-2330) on a 100/200 mesh Chromosorb WAW support (Supelco, Inc., Bellefonte, PA) are used. Helium at a flow rate of 30 mL/min is the carrier gas. The injection port and detector temperatures are 225 and 235°C, respectively.

For the separation of FAMEs (Fig. 6), the initial column temperature of 185°C during injection of sample is held for 15 min and then raised to 210°C at a rate of 7.5°C/min. After 5 min at 210°C, the

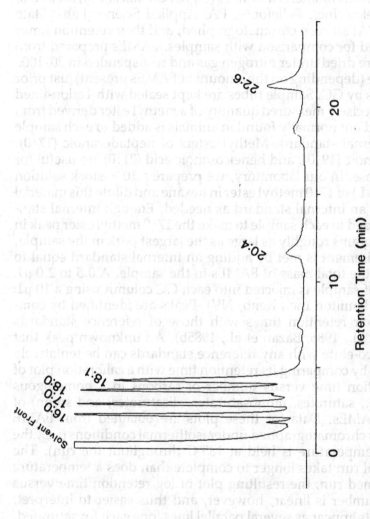

Fig. 6. Separation of FAMEs by open column gas-liquid chromatography. Sample from phospholipids of Rhesus monkey thalamus, 2 μg injected onto column. Column measures 1.8 m × 2 mm (id) and is packed with 10% SP-2330 on a 100/200 mesh Chromosorb WAW support (Supelco Inc). Carrier is helium gas at a flow rate of 30 mL/min. FID sensitivity is 1×10^{-12} afs.

column temperature is raised to 230°C at 20°C/min to purge the column in preparation for the next run. Authentic FAME standards obtained from commercial sources (Sigma Chemical Co., St. Louis, MO; Supelco, Inc., Bellefonte, PA; Applied Science Labs, State College, PA) are also chromatographed, and their retention times determined for comparison with samples. FAMEs prepared from samples are dried under nitrogen gas and resuspended in 20–1000 μL hexane (depending on the amount of FAMEs present) just prior to analysis by GC. Sample tubes are kept sealed with Teflon-lined caps. A precisely measured quantity of a methyl ester derived from a fatty acid not normally found in animals is added to each sample as an internal standard. Methyl esters of heptadecanoic (17:0), nonadecanoic (19:0), and heneicosanoic acid (21:0) are useful for this purpose. In our laboratory, we prepare a 10 × stock solution (50 nmol/μL) of 17:0 methyl ester in hexane and dilute this material for use as an internal standard as needed. Enough internal standard is added to each sample to make the 17:0 methyl ester peak in chromatograms roughly as large as the largest peak in the sample. This requirement is met by adding an internal standard equal to 5–10% of the total mass of FAMEs in the sample. A 0.5 to 2.0 μL (1–5 μg) of sample is injected into each GC column using a 10 μL syringe (Hamilton Inc., Reno, NV). Peaks are identified by comparing their retention times with those of reference standards (Reddy et al., 1985; Bazan et al., 1985b). An unknown peak that does not co-elute with any reference standards can be tentatively identified by comparing its retention time with a calibration plot of log retention time versus number of carbons in a homologous series (i.e., saturates, monosaturates, disaturates, and so on) of known FAMEs. Data for these plots are obtained from FAME standards chromatographed under isothermal conditions (i.e., the column temperature is held at 185°C throughout the run). The isothermal run takes longer to complete than does a temperature programmed run; the resulting plot of log retention time versus carbon number is linear, however, and thus easier to interpret. These plots appear as several parallel lines, one each for saturated, mono-, di-, tri-, tetraenoic, and so on, FAMEs.

A complementary approach to the identification of unknown peaks is to rechromatograph the unknown sample after subjecting it to catalytic hydrogenation, a process that reduces double bonds in the acyl chain to single bonds. To carry out hydrogenation, a sample is first dried under nitrogen and redissolved in 0.5 mL of methanol. A pinch of palladium black is added to the sample, and hydrogen gas bubbled through for 30 min. Samples are sealed and placed in a water bath at 50°C for 50 min, and the resulting hydrogenated FAMEs extracted twice into hexane (Aveldano and Bazan, 1973). If hydrogenation fails to alter the retention time of an unknown peak, the peak probably represents a saturated molecule. If an unknown peak migrates (its retention time changes) after hydrogenation, the unknown must be unsaturated. Often the carbon number of an unknown, unsaturated FAME can be inferred by comparing the integrated areas of peaks obtained before, versus after, hydrogenation; for example, an unsaturated, 24-carbon FAME producing a 50,000 unit area peak *before* catalytic hydrogenation would be expected to add 50,000 unit areas to the 24:0 methyl ester peak *after* hydrogenation.

The number of double bonds present in an unknown FAME can be determined by subjecting the sample to argentation TLC prior to GC. Argentation TLC exploits the weak attraction that exists between Ag^+ and double bonds, thereby separating FAMEs according to the number of double bonds present. Plates are prepared with silica gel containing 12% $AgNO_3$ in 30% NH_4OH, spotted with FAME samples, and developed with hexane:ether (80:20 by vol; Aveldano and Bazan, 1983). Bands, each containing FAMEs with the same number of double bonds, are visualized as described above with dichlorofluorescein under UV light and scraped into tubes. FAMEs are eluted from the silica gel with hexane and resolved by GC.

Most of the PE present in brain, especially in white matter PE, is in the ether form, in which one of the acyl groups esterified to glycerol in a typical diacyl phospholipid is substituted by an alkyl or alkenyl group in ether linkage with glycerol. Alkenyl groups account for up to 69% of the long-chain hydrocarbons present in PE of white matter (Nakagawa and Horrocks, 1983) and alkyl groups, for about 4% (Nakagawa and Horrocks, 1983). Ethanolamine and choline plasmalogens co-migrate with the respective diacyl phospholipid (PE or PC) during TLC. But, whereas methanolysis converts acyl groups of phospholipids to FAMEs, the alkyl and alkenyl groups of plasmalogens are converted to dimethyl acetals, which are susceptible to acid hydrolysis and behave somewhat differently during GC. Therefore, when working with plasmalogens, it becomes necessary to modify the procedure used for extracting the products of methanolysis. Instead of adding 3 drops of $3N$ HCl after methanolysis, a pinch of Na_2CO_3 is added, in order to neutralize fluoride and prevent acid hydrolysis of dimethyl acetals (Marta

Aveldano, personal communication). FAMEs and dimethyl acetals are then extracted into hexane and separated from each other by the following TLC method. A silica gel G plate is prewashed by development with 2:1 chloroform:methanol to remove any organic contaminants. The top of the plate is marked to assure that subsequent runs are carried out in the same direction. The plate is dried completely, activated for 30 min in a 120°C oven, and spotted with the extracted methanolysis products. Adjacent lanes are spotted with fatty acid methyl ester and dimethyl acetal standards. Plates are developed with 95:5 hexane:diethyl ether and visualized with dichlorofluorescein as described. Methyl esters have an approximate R_f of 0.67 using this system, whereas dimethyl acetals remain near the origin. Silica containing dimethyl acetals and FAMEs are scraped into separate 5 mL tubes and the derivatized lipids are eluted from the silica (modified method of Arvidson, 1968) by extracting three times with 4 mL of chloroform:methanol:acetic acid:water (50:39:1:10 by vol). The collected eluate is brought to a final volume of 12 mL, 4 mL of 1N NH_4OH added, and the solution vortexed and centrifuged. The upper phase is discarded and the lower phase mixed with 4 mL of 1:1 methanol:water. The new lower phase is collected, dried under nitrogen gas, and resuspended in a known volume of hexane for injection into the GLC.

7. Separation of Fatty Acid Methyl Esters by Capillary Gas-Liquid Chromatography

As neurochemical studies become progressively more focused upon specific brain regions, cultured neurons, and glia, as opposed to whole brain, it has become increasingly important to develop more sensitive methods for the detection and quantitation of fatty acyl groups in small portions of tissue. One method developed in response to this need is capillary GC.

By using capillary GC, one can improve the sensitivity of FAME detection by a factor of at least 10-fold over that obtained with a packed GC column containing the same stationary phase. For our FAME separations (Fig 7; Bazan et al., 1986), we employ a 30-m fused-silica capillary column containing a 0.20-μm coating of SP-2330, a highly polar stationary phase consisting of 90% biscyanopropyl:10% phenyl cyanopropyl polysiloxane (Supelco Inc., Bellefonte, PA). The column has an inner diameter of 0.25 mm and

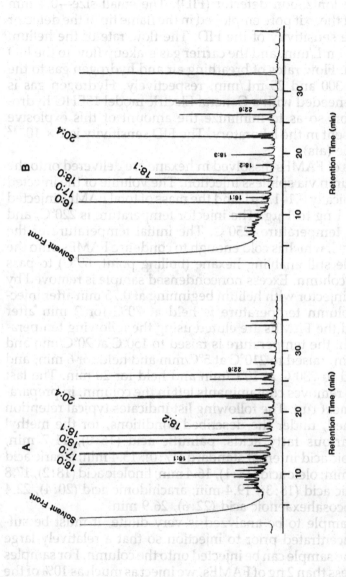

Fig. 7. Separation of FAMEs by capillary gas-liquid chromatography. (A) free fatty acid and (B) diacylglycerol fractions isolated from hippocampus. Sample size, 500 ng. Column is a 30 m × 0.25 mm id fused-silica capillary column containing a 0.20 μm-thick coating of SP-2330 stationary phase (Supelco Inc). Temperature programming is as described in the text. Carrier is helium gas at a flow rate of 1.0 mL/min. FID sensitivity is 1×10^{-12} afs.

is installed in a Varian 6000 gas-liquid chromatograph equipped with a flame ionization detector (FID). The small size—0.1 mm diameter—of the exit pore employed in the flame tip of the detector enhances the sensitivity of the FID. The flow rate of the helium carrier gas is 1 mL/min, and the carrier gas makeup flow to the FID is 30 mL/min. Flow rates of breathing air and hydrogen gas to the detector are 300 and 25 mL/min, respectively. Hydrogen gas is produced as needed with a General Electric model 15EHG hydrogen generator, so as to minimize the amount of this explosive material present in the laboratory. The FID sensitivity is 1×10^{-12} amp full scale (afs).

Aliquots of FAMEs dissolved in hexane are delivered onto the capillary column via splitless injection. The volume of the injected sample is typically 0.1–1.0 μL and the mass of total FAMEs injected ranges from 1 ng to 1 μg. The injector temperature is 220°C, and the detector temperature, 250°C. The initial temperature of the column is 70°C, which is cold enough to condense FAMEs onto the column while still enabling hexane (boiling point, 69°C) to pass through the column. Excess noncondensed sample is removed by purging the injector with helium beginning at 0.75 min after injection. The column temperature is held at 70°C for 2 min after injection and the FAMEs are eluted using the following temperature program: the temperature is raised to 150°C at 20°C/min and held for 8 min, raised to 210°C at 5°C/min and held for 6 min, and finally raised to 230°C at 10°C/min and held for 20 min. The last heating step removes contaminants left in the column, in preparation for the next run. The following list indicates typical retention times obtained, under the described conditions, for the methyl esters of various fatty acids: palmitic acid (16:0), 11.7 min; heptadecanoic acid internal standard (17:0), 13.4 min; stearic acid (18:0), 15.5 min; oleic acid (18:1), 16.4 min; linoleic acid (18:2), 17.8 min; linolenic acid (18:3), 19.4 min; arachidonic acid (20:4), 22.4 min; and docosahexaenoic acid (22:6), 26.9 min.

If the sample to be analyzed is very dilute, it must be sufficiently concentrated prior to injection so that a relatively large fraction of the sample can be injected onto the column. For samples containing less than 2 ng of FAMEs, we inject as much as 10% of the total sample onto the column, by injecting 1 μL of the total sample dissolved in as little as 10 μL of hexane. The minimum mass that can be detected by our system is about 100 pg in a single peak. Therefore, as little as 1 ng (10 × 100 pg) of a single FAME in an entire sample can be detected, provided that blank values are sufficiently low. To minimize potential contamination of highly concentrated blanks or samples, one must take great care throughout all preparative steps (lipid extraction, TLC, methanolysis, and so on) to assure that the glassware, solvents, and reagents used are free of fatty acids and other contaminants. The importance of this step to the accurate measurement of trace acyl groups in a dilute sample cannot be over emphasized.

8. Separation of Fatty Acid Methyl Esters by High Performance Liquid Chromatography

In our laboratory, we are particularly interested in the metabolism of the n-3 essential fatty acid family during development. (Note: "n-3" refers to the existence of a double bond at the third carbon from the methyl end of the fatty acid.) In general, n-3 fatty acids enter mammalian tissues as the dietary precursor linolenic acid (18:3, n-3). Subsequent elongation and desaturation of the precursor produces docosahexaenoic acid (22:6, n-3), a long-chain, polyunsaturated fatty acid that is highly enriched within synaptic membranes and retinal photoreceptor cells. The large concentration of docosahexaenoate found in neural tissue suggests that it plays an important role in normal functioning of excitable membranes.

The metabolism of n-3 fatty acids can be studied in vivo by determining the identity and distribution of radiolabeled products subsequent to intraperitoneal injection of [1-^{14}C]linolenic acid. In our studies, we use C57BL/6J mice, focusing on the period of early postnatal development. [1-^{14}C]Linolenic acid (specific activity, 55 Ci/mol; New England Nuclear, Inc.) is prepared for intraperitoneal injection in the following way: an aliquot of radiolabeled linolenic acid is dried under a stream of nitrogen gas. One microliter of 50 mM $NaHCO_3$ is added per μCi of radioactivity, and the mixture is vigorously sonicated and vortexed until a homogenous, usually milky appearing suspension of the sodium salt of the fatty acid is obtained. The radioactivity of a 1-μL aliquot of the material is measured by scintillation counting just prior to injection into animals. We normally inject 5 μCi of radiolabeled fatty acid in 5 μL of solution into each mouse pup; pups range in weight from 1.5 g at birth to 7 g at 14 d of postnatal age. Animals are sacrificed from 2 h to 10 d postinjection, and target organs dissected out. Lipids are extracted from these tissues with 2:1 chloroform:methanol, following the method of Folch et al. (1957), as previously described. Lipid phosphorus is measured in the extracts by the method of Rouser et al. (1970).

Acyl groups in the lipid extracts are converted to FAMEs, as described, by the method of Morrison and Smith (1964). Samples are kept in an inert, nitrogen atmosphere as much as possible to prevent oxidation of the fatty acids.

In order to prepare the FAME samples for separation by reverse phase HPLC, they must be redissolved in a solvent more polar than hexane. This solvent change is accomplished by evaporating the hexane from the sample under N_2 prior to redissolving it in the desired solvent. Usually, we dissolve the FAME samples in 100% acetonitrile. Often, however, concentrated FAME samples require a somewhat less polar solvent system in order for complete solubilization to occur. We find that, under our HPLC conditions, a mixture of 6:2:1 (by vol) acetonitrile:chloroform:methanol efficiently dissolved the more concentrated samples without altering the retention times of individual FAMEs as separated by reverse phase HPLC (see below).

The redissolved samples are subjected to scintillation counting to determine the injection volume required for successful detection of radioactive peaks during HPLC. In our hands, a minimum radioactivity of 2000 dpm per sample is required to yield one or two easily resolvable HPLC peaks.

FAMEs are separated by a method based upon the work of Aveldano et al. (1983); we have fine tuned this method for the separation of n-3 FAMEs. Individual FAMEs are resolved using an 8 mm × 10 cm Radial Pak liquid chromatography cartridge containing μBondapak C18 (Waters Associates, Inc., Milford, MA), and the column is protected with a Guard Pak precolumn insert also containing μBondapak C18. The nominal particle size of the packing material is 10 μm. Sample size is typically 200 μL of radiolabeled FAMEs. Samples are introduced onto the column via a Waters WISP 710B automatic injection unit, and individual FAMEs are eluted with acetonitrile:water mixtures. The acetonitrile concentration is initially held at 70% (by vol) for 55 min, then linearly increased to 100% during 10 min and held for 15 min longer. After each run, the solvents are returned to initial conditions (70% acetonitrile) and held for 10 min, in order to equilibrate the column in preparation for the next injection. The water employed for HPLC undergoes a primary cycle of deionization and distillation (Corning Mega-Pure water purifier; Corning, NY), followed by a second cycle of deionization (Gelman Water I, Ann Arbor, MI) and filtration (Sybron/Barnstead Organic Pure, Boston, MA). Radioactive peaks are resolved with a Flo-One Beta Radioactive Flow Detector (Radiomatic Instruments, Tampa, FL). Eluted peaks are monitored for UV absorption at 205 nm with a Waters Lambda-Max Model 480 Liquid Chromatography Spectrophotometer. Selection of 205 nm as the monitored wavelength avoids significant interference by acetonitrile in the mobile phase, yet permits excellent detection of unsaturated FAMEs, as well as moderately sensitive detection of saturated FAMEs.

For the separation of individual FAMEs by HPLC, we employ flow rates of 2 mL/min for the mobile phase and 6 mL/min for the scintillation cocktail (Beckman Ready-Solv EP), yielding a total flow rate of 8 mL/min. The relatively high cocktail flow rate is required to assure that the mobile phase and scintillation cocktail are mixed sufficiently well to form a single phase system, which is essential for reliable scintillation counting to be achieved. We find that individual peaks containing a minimum of 500–1000 dpm (or about 2000 dpm per sample) are resolved by the flow scintillation detector. Under the described conditions, we obtain the following retention times for various FAMEs (Fig. 8): 20:5, n-3, 25 min; 18:3, n-3, 28 min; 22:6, n-3, 32 min; 22:5, n-3, 40 min; and 16:0, 63 min. Retention times are sometimes observed to shift by several min from run to run; however, the order of elution of FAMEs does not change. Because of this variation, it is recommended that a set of FAME standards be run with each batch of samples to aid in the identification of the various eluted peaks.

9. Conclusion

In this chapter, we have presented a detailed account of current methodology for the analysis and separation of lipids. Accurate determination of the fatty acyl composition of brain requires care throughout a multi-step procedure involving: 1) tissue fixation, 2) lipid extraction, 3) isolation of desired lipid classes, 4) transmethylation of acyl groups, and 5) separation and quantitation of methylated acyl groups. Tissue fixation (Step 1) is usually accomplished by homogenization in the solvent system used for lipid extraction (Step 2). However, fixation by high power head-focussed microwave (6.5 KW) irradiation followed by homogenization in the lipid extraction solvent system is preferred in the analy-

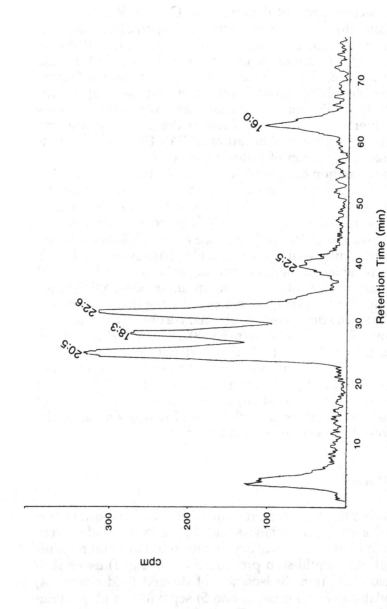

Fig. 8. Separation of radiolabeled fatty acid methyl ester standards by reverse phase high performance liquid chromatography. (Waters Inc. system is described in the text.) Sample was eluted with 70% acetonitrile (in water) for 55 min. The acetonitrile concentration was then linearly increased to 100% during 10 min and held for 15 min. Sample volume = 200 μL, total dpm = 10000. Solvent flow rate = 2 mL/min. Scintillation cocktail (Beckman EP) flow rate = 6 mL/min. The peak at 4 min arises from polar contaminants.

sis of highly labile lipid pools, because ischemia can affect extremely fast changes in lipid composition. Where appropriate, tissues from sham-operated animals should be processed in parallel with tissues obtained from experimental animals. Isolation of the lipid classes of interest (Step 3) is conveniently achieved by mono- or two-dimensional thin-layer chromatography on silica gel plates using appropriate solvent systems. Acyl groups are cleaved from complex lipids and converted to the corresponding fatty acid methyl esters (Step 4) by treating the isolated lipid classes with boron trifluoride. Finally, individual fatty acid methyl esters are resolved and measured (Step 5) by gas-liquid chromatography or by reverse phase high performance-liquid chromatography. Gas-liquid chromatography on a narrow-bore capillary column can detect as little as 100 pg of a fatty acid methyl ester in a single peak.

Throughout all of the steps of the fatty acid analysis, care must be taken to prevent oxidation and contamination of lipid extracts. High purity solvents should be employed in the extraction steps and samples should be maintained in an inert atmosphere (i.e. pure nitrogen gas), as much as possible. Oxidation of lipid extracts is further reduced by storing them cold ($< -20°C$) and protected from ultraviolet light. Finally, it is important that blank extracts (no tissue) are carried through all of the steps of the analysis so that sample data can be properly corrected. This step is made necessary in part because HPLC plates, solvents, glassware and derivatization reagents can contain variable amounts of lipids and other contaminants which may become concentrated during sample processing.

Acknowledgment

This work was supported by research grant NS-23002 from the NIH.

References

Arvidson G. A. E. (1968) Structural and metabolic heterogeneity of rat liver glycerophosphatides. *Eur. J. Biochem.* **4**, 478–486.

Aveldano M. I. and Bazan N. G. (1973) Fatty acid composition and level of diacylglycerols and phosphoglycerides in brain and retina. *Biochim. Biophys. Acta* **296**, 1–19.

Aveldano M. I. and Bazan N. G. (1975a) Differential lipid deacylation during brain ischemia in a homeotherm and a poikilotherm. Content and composition of free fatty acids and triacylglycerols. *Brain Res.* **100,** 99–110.

Aveldano M. I. and Bazan N. G. (1975b) Rapid production of diacylglycerols enriched in arachidonate and stearate during early brain ischemia. *J. Neurochem.* **25,** 919–920.

Aveldano M. I. and Bazan N. G. (1983) Molecular species of phosphatidylcholine, -ethanolamine, -serine and -inositol in microsomal and photoreceptor membranes of bovine retina. *J. Lipid Res.* **24,** 620–627.

Aveldano M. I., Van Rollins M., and Horrocks L. A. (1983) Separation and quantitation of free fatty acids and fatty acid methyl esters by reverse phase high pressure liquid chromatography. *J. Lipid Res.* **24,** 83–93.

Banschbach M. W. and Geison R. L. (1974) Post-mortem increase in rat cerebral hemisphere diglyceride pool size. *J. Neurochem.* **23,** 875–877.

Bazan H. E. P. and Bazan N. G. (1984) Composition of phospholipids and free fatty acids and incorporation of labeled arachidonic acid in rabbit cornea. Comparison of epithelium, stroma and endothelium. *Curr. Eye Res.* **3,** 1313–1319.

Bazan H. E. P., Sprecher H., and Bazan N. G. (1984) De novo biosynthesis of docosahexaenoyl phosphatidic acid in bovine retinal microsomes. *Biochim. Biophys. Acta* **796,** 11–19.

Bazan N. G. (1970) Effects of ischemia and electroconvulsive shock on free fatty acid pool in the brain. *Biochim. Biophys. Acta* **218,** 1–10.

Bazan N. G. (1971) Changes in free fatty acids of brain by drug-induced convulsions, electroshock and anesthesia. *J. Neurochem.* **18,** 1379–1385.

Bazan N. G. (1982) Metabolism of phospholipids in the retina. *Vision Res.* **22,** 1539–1548.

Bazan N. G. (1983) Metabolism of Phosphatidic Acid, in *Handbook of Neurochemistry* 2nd Edn. vol. 3 (Lajtha A., ed.) Plenum New York.

Bazan N. G. and Bazan H. E. P. (1975) Analysis of Free and Esterified Fatty Acids in Neural Tissues Using Gradient-Thickness Thin-Layer Chromatography, in *Research Methods in Neurochemistry* vol. III (Marks N. and R Rodnight R., eds.) Plenum, New York.

Bazan N. G. and Cellik S. (1972) Improved separation and quantification of free fatty acids and other tissue lipids by gradient-thickness thin-layer chromatography. *Analyt. Biochem.* **45,** 309–314.

Bazan N. G. and Joel C. D. (1970) Gradient-thickness thin-layer chromatography for the isolation and analysis of trace amounts of free fatty acids in large lipid samples. *J. Lipid Res.* **11,** 42–47.

Bazan N. G., Bazan H. E. P., Kennedy W. G., and Joel C. D., (1971) Regional distribution and rate of production of free fatty acids in rat brain. *J. Neurochem.* **18,** 1387–1393.

Bazan N. G., Birkle D. L., and Reddy T. S. (1985a) Biochemical and Nutritional Aspects of the Metabolism of Polyunsaturated Fatty Acids and Phospholipids in Experimental Models of Retinal Degeneration, in *Retinal Degeneration: Contemporary Experimental and Clincal Studies* (LaVail M. M., Anderson G., and Hollyfield J., eds.) Alan R. Liss, New York.

Bazan N. G., Reddy T. S., Redmond T. M., Wiggert B., and Chader G. J. (1985b) Endogenous fatty acids are covalently and non-covalently bound to interphotoreceptor retinoid-binding protein in the monkey retina. *J. Biol. Chem.* **260**, 13677–13680.

Bazan N. G., Morelli de Liberti S. G., Rodriguez de Turco E. B., and Pediconi M. F. (1983) Free Arachidonic and Docosahexaenoic Acid Accumulation in the Central Nervous System During Stimulation, in *Neural Membranes* (Sun G. Y., Bazan N. G., Wu J., Porcellati G., and Sun A. Y., eds.) Humana, Clifton, New Jersey.

Bazan N. G., Scott B. L., Reddy T. S., and Pelias M. Z. (1986) Decreased content of docosahexaenoate and arachidonate in plasma phospholipids in Usher's syndrome. *Biophys. Biochem. Res. Commun.* **141**, 600–604.

Cenedella R. J., Galli C., and Paoletti R. (1975) Brain free fatty acid levels in rats sacrificed by decapitation versus focused microwave-irradiation. *Lipids* **10**, 290–293.

Cruess R. L. and Seguin F. W. (1965) Box for application of samples to thin-layer chromatograms under nitrogen. *J. Lipid Res.* **6**, 441–442.

Eichberg J. and Hauser G. (1967) Concentrations and disappearance post-mortem of polyphosphoinositides in developing rat brain. *Biochim. Biophys. Acta* **14**, 415–422.

Fewster M. E., Burns B. J., and Mead J. F. (1969) Quantitative densitometric thin-layer chromatography of lipids using copper acetate reagent. *J. Chromatogr.* **43**, 120–126.

Folch J., Lees M., and Sloane Stanley G. H. (1957) A simple method for the isolation and purification of total lipides from animal tissues. *J. Biol. Chem.* **226**, 497–509.

Galli C. and Spagnuolo C. (1976) The release of brain free fatty acids during ischemia in essential fatty acid-deficient rats. *J. Neurochem.* **26**, 401–404.

Hauser G. and Eichberg J. (1973) Improved conditions for the preservation and extraction of polyphosphoinositides. *Biochim. Biophys. Acta* **326**, 201–209.

Hauser G., Eichberg J., and Gonzalez-Sastre F. (1971) Regional distribution of polyphosphoinositides in rat brain. *Biochim. Biophys. Acta* **248**, 87–95.

Horrocks L. A. and Ansell G. B. (1967) The incorporation of ethanolamine into ether-containing lipids in rat brain. *Lipids* **2**, 329–333.

Levy D. E. and Duffy T. E. (1975) Effect of ischemia on energy metabolism in the gerbil cerebral cortex. *J. Neurochem.* **24,** 1287–1289.

Morrison W. R. and Smith L. M. (1964) Preparation of fatty methyl esters and dimethylacetals from lipids with boron fluoride-methanol. *J. Lipid Res.* **5,** 600–608.

Nakagawa Y. and Horrocks L. A. (1983) Separation of alkenylacyl, alkylacyl and diacyl analogues and their molecular species by high performance liquid chromatography. *J. Lipid Res.* **24,** 1268–1275.

Nishihara M. and Keenan R. W. (1985) Inositol phospholipid levels of rat forebrain obtained by freeze-blowing method. *Biochim. Biophys. Acta* **835,** 415–518.

Ponten V., Ratcheson R. A., Salford L. G., and Siesjö B. K. (1973) Optimal freezing conditions for cerebral metabolites in rats. *J. Neurochem.* **21,** 1127–1138.

Reddy T. S., Birkle D. L., Armstrong D., and Bazan N. G. (1985) Change in content, incorporation and lipoxygenation of docosahexaenoic acid in retina and retinal pigment epithelium in canine ceroid lipofuscinosis. *Neurosci. Lett.* **59,** 67–72.

Rodriguez de Turco E. B. and Bazan N. G. (1977) Simple preparative and analytical thin-layer chromatographic method for the rapid isolation of phosphatidic acid from tissue lipid extracts. *J. Chromatogr.* **137,** 194–197.

Roughan P., Holland R., and Slack G. (1979) On the control of long chain fatty acid synthesis in isolated spinach *(Spinacia oleracea)* chloroplasts. *Biochem. J.* **184,** 193–199.

Rouser G., Fleischer S., and Yamamoto A. (1970) Two dimensional thin-layer chromatographic separation of polar lipids and determination of phospholipids by phosphorus analysis of spots. *Lipids* **5,** 494–496.

Steroids and Related Isoprenoids

**Thomas J. Langan, Robert S. Rust,
and Joseph J. Volpe**

1. Introduction

The principal nonsaponifiable lipids of mammalian cells are derived from multiple units of the five-carbon hydrocarbon isoprene (2-methyl-1,3-butadiene) (Fig. 1A), and may consequently be termed isoprenoids. These compounds are further subcategorized as *terpenes* and *steroids* (for reviews, *see* Heftmann, 1969; Brown and Goldstein, 1980; Poulter and Rilling, 1981). Steroids have in common a 19-carbon saturated tetracyclic structure known as perhydrocyclopentanophenanthrene, which may thus be considered the basic steroid nucleus (Fig. 1B). Terpenes consist of either linear or cyclic arrangements of multiple isoprene units; since the steroid nucleus is derived from cyclization of the linear terpene squalene, it is evident that steroids, in fact, represent a subclass of terpenoids. Cholesterol is quantitatively the major steroid synthesized within neural tissue (Kabara, 1973; Bhat and Volpe, 1983), and methodologies relating to the metabolism of cholesterol will consequently be a major focus of this review.

The biogenesis of cholesterol is of considerable interest to neurobiologists, since the relative proportion of cholesterol in brain exceeds that in extraneural tissues (Jones et al., 1975; Norton and Poduslo, 1973; Wells and Dittmer, 1967). Moreover, endogenous sterol synthesis in developing brain is extremely active, with this cholesterologenic activity capacity being located predominantly in glial cells (Jones et al., 1975; Volpe and Hennessy, 1977; Volpe et al., 1978). This conspicuous contribution of sterol metabolism to brain ontogenesis relates in part to the multiplicity of cell membranes and cell processes resulting from the extensive arborization of neuronal and glial cells (Cuzner and Davison, 1968; Kishimoto et al., 1965), but also to the fact that the elaborations of neural membranes, specifically myelin, are particularly enriched in sterol (Wells and Dittmer, 1967; Kabara, 1973; Norton and Poduslo, 1973).

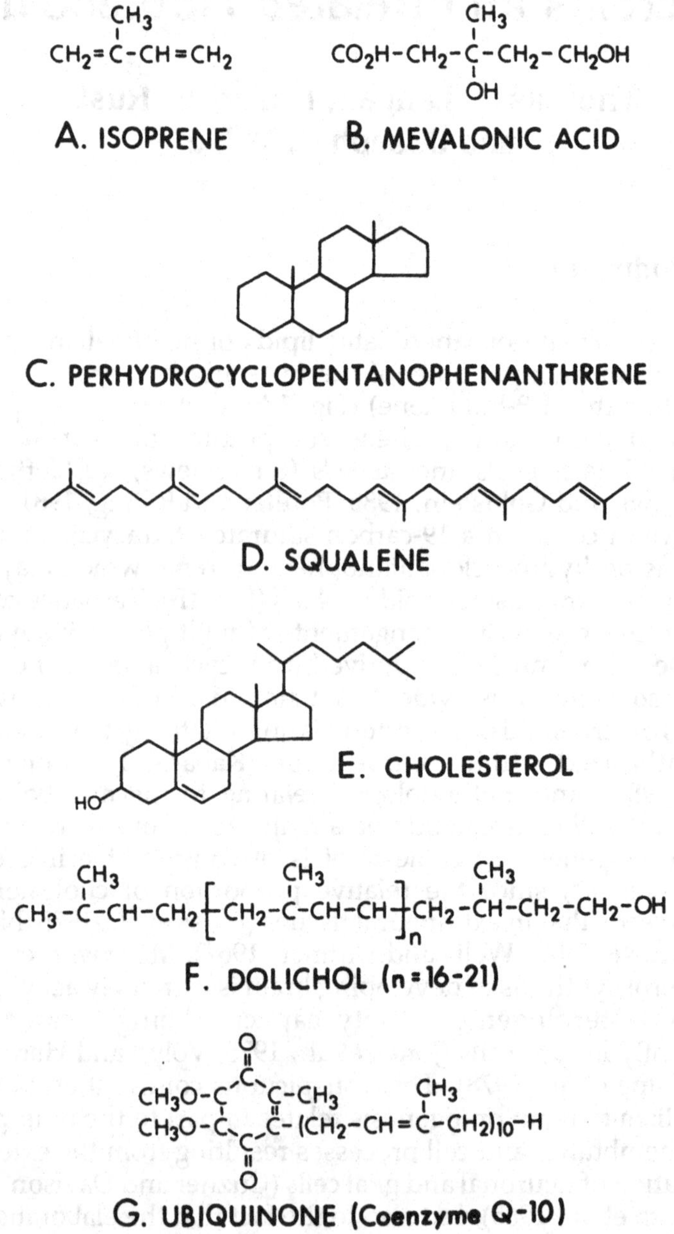

$$CH_2=\overset{\overset{\displaystyle CH_3}{|}}{C}-CH=CH_2$$

A. ISOPRENE

$$CO_2H-CH_2-\overset{\overset{\displaystyle CH_3}{|}}{\underset{\underset{\displaystyle OH}{|}}{C}}-CH_2-CH_2OH$$

B. MEVALONIC ACID

C. PERHYDROCYCLOPENTANOPHENANTHRENE

D. SQUALENE

E. CHOLESTEROL

$$CH_3-\overset{\overset{\displaystyle CH_3}{|}}{C}-CH-CH_2\left[CH_2-\overset{\overset{\displaystyle CH_3}{|}}{C}-CH-CH\right]_nCH_2-\overset{\overset{\displaystyle CH_3}{|}}{CH}-CH_2-CH_2-OH$$

F. DOLICHOL (n = 16-21)

$$\begin{array}{c} CH_3O-\overset{\displaystyle O}{\overset{\displaystyle \|}{C}}\diagdown C-CH_3 \\ CH_3O-\underset{\displaystyle \underset{O}{\|}}{C}\diagup C-(CH_2-CH=\overset{\overset{\displaystyle CH_3}{|}}{C}-CH_2)_{10}-H \end{array}$$

G. UBIQUINONE (Coenzyme Q-10)

Fig. 1. Chemical structure of isoprene and major isoprenoids of mammalian cells.

In addition to its role as a source of the steroid nucleus via cyclization of squalene, the sterol biosynthetic pathway (Fig. 2) provides several isoprenoid intermediates that have an impressive range of functions in intermediary metabolism (for reviews, *see* Brown and Goldstein, 1980; Lennarz, 1983). These nonsterol derivates include dolichol, a long-chain polyisoprenoid alcohol that functions as a carrier of sugars during the synthesis of N-linked glycoproteins (Waechter and Lennarz, 1976; Carson and Lennarz, 1981; Lennarz, 1983). A polyisoprenoid side chain contributes also to the structure of ubiquinone, a constituent of the electron transport chain (Faust et al., 1979; Nambudiri et al., 1980). Farnesyl pyrophosphate is a precursor of both dolichol and ubiquinone, in addition to cholesterol, and therefore serves as a major branch point of the sterol biosynthetic pathway (Fig. 2). Thus, this pathway, by providing sterol and nonsterol derivatives with key functions in a variety of important biochemical pathways, potentially represents a mechanism involved in coordination of intracellular metabolism (Brown and Goldstein, 1980; Chen, 1984). In addition, these sterol and nonsterol intermediates have been demonstrated to modulate the growth and differentiation of both extraneural (Chen et al., 1975; Habenicht et al., 1980; Schmidt et al., 1982; Quesney-Huneeus et al., 1983; Fairbanks et al., 1984; Yachnin et al., 1984) and neural cells (Maltese et al., 1981; Volpe and Obert, 1982; Bhat and Volpe, 1984; Maltese and Sheridan, 1985; Langan and Volpe, 1986).

2. Neurochemical Strategies for Examination of Steroid Metabolism

Techniques applicable to the study of lipid metabolism in neural tissue have progressed substantially during the past several decades. The classical approach of quantitative extraction of lipids from homogenates of either whole brain or from dissected brain regions (Folch et al., 1957; Johnson, 1979) or from subcellular fractions (Ramsey et al., 1971) remains useful for several specific purposes, such as the determination of developmental profiles of accumulation of lipid metabolites in vivo (Wells and Dittmer, 1967; Eto and Suzuki, 1972; Ramsey et al., 1971; Mitzen and Koeppen, 1984; Sakakihara and Volpe, 1984, 1985a,b) or the performance of procedures that require large amounts of cerebral tissue (Freshney,

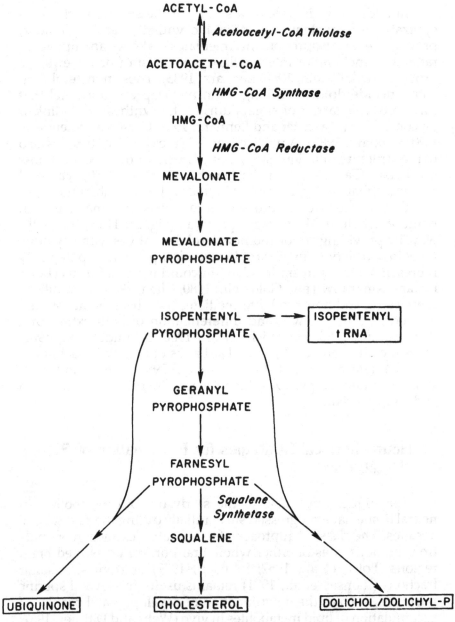

Fig. 2. Cholesterol biosynthetic pathway and its major branched pathways (adapted from Brown and Goldstein, 1980).

1983). Employing brain homogenates for analytical techniques related to lipid biochemistry has distinct disadvantages, however. A major short-coming results from the fact that mammalian brain, as a result of the incompletely understood divergence of its immature neural epithelial precursor cells along distinct pathways of cellular differentiation, is extremely heterogeneous with respect to its cellular composition (Dobbing and Sands, 1973, 1979; Federoff et al., 1977, 1985; Raff et al., 1983; Martin, 1985). Moreover, specific biochemical functions vary markedly among cells of differentiated phenotypes as well as among different brain regions (Hertz et al., 1982; DeVellis et al., 1983; Hertz et al., 1985). Particularly relevant to the current discussion is the demonstration that cells differentiated along glial and neuronal directions differ with respect to the details of the regulation of the synthesis of the steroid nucleus (Ramsey et al., 1971; Jones et al., 1975; Volpe and Hennessy, 1977; Volpe et al., 1978; Volpe and Obert, 1982; Bhat and Volpe, 1983).

Thus, the necessity for examining the biochemistry of specific cell types has resulted in the development of techniques for deriving relatively homogeneous cellular preparations. Such techniques include utilization of established transformed cell lines (Benda et al., 1968; Schein et al., 1970; Federoff, 1977), or of primary cultures derived from brain explants (Murray, 1971), or from dissociated developing brain (Booher and Sensenbrenner, 1972; Raff et al., 1979; Hertz et al., 1982; DeVellis et al., 1983). The technical details involved in employing such strategies, as well as the many factors that determine which strategy is appropriate in a given experimental situation, have been reviewed recently (Hertz et al., 1985).

Studies of isoprenoid metabolism in neural cells conducted in our laboratory have incorporated several such neurochemical strategies, including analysis of brain homogenates (Sakakihara and Volpe, 1984, 1985a,b), utilization of established glial (Volpe and Hennessy, 1977; Maltese et al., 1980; Volpe and Obert, 1982; Volpe and Goldberg, 1983; Volpe et al., 1985; Bhat and Volpe, 1984) and neuronal (Volpe and Hennessy, 1977; Maltese et al., 1981) cell lines, and derivation of primary cultures from the dissociation of fetal (Volpe et al., 1985) and neonatal (Wernicke and Volpe, 1986; Langan and Volpe, 1986, 1987) rodent brain. The specific procedures that will be described include enzymologic and chromatographic techniques for examination of the synthesis of cholesterol and its major precursors, as well as the metabolism of the important nonsterol isoprenoids ubiquinone and dolichol. Although some of the methods that will be presented were developed by investigators interested in isoprenoid metabolism in extraneural tissues, we have found that these procedures are generally applicable to the various neurochemical strategies that have been mentioned in this section.

3. Procedures Related to the Biosynthesis of the Steroid Nucleus

3.1. Biosynthesis of Sterols from [14C]Acetate

The determination of the net rate of synthesis of the steroid nucleus may be readily determined in cultured neural cells as incorporation of radiolabeled percursors into sterols. A modification (Volpe and Hennessy, 1977) of the method of Popjak (1969) exploits the fact that digitonin quantitatively precipitates 3-β-hydroxysterols. The procedure is extremely reproducible and in our hands enables recovery of >90% of sterol added as internal standard.

3.1.1. [14C]Acetate Pulse

Cell cultures are exposed to [14C]acetic acid by addition of the isotope to the culture medium. The conditions of the acetate pulse need to be adjusted depending on such factors as type of cell and cell density. Consistent results are usually yielded by addition of the [14C]acetate at a final concentration of 1.0–2.0 μCi/mL of culture medium and by a pulse duration of 1–2 h. The pulse is ended by removal of the culture medium by gentle suction, followed by washing with isotonic Tris-buffered saline. The adherent cells are then harvested in 0.5N KOH. In 25-cm^2 screw-cap tissue culture flasks, a total of two 1.0-mL washes of KOH is usually sufficient to harvest the cells. The cellular extract in KOH may be maintained at −20°C until assayed. Storage of the frozen extract for several weeks results in no impairment of recovery of the sterol fraction after thawing.

3.1.2. Digitonin Precipitation

Quantification of sterol synthesis from the administered [14C]acetate is achieved by extraction of nonsaponifiable lipids followed by addition of digitonin. Thus, 1 mL of the thawed KOH extract is saponified by heating at 70°C for 1 h after addition of 1 mL of 20% KOH in methanol, 500 μg of nonradioactive cholesterol, and 10^5 cpm of [^{3}H]cholesterol. The latter two additives serve, respectively, as a carrier to enhance precipitation with digitonin and as an internal standard. The nonsaponifiable lipids are extracted into light ether, and the ether extracts are dried under a stream of nitrogen. Resuspension of the dried residue in 2 mL of acetone/ethanol (1:1) is followed by the addition of 0.2 mL of 20% acetic acid and 1.0 mL of a solution of 0.5% digitonin and 50% ethanol.

Incubation with digitonin at 37°C for at least 3 h results in formation of a solid digitonin pellet in which sterols are complexed. The pellet is washed three times with diethyl ether and dissolved in 1.5 mL of methanol. It is frequently necessary to heat the tubes containing the methanol and sterol digitonide at 37°C in order to promote full solubilization of the pellet. The dissolved digitonides are added directly to scintillation vials along with an appropriate counting cocktail. We utilize 3a70B (Research Products International, Elk Grove Village, IL USA). After correction for recovery of the [^{3}H]cholesterol internal standard, the synthetic rates of digitonin-precipitated sterols are expressed conveniently as cpm/mg protein for the duration of the pulse.

3.2. Biosynthesis of Sterols from Precursors Other than [^{14}C]Acetate

Sterol biosynthetic rates may also be determined with radiolabeled acetoacetate, β-hydroxybutyrate, mevalonate, pyruvate, amino acids, and water (for review, *see* Bhat and Volpe, 1983). There are two examples of experimental circumstances in which use of precursors other than acetate provides particularly useful information regarding regulation of the pathway that elaborates the steroid nucleus. In the first example, it is necessary to exclude the possibility that a measured alteration of the rate of sterol synthesis as determined by acetate incorporation is actually caused by a change in the acetate pool size. If the observed change in incorporation of [^{14}C]acetate into digitonin-precipitatable steroids is caused by such a change in pool size, then the alteration in the apparent rate of sterol synthesis will not be observed after pulsing with [^{3}H]water (Volpe and Obert, 1981).

The second circumstance in which determining sterol synthesis by measuring incorporation of a precursor other than [^{14}C]acetate is particularly informative occurs when it is desirable

Table 1

Interpretation of Studies of Incorporation of Radioactivity into Sterols
with Major Precursors

Change in incorporation of precursor into sterols[a]			
[^{14}C]Acetate	[^{3}H]Mevalonate	[^{3}H]Water	Interpretation
+	−	+	Regulation of sterol biosynthesis prior to mevalonate formation
+	−	−	Change in acetate pool size
+	+	+	Regulation of sterol biosynthesis subsequent to mevalonate formation
−	+	−	Change in mevalonate pool size

[a]+, Change in incorporation with experimental manipulation; −, no change in incorporation with same experimental manipulation.

to localize the regulatory sites in the sterol biosynthetic pathway that are responsible for an observed alteration in net sterol synthesis. Thus the rate of [^{14}C]acetate incorporation into sterols is compared to the rate of sterol synthesis as determined by incorporation of radiolabeled mevalonate (Maltese et al., 1980; Volpe and Obert, 1981; Volpe and Goldberg, 1983; Yachnin et al., 1984). If the regulatory site accounting for a net change in synthesis is proximal to the formation of mevalonate [e.g., 3-hydroxy-3-methylglutaryl coenzyme A (HMG-CoA) reductase] (Fig. 2), then the change in synthesis will be apparent after pulsing with [^{14}C]acetate, but not with radioactive mevalonate. If, on the other hand, the regulatory site occurs distal to the formation of mevalonate, i.e., after the HMG-CoA reductase reaction, the phenomenon will be demonstrable with either mevalonate or acetate as radioactive precursor. The most common combinations of major changes in incorporation of precursors into sterols and the interpretation of these combinations *vis-a-vis* sterol biosynthesis are illustrated in Table 1.

The specific procedure for determining sterol biosynthesis from these alternative percursors is modified only slightly from that just described for [^{14}C]acetate (section 3.1). We have obtained satisfactory results with [^{3}H]mevalonalactone and ^{3}H$_2$O. The pulse consists of 5 μCi/mL of either isotope in the medium containing the cultured cells with [^{14}C]cholesterol utilized as an internal standard in the digitonin precipitation.

3.3. HMG-CoA Reductase

Considerable evidence indicates that the enzyme HMG-CoA reductase is the major regulatory site of the sterol biosynthetic pathway in both extraneural and neural tissues (for reviews, *see* Brown et al., 1978; Brown and Goldstein, 1980; Bhat and Volpe, 1983; Volpe et al., 1985). Although appropriate studies of neural systems have not yet been undertaken, the data in extraneural systems indicate that most regulatory effects on HMG-CoA reductase are regulated at the level of enzyme synthesis and/or degradation (Goldstein and Brown, 1984). Evidence from several in vivo and in vitro systems, however, indicates that HMG-CoA reductase activity may be catalytically modified by a phosphorylation-dephosphorylation mechanism (Saucier and Kandutsch, 1979; Nordstrom et al., 1977; Brown and Goldstein, 1980; Shah, 1981; Bhat and Volpe, 1983; Kennelly and Rodwell, 1985). The following discussion will describe the assay of total HMG-CoA reductase activity and a modification of this basic assay that allows determination of the proportion of total enzyme that is in a latent (i.e., phosphorylated) form.

3.3.1. Assay of Total HMG-CoA Reductase

The procedure that will be described (Volpe and Hennessy, 1977; Volpe, 1979) is a modification of earlier methods based on isolation of [^{3}H]mevalonate after catalytic formation from [^{3}H]HMG-CoA by HMG-CoA reductase (Brown et al., 1974; Avigan et al., 1975). The assay is performed on cellular pellet or tissue homogenates that usually have been stored at −80°C. Enzyme activity in neural tissue is stable for up to 2 mo under these conditions (Maltese and Volpe, 1979). The cellular material is thoroughly disrupted by vortexing it in a hypotonic buffer consisting of 10 mM Tris-buffered saline, 10 mM KCl, and 5 mM MgCl$_2$ at pH 7.4 and then homogenizing the suspension in a final volume of 150–300 μL. After the initial homogenization, one half volume of a hypertonic buffer (30 mM β-mercaptoethanol, 90 mM EDTA, and 210 mM NaCl at pH 7.4) is added, and homogenization is repeated. The volume in each reaction tube is adjusted to 163 μL with buffer, and the assay mixture is completed by addition of 37 μL of a freshly prepared solution of 2.5 mM NADPH in 100 mM K$_2$HPO$_4$ and 5 mM dithiothreitol (pH 7.5). After 10 min of preincubation at 37°C, the reaction is started by addition of [^{3}H]HMG-CoA substrate (final concentration of HMG-CoA in the mixture, 75 mM, specific activity 10 mCi/mmol). The reaction is terminated after 1 h by the addition of 20 μL of 5N HCl. [^{14}C]Mevalonate is then added as an internal standard, and the samples are allowed to incubate at 37°C for an additional 30 min in order to promote lactonization of mevalonate.

The final separation of the tritiated reaction product ([^{3}H]mevalonolactone) and the unreacted substrate ([^{3}H]HMG-CoA) is accomplished by use of the formate form of an anion exchange resin (Avigan et al., 1975). The commercially available resin (AG1-X8, BioRad, Richmond, CA) is pretreated by washing in a 20-fold volume of 10 mM formate. A 75% slurry of the resin is transferred to disposable centrifuge tubes. The reaction mixture is added to this slurry, which then is thoroughly mixed by vortexing and inverting the tubes. After centrifugation in a table-top centrifuge, the supernatant solution is carefully removed with a Pasteur pipet and transferred to a scintilation vial for counting. The supernatant solution contains the [^{3}H]mevalonolactone, and the [^{3}H]HMG-CoA is retained on the resin. Enzyme activity is calculated from recovery of the internal standard and from the concentration and radiochemical specific activity of the [^{3}H]HMG-CoA. Specific activity is expressed as nmol of mevalonate formed/min/mg protein.

3.3.2. Assay of Latent Activity of HMG-CoA Reductase

The assay of latent activity of HMG-CoA reductase is essentially that of Saucier and Kandutsch (1979). Cellular material is homogenized, and for purposes of comparison, separate aliquots are preincubated for 30 min at 37°C in buffers either containing or lacking 50 mM sodium fluoride in addition to 50 mM K$_2$HPO$_4$, 5 mM dithiothreitol, and 1 mM EDTA. After estimation of cellular protein, a volume of each homogenate containing 100 μg of protein is brought to 100 μL with buffer and added to separate tubes containing 95 μL of a reaction mixture containing 5 mM NaDPH and 160 mM K$_2$HPO$_4$. Homogenates that have been preincubated in the presence of fluoride should also have fluoride added to the reaction mixture at a final concentration of 50 mM. The reaction is commenced with the addition of [^{3}H]HMG-CoA, and separation of the reaction product from the substrate is perfomed as described in section 3.3.1.

Under these conditions, the fluoride ion functions as an inhibitor of phosphatases, which would otherwise convert the inactive phosphorylated reductase to the active dephosphorylated form during such procedures as homogenization and centrifugation (Nordstrom et al., 1977; Beg et al., 1978; Ness, 1983). Thus comparison of the reductase-specific activities in cells or tissue homogenized and assayed in the presence and absence of fluoride enables determination of the portion of reductase activity that is in the latent or phorphorylated form prior to cellular disruption.

3.4. Assay of HMG-CoA Synthase

Even though, as noted in the preceding section, HMG-CoA reductase is the major regulatory site for synthesis of the steroid nucleus in brain, several studies have provided evidence that other potential regulatory sites exist (for review, *see* Bhat and Volpe, 1983). Moreover, it has been established that the enzymes preceding HMG-CoA reductase in the sterol biosynthetic pathway have major regulatory actions in extraneural cells under specific experimental conditions, such as treatment of Hela cells with glucocorticoids (Cavenee et al., 1978; Cavenee and Melnykovych, 1977) and baby hamster kidney cells with oxygenated sterols (Chang and Limanek, 1980). Similarly, determinations of sterol synthesis after cholesterol feeding in both avian liver (Clinkenbeard et al., 1973) and in rat adrenal gland (Balasubramaniam et al., 1977) suggest such regulatory sites prior to the reductase as HMG-CoA synthase. HMG-CoA synthase, the cytosolic enzyme immediately preceding the reductase in the biosynthetic pathway (Fig. 2), is of particular significance since it has been demonstrated that the synthase may be regulated independently of HMG-CoA reductase in neural cells. Thus results from our laboratory established that sterol synthesis in C-6 glial cells, when maintained in serum-free medium supplemented with polyunsaturated fatty acids, demonstrated a coordinate regulation with HMG-CoA synthase, but not with HMG-CoA reductase (Volpe and Obert, 1981).

The procedure for assay of HMG-CoA synthase is modified from Balasubramaniam et al. (1977). A cytosolic fraction is derived from either cerebral tissue or from cultured neural cells and homogenized in a buffer containing, at the final concentrations indicated, 20 mM K₂HPO₄, 1 mM EDTA (pH 7.0). The homogenate is centrifuged at 100,000 g for 1 h at 4°C, and the resulting cytosolic fraction is dialyzed overnight in the homogenization buffer. The dialyzed extract is added to the reaction mixture, consisting of 100 mM Tris-HCl (pH 8.0), 0.1 mM EDTA, 20 mM MgCl₂, 10 μM acetoacetyl-CoA, and 0.6 mM [¹⁴C]acetyl-CoA at the final molalities indicated. The reaction is started at 30°C by addition of the [¹⁴C]acetyl-CoA (2 mCi/mmol) and is terminated after 10 min by addition of 6N HCl. The acidified final mixture is added to glass scintilation vials that are incubated at 90°C for 2 h. This procedure enables isolation of [¹⁴C]HMG-CoA as the nonvolatile radioactivity remaining in the vials (Balasubramaniam et al., 1977; Volpe and Obert, 1981). The radiochemical specific activity of the substrate [¹⁴C]acetyl-CoA is used to calculate synthase specific activity as nmol of HMG-CoA formed/min/mg protein.

3.5. Assay of Cholesterol

Several analytical techniques have been used to isolate and quantify cholesterol in neural tissue (Wells and Dittmer, 1967; Jones et al., 1975; Maltese et al., 1981; Volpe and Obert, 1982; Sakakihara and Volpe, 1984). Thus, as detailed subsequently in this review (section 4.1.1), it is possible to separate and qualitatively estimate the amounts of cholesterol and other isoprenoids by comparing their migration to that of known standards in thin layer chromatographic preparations (Choi and Suzuki, 1978; Jagannatha and Sastry, 1981). In addition, application of high performance liquid chromatography (see below, sections 5.1.1 and 5.2.2.) enables the identification and precise quantification of sterols and nonsterol isoprenoids (Palmer et al., 1984; Sakakihara and Volpe, 1984).

The specificity of chromatographic techniques makes them essential tools for quantitative biochemical analyses. Several colorimetric techniques have been described for a more rapid estimation of cholesterol content, however (reviewed by Johnson, 1979), and we have found such techniques to be particularly useful in experiments utilizing a large number of samples derived from neural cell cultures (Volpe and Obert, 1982; Volpe et al., 1985, 1986). Available data indicate that quantification of cholesterol by a method using cholesterol oxidase yields results that compare favorably with chromatographic analyses in terms of both sensitivity and specificity (Johnson, 1979; Ott et al., 1982). We consequently use a cholesterol oxidase technique routinely for determining cholesterol content of extracts from neural tissue.

Thus, lipids are extracted by a standard technique of partitioning into chloroform:methanol:water (see below, section 4.1.1) after homogenization in 0.15M NaCl. The organic phase is dried down under nitrogen and then resuspended in 20% (w/v) KOH in methanol. Saponification (70°C for 30 min) is followed by three sequential extractions into petroleum ether. The combined ether extracts are dried under nitrogen and resuspended in isopropanol. A rapid and extremely sensitive cholesterol oxidase assay (Ott et al., 1982) is then performed by using a commercially available kit (Boehringer Monotest, Boehringer Mannheim Biochemicals). In this assay, an aliquot of the isopropanol suspension is added to a solution containing catalase (2 U/mL), ammonium phosphate (0.15M), and acetylacetone (0.015M) in methanol:water [1:3 (v/v)]. The reaction is started by addition of cholesterol oxidase at a final concentration of 0.10 U/mL. The colorimetric reaction develops after 60 min at 37°C as a consequence of the oxidative conversion of cholesterol to 4-cholesten-3-one and hydrogen peroxide (Johnson, 1979). A standard curve is prepared by assaying samples containing known amounts of authentic cholesterol, and cholesterol is determined as a function of optical absorbance at 405 nm (Ott et al., 1982).

4. Further Metabolism of Steroid Compounds

Following its formation via the sterol biosynthetic pathway, the tetracyclic steroid nucleus becomes a major structural component of neural membranes and an important determinant of the physiologic properties of neural cells (Kabara, 1973; Jones et al., 1975; Kishimoto et al., 1965). In extraneural cells cholesterol contributes to the synthesis of a wide array of cyclic isoprenoid substances, including both adrenocortical and gonadal steroid hormones, vitamins, and bile acids (Heftmann, 1969). In brain it has been demonstrated that specific receptor sites for androgenic, estrogenic, and adrenocortical steroid hormones are distributed in a number of anatomic regions and also that there is active uptake of these steroid hormones into neural cells (Leiberburg and McEwen, 1975; Denef et al., 1973; McEwen, 1980; Hutchison and Schumacher, 1986). Our current understanding of the further metabol-

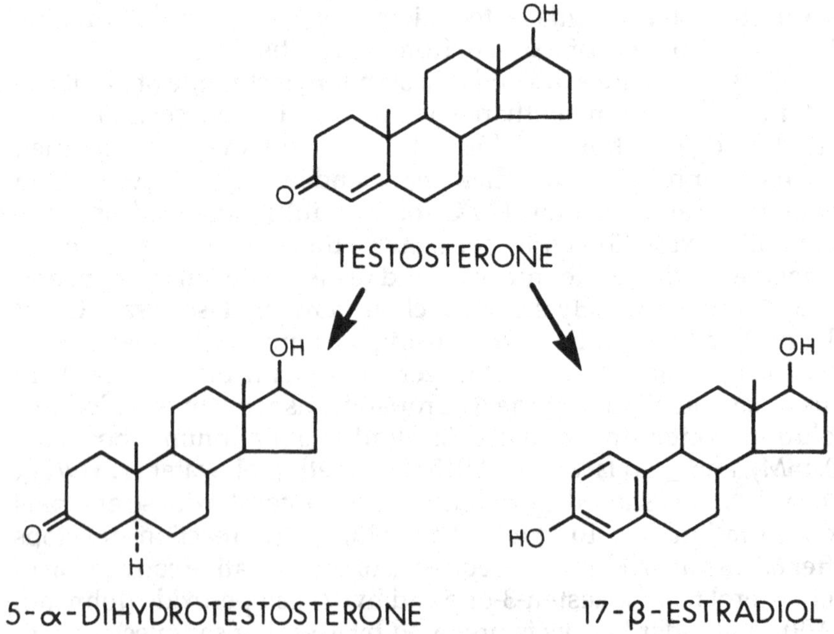

TESTOSTERONE

5-α-DIHYDROTESTOSTERONE   17-β-ESTRADIOL

Fig. 3. Metabolism of testosterone in neural tissue.

ism of steroid compounds in brain is limited to two principal biochemical processes, however: (1) esterification of cholesterol with fatty acid, especially during brain development (Eto and Suzuki, 1972; Choi and Suzuki 1978; Jagannatha and Sastry, 1981) and (2) conversion of testosterone both to 5-α-dihydrotestosterone and to 17-β-estradiol (Fig. 3) (Lieberburg and McEwen, 1975; Naftolin et al., 1975; Hutchison and Shumacher, 1986).

4.1. Esterification of Cholesterol

Eto and Suzuki (1972) originally reported that the accumulation of cholesterol esters in rodent brain demonstrates a distinct developmental profile, with a gradual decline from relatively high newborn levels to minute adult values over the first 30 d of life, upon which is superimposed a transient increase corresponding to the period of active myelination. Subsequent investigations (Choi and Suzuki, 1978; Jagannatha and Sastry, 1981) demonstrated that elaboration of cholesterol esters in developing brain was caused by the contributions of two distinct enzyme activities. One esterifying enzyme has a pH optimum of 5.2, does not require cofactors, is distributed among several particulate subcellular fractions, and remains at approximately constant specific activity in rat brain after the sixth day of life (Jagannatha and Sastry, 1981). The other esterifying enzyme has a pH optimum of 7.4, requires the presence of ATP and CoA, is particularly enriched in the microsomal fraction, and has a peak of activity at day 15 in rat brain (Jagannatha and Sastry, 1981).

4.1.1. Identification of Sterol Esters (and Sterols) by Thin Layer Chromatography

Lipids are extracted essentially according to Bligh and Dyer (1959). Thus the starting material, consisting of either cell pellets, cerebral tissue, or subcellular fractions, is homogenized in 0.15M NaCl, then extracted with chloroform:methanol:water [4:2:1 (v/v/v)] mixture. The upper phase is discarded after centrifugation in a table-top centrifuge, and the remaining organic phase is dried under nitrogen and resuspended in chloroform. Phospholipids are removed at this point by passing each sample through a silicic acid column that is prepared by adding 0.3 g of silicic acid to a Pastuer pipet that has been filled with glass wool. The lipids are eluted from the column with chloroform, which is again dried under nitrogen. The resulting residue is dissolved in approximately 200 μL of hexane and then spotted, along with appropriate authentic standards, onto silica-G thin layer plates that have been activated by heating at 100°C for 30 min. The solvent for the chromatographic separation of sterols and sterol esters is heptane:diethyl ether:acetic acid [127.5:22.5:3.0 (v/v/v)]. Spots containing lipid components are visualized by exposure to iodine vapor and identified by comparison with migration of authentic standards.

4.1.2. Biosynthesis of Cholesterol Esters In Vivo

Cholesterol ester synthesis has been studied in intact tissue culture cells most commonly by measuring the incorporation of radioactivity from [^{14}C]oleate added to the medium or from pre-labeled intracellular [^{3}H]cholesterol (Volpe et al., 1978). When [^{14}C]oleate is utilized as the precursor for cholesterol ester formation, the fatty acid is complexed to bovine albumin in 0.15M NaCl. In studies of C-6 glioma cells, the optimal concentration of [^{14}C]oleate was 0.1 mM, and the appropriate amount of radioactivity was found to be 1 μCi/mL of culture medium. Duration of the pulse with [^{14}C]oleate was 1 h. Cholesterol esters are isolated essentially as described in the immediately preceding section. When authentic cholesterol [^{14}C]oleate is carried through this procedure, recoveries are consistently 85–95%. In representative experiments in which [^{14}C]oleate is the precursor for cholesterol ester formation, the isolated cholesterol fraction has been subjected to hydrolysis in ethanolic KOH (Volpe et al., 1978). The products have been subjected to thin layer chromatography, the plates developed in light petroleum:ethyl ether:acetic acid [75:25:4 (v/v/v)], and the cholesterol and free fatty acid spots scraped into 3a70B scintillation fluid. In such experiments, 90–95% of the radioactivity in the sample has been recovered in the free fatty acid spot (i.e., same R_f as authentic oleic acid).

In experiments designed to measure esterification of *endogenous* cholesterol, the cellular cholesterol is labeled by incubating the cells for 18 h in medium containing [1,2-^{3}H]cholesterol, 10 μCi/flask. After labeling, the cells are washed three times with an excess of Tris-NaCl buffer before addition of nonradioactive oleate. The cholesterol ester fraction is then isolated and the amount of tritium in the cholesterol fraction determined as described immediately above. When this material is subjected to alkaline hydrolysis, 90–95% of the radioactivity is recovered in the cholesterol spot.

4.1.3. Biosynthesis of Cholesterol Esters In Vitro

This procedure is essentially a modification of those described by Choi and Suzuki (1978) and Jagannatha and Sastry (1981). It should be noted that the content of cofactors and the pH of the reaction mixture can be adjusted in order to determine activities of the different esterifying enzymes that were described in the preceding section. In addition, since it may be difficult to isolate a microsomal fraction from the amount of cells usually available in a cell culture, it may be necessary to assay the whole homogenate from a cell pellet derived from such a culture.

This final incubation volume should be 400 μL. The reaction mixture for assay of the pH 7.4 enzyme consists of the following reagents at the final concentrations designated: 25 mM NaH$_2$PO$_4$ adjusted to pH 7.4, 20 mM ATP and 250 μM coenzyme A. For the pH 5.2 enzyme, the buffer is sodium citrate 25 mM (pH 5.2), and both ATP and CoA are omitted. For either enzyme the additional constituents of the reaction mixture include 20 mM MgCl$_2$, 725 μM of unlabeled oleic acid, and 625 μM of unlabeled cholesterol that is added from an ethanolic stock solution. The reaction is started at 37°C by addition of 40 nmol of [^{14}C]oleic acid dissolved in 10 μL of ethanol and terminated after 1 h by addition of 1.5 mL of chloroform:methanol [2:1 (v/v)]. Lipids are extracted and subjected to thin layer chromatography as described in section 4.1.1. Radioactivity from the appropriate spots on the chromatogram is scraped into counting vials, and radioactivity is determined. The rate of formation of cholesterol esters is determined from the radiochemical specific activity and concentration of the precursor [^{14}C]oleic acid, and enzyme specific activity is expressed as nmol/h/mg protein (Jagannatha and Sastry, 1981).

4.2. Metabolism of Testosterone

After entering neural tissue and interacting with specific receptors, testosterone is metabolized further by two major mechanisms (for review, *see* McEwen, 1980): reduction may occur at the 5 position (Fig. 3), forming the androgenically active 5-α-dihydrotestosterone or, alternatively, aromatization may occur to produce 17-β-estradiol (Fig. 3). It is of interest that these reactions have been identified in brain regions usually associated with neuroendocrine functions, such as hypothalamus and pituitary, but also in numerous additional regions, including thalamus, midbrain, and cerebral cortex (Naftolin et al., 1975; McEwen, 1980).

4.2.1. Assay of Conversion of Testerone to Androgenic and Estrogenic Steroids

The procedure to be described is a microassay (Hutchison and Schumacher, 1986) that is based on earlier techniques (Naftolin et al., 1975). The procedure has been used to determine synthetic rates of testosterone derivatives by homogenates derived from both whole brain and from specific anatomic regions; the sensitivity of this microassay presumably should enable its adaptation for use with cell cultures (Hutchinson and Schumacher, 1986).

Samples are homogenized in a volume of an ice-cold buffer consisting of 250 mM sucrose, 50 mM Tris-HCl (pH 7.4), 5 mM MgCl$_2$, and 12.8 mM 2-mercaptoethanol (final concentrations) and sufficient to yield between 1 and 4 mg of protein in 100 μL. This homogenate may be assayed immediately or may be stored at −80°C for up to 2 wk (Hutchinson and Schumacher, 1986). The reaction is started at 41°C by adding equal volumes of the homogenate and a reaction mixture consisting of the original homogenization buffer in addition to 2.4 mM NADPH and 75 nM (1-α, 2-α-^{3}H)testosterone (specific activity, 60 Ci/mmol). The reaction is terminated after 15 min by freezing the samples at −20°C, and [^{14}C] derivatives of 17-β-estradiol and 5-α-dihydrotestosterone are added as internal standards. Steroids are extracted from the incubates by three extractions with diethyl ether. After the residue is dried under nitrogen, it is resuspended in $0.1M$ NaOH for extraction of androgens by a mixture of cyclohexane : toluene [1:1 (v/v)]. For the extraction of phenolic steroids, which include 17-β-estradiol, the remaining aqueous phase is neutralized with $1M$ HCl, then extracted a final time with diethyl ether. Androgenic and estrogenic metabolites of [³H]testosterone are separated by thin layer chromatography on silica gel in a mixture of dichloromethane : ether [85:15 (v/v)]. Estrogenic spots are visualized with iodine vapor, and androgens are visualized with H_2SO_4 : ethanol [2:3 (v/v)]. Synthetic rates of 5-α-dihydrotestosterone and 17-β-estradiol are calculated from the concentration and radiochemical specific activity of the precursor [1-α,2-α-³H]testosterone.

5. Procedures Related to the Metabolism of Nonsteroid Isoprenoids

5.1. Ubiquinone

Ubiquinone (coenzyme Q) is an important electron carrier in the mitochondrial respiratory chain and consists of a polyisoprenoid side chain attached to a benzoquinone moiety (Fig. 2). The isoprenoid side chain is derived from mevalonate, indicating a relationship between the sterol biosynthetic pathway and metabolism of ubiquinone (for review, see Brown and Goldstein, 1980). Thus studies of cultured fibroblasts (Faust et al., 1979; Nambudiri et al., 1980) and C-6 glioma cells (Volpe and Obert, 1982) demonstrated an apparent enhancement of ubiquinone synthesis under several conditions that included suppression of sterol synthesis from mevalonate. Later studies of cultured fibroblasts (Ranganathan et al., 1981) and intestinal epithelial cells (Sexton et al., 1983) demonstrated coordinate inhibition of ubiquinone and sterol syntheses, however. Inhibition of synthesis of ubiquinone following inhibition of HMG-CoA reductase has been reported recently in neuroblastoma cells (Maltese and Aprille, 1985). Possible reasons for the differences in some of the reported results include cell-specific metabolic differences, alterations in observed rates of syntheses caused by changes in the specific activity of the tracer, [³H]mevalonate secondary to a decrease in the intracellular mevalonate pool size (Ranganathan et al., 1981; Sexton et al., 1983), or accumulation of labeled precursors in contaminating epoxides of squalene that are not separated from ubiquinone by one-dimensional thin layer chromatography (Sexton, 1983). Hence the procedures to be described here will include identification of ubiquinone by either two-dimensional thin layer chromatography or by high performance liquid chromatography (HPLC).

5.1.1. Isolation of Ubiquinone by Two-Dimensional Thin Layer Chromatography

This two-dimensional system is modified from those reported by Sexton et al. (1983) and by Maltese and Aprille (1985). Cellular material from cell cultures, from intact neural tissue, or from isolated mitochondria may be used for this analysis. Each sample is homogenized in 4 vol of a buffer consisting of 250 mM sucrose and 10 mM Tris-HCl (pH 7.5). Lipids are extracted by addition of a mixture of chloroform : methanol such that the final proportion of chloroform : methanol : water is 4 : 2 : 1 (v/v/v) as described in section 4.1.1. The lower phase is evaporated under nitrogen and resuspended in chloroform, and chromatography is performed by spotting 100 μL of the chloroform suspension, along with authentic standards, onto silica-G plates that have been activated at 100°C for 30 min. The solvent for the first dimension is petroleum ether : acetone [90 : 10 (v/v)]. For the second dimension, a reverse phase separation is achieved by first impregnating the plate with paraffin, which is accomplished by dipping the plates in 5% (v/v) paraffin oil and petroleum ether and then developing them in the second dimension with acetone : paraffin-saturated water [85 : 15 (v/v)] as the solvent. Ubiquinone is then visualized with iodine vapor. This technique enables satisfactory separation of ubiquinone from the squalene epoxides that may contaminate the ubiquinone spot on one-dimensional chromatograms (Sexton et al., 1983).

5.1.2. Identification of Ubiquinone by HPLC

Although less widely available HPLC represents a valuable alternative to TLC for separation of ubiquinone. The principal method is that of Palmer et al. (1984). It is suitable for quantities of cellular material in the range of 1–2 mg, and it may consequently be used to analyze cells derived from cultures as well as homogenized cerebral tissue.

The cellular material is homogenized and extracted with 2 : 1 (v/v) chloroform : methanol as described in the preceding section.

After drying of the organic phase under nitrogen, the extracted lipids from each sample are redissolved in cyclohexane (100 μL). The cyclohexane suspension is added to 0.1 mL of ice-cold acetone containing 0.25% $MgCl_2$ and then placed on ice for 30 min. The purpose of the last step is to precipitate phospholipids, since the neutral lipids are soluble in acetone and may be removed after centrifugation. The supernatant solutions are dried down, and the neutral lipids are dissolved in 200 μL of 0.05% (v/v) isopropanol:hexane. Aliquots of 50 μL are injected into the HPLC. The solvent is 0.05% isopropanol:hexane, delivered at 1 mL/min. The liquid chromatography apparatus must be fitted with a cyanopropyl column, since the quantitative separation of ubiquinone cannot be achieved by using a standard silica gel column (Palmer et al., 1984).

5.1.3. Biosynthesis of Ubiquinone

Synthetic rates of ubiquinone have been determined in both extraneural cells (Faust et al., 1979; Nambudiri, 1980; Sexton, 1983) and neural cells (Volpe and Obert, 1982; Maltese and Aprille, 1985). The procedure consists of adding appropriately radiolabeled precursors, such as [^{14}C]acetate, [^{3}H]mevalonate, and [^{14}C]tyrosine to the culture medium. Satisfactory results are obtained by using concentrations of radioisotopes in the range of between 1 and 10 μCi/mL of medium. A pulse duration of 2 h is usually appropriate, i.e., rates of incorporation are linear with respect to time, and the incorporated radioactivity is appreciable. After removal of the media containing the radioisotope, the cultures are gently washed three times with $0.15M$ NaCl:Tris-HCl, and ubiquinone is isolated according to either of the chromatographic procedures described in the preceding sections.

5.2. Dolichol

Dolichol consists of a series of long-chain isoprenoid alcohols that are derived from the cholesterol biosynthetic pathway and play a critical role in biosynthesis of glycoproteins (Lennarz, 1983). Thus dolichol in its phosphorylated form serves as the carrier of saccharide moieties in the transfer of sugars to the amino nitrogen of the asparagine of the polypeptides. Consistent with this role of dolichol in N-linked glycoprotein synthesis and the importance of glycoproteins in tissue differentiation, synthesis of dolichol has been shown to be particularly active in developing sea urchin

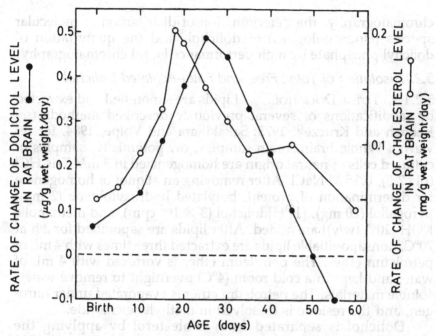

Fig. 4. Rate of increase of total dolichol (solid circles) and total cholesterol (open circles) in rat brain as a function of age. Values were obtained from the developmental curves of brain dolichol (*see* Fig. 5) and cholesterol content (personal data) from the slopes of the tangents at each age.

embryo (Carson and Lennarz, 1981; Lennarz, 1983), erythropoietic spleen (Potter et al., 1981), spermatogenic testis (James and Kandutsch, 1980), and developing brain in vivo (Waechter and Scher, 1981). Because glycoproteins are involved in several aspects of mamalian brain development, important roles for dolichol might be anticipated in this tissue during development. Studies of brain dolichol thus far have focused on the sharp increase in tissue levels with aging, a sharp developmental rise during the first 25 d postnatal that is separable from the developmental increase in brain cholesterol (Fig. 4), and the elevated brain content in certain human neurological disorders, especially Alzheimer's disease and neuronal ceroid lipofuscinosis (Pullarkat and Rehe, 1982; Ng Ying Kin et al., 1983; Wolfe et al., 1982, 1983; Sakakihara and Volpe, 1984).

In this section we will review the isolation of total and free dolichol, the quantitation of dolichol by high performance liquid chromatography, the determination of distribution of molecular species (isoprenologues) of dolichol, and the quantitation of dolichyl phosphate by high performance liquid chromatography.

5.2.1. Isolation of Total, Free, and Fatty-Acylated Dolichol

5.2.1.1. TOTAL DOLICHOL.   Lipids are saponified and extracted by modifications of several previously described methods (see Keenan and Kruczek, 1975; Sakakihara and Volpe, 1984, for reviews). Whole brain, brain samples, or, potentially, samples of cultured cells of neural origin are homogenized in 3 mM Tris-HCl (pH 7.5), 0.15 M NaCl. After removing an aliquot of homogenate for determination of protein, butylated hydroxytoluene (2 mg), pyrogallol (20 mg), [1-³H]dolichol (3 × 10⁴ cpm), and methanolic KOH [20% (w/v)] are added. After lipids are saponified for 2 h at 70°C, nonsaponifiable lipids are extracted three times with 4 mL of petroleum ether. The petroleum ether is vortexed with 4 mL of water and kept in a cold room (4°C) overnight to remove water-soluble materials. The petroleum ether is evaporated under nitrogen, and the residue is dissolved in methylene chloride.

Dolichol is separated from cholesterol by applying the methylene chloride solution to a disposable cartridge-type C18 column (J. T. Baker, Phillipsburg, New Jersey) that has been equilibrated with methanol prior to use. The column is eluted first with 10 mL of methanol, which recovers approximately 95% of cholesterol (assessed in separate experiments with [³H]cholesterol), and next with 8 mL of acetone, which recovers approximately 85% of the dolichol (assessed in separate experiments with [³H]dolichol). The acetone is evaporated under nitrogen and the residue disolved in methanol:isopropanol:hexane [2:2:1 (v/v/v)] for high performance liquid chromatography. An aliquot of this solution is also utilized to determine recovery of the internal standard ([³H]dolichol), which routinely averaged approximately 75–80%.

5.2.1.2. FREE DOLICHOL.   For extraction of free dolichol, homogenates are prepared exactly as described above for total dolichol, but saponification is not carried out. Free dolichol is extracted with chloroform:methanol [2:1 (v/v)]. The combined chloroform extracts are evaporated under nitrogen, dissolved in methylene chloride, and applied to the disposable C18 column as described in the previous section. The remainder of the procedure is identical to that described for the isolation of total dolichol. The recovery of [^{3}H]dolichol in this procedure routinely averaged 75–80%.

5.2.1.3. FATTY-ACYLATED DOLICHOL. Fatty-acylated dolichol is calculated as the difference between the determinations of total and free dolichol as just described. Fatty-acylated dolichol accounts for a relatively small proportion of total brain dolichol, except with aging, as shown in Fig. 5.

5.2.2. Quantitation of Dolichol by High Performance Liquid Chromatography

High performance liquid chromatography is carried out with a standard system. The system utilized in this laboratory consists of a Tracor 955 liquid chromatography pump apparatus, equipped with a 970A variable wave length detector and a Hewlett-Packard 3390A recorder-integrater. Elution is monitored at 210 nm. A 5-μm Ultrasphere (Beckman, Irvine, CA) 15 × 0.46 cm column or a 3-μm Microsorb (Rainin, Woolburn, MA) 10 × 0.46 cm column is used with methanol:isopropanol:hexane [2:2:1 (v/v/v)] as mobile phase, at a flow rate of 1 mL/min or 0.5 mL/min, respectively. In a typical chromatogram from mammalian tissues (Fig. 6), approximately six discrete peaks are noted. Identification of peaks of dolichol from samples of brain is accomplished by comparison with simultaneously chromatographed tritiated pig liver dolichol.

The amount of dolichol in the rat brain samples is quantitated by calculating peak areas and comparing these areas with those obtained with authentic dolichol. The standard curve for authentic dolichol is linear from 50 to 500 ng. Final calculation of the brain quantity of dolichol is made by correcting the value determined by liquid chromatography according to recovery of the internal standard.

5.2.3. Determination of the Distribution of Molecular Species (Isoprenologs) of Dolichol

Determination of the number of isoprene units in each isoprenolog of dolichol in rat brain samples is based on the linear relationship between log of the retention time and the number of isoprene units (Fig. 7). Utilizing this relationship, we have observed that the most prominent isoprenologs for developing rat brain are as shown in Fig. 8. Quantitation of the distribution of molecular species is accomplished by calculating the proportion of each individual peak areas to the total peak area for dolichol.

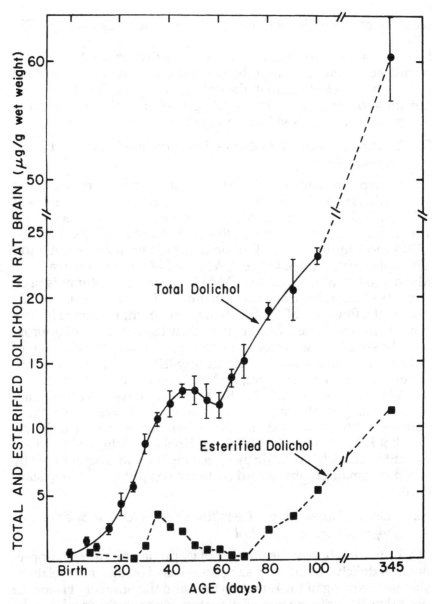

Fig. 5. Content of total and fatty-acylated dolichol in rat brain as a function of age. Analyses were carried out as described in the text. Values are means obtained from determinations of individual whole brain samples from 3 to 6 rats. (Note the break between the 100-d and 345-d points.)

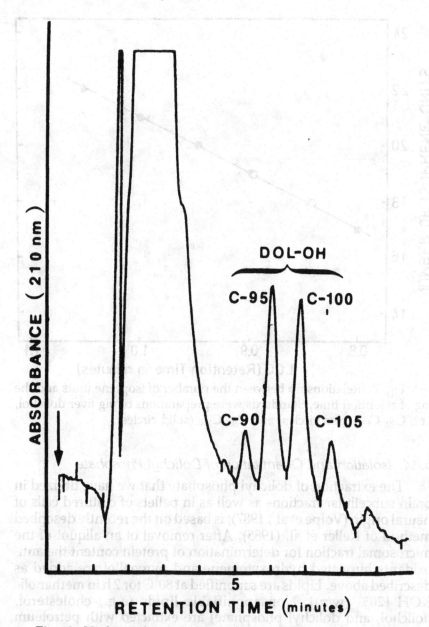

Fig. 6. High performance liquid chromatography of rat brain dolichol, carried out as described in the text.

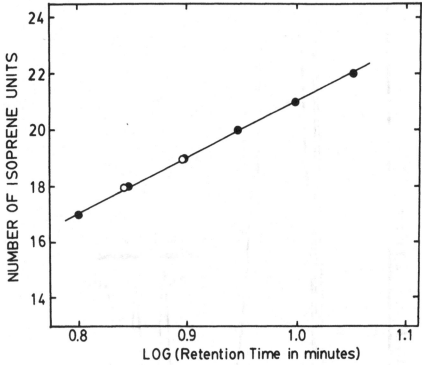

Fig. 7. Relationship between the number of isoprene units and the log of retention time. Standards were preparations of pig liver dolichol, i.e., C_{90}–C_{95} (open circles) and C_{85}–C_{110} (solid circles).

5.2.4. Isolation and Quantitation of Dolichyl Phosphate

The extraction of dolichyl phosphate that we have utilized in brain subcellular fractions as well as in pellets of cultured cells of neural origin (Volpe et al., 1987), is based on the recently described method of Keller et al. (1985). After removal of an aliquot of the microsomal fraction for determination of protein content the antioxidants butylated hydroxytoluene and pyrogallol are added as described above. Lipids are saponified at 80°C for 2 h in methanolic KOH [20% (w/v)]. Nonsaponifiable lipids (e.g., cholesterol, dolichol, and dolichyl phosphate) are extracted with petroleum ether. The petroleum ether is vortexed with water and kept in a cold room (4°C) overnight to remove water-soluble materials. The petroleum ether is evaporated under nitrogen, and the residue is dissolved in chloroform:methanol [2:1 (v/v)]. In separate experiments, recoveries of internal standards added to microsomal

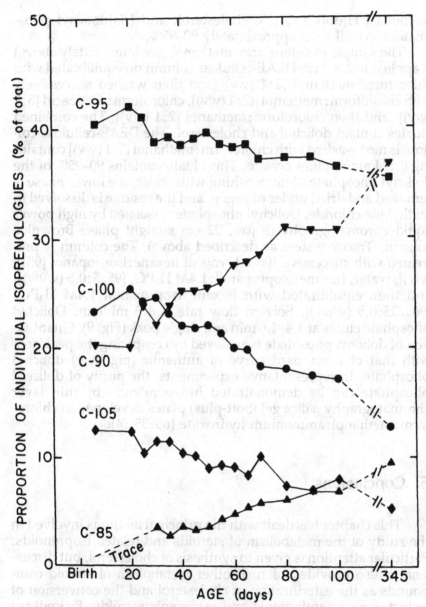

Fig. 8. Relative proportions of individual isoprenologues of total dolichol of rat brain as a function of age.

fractions ([3H]dolichol, [14C]cholesterol, and [3H]dolichyl phosphate) have all been approximately 90–95%.

The sample in chloroform:methanol (see immediately above) is applied to 2 × 1 cm DEAE-cellulose column preequilibrated with chloroform:methanol [2:1 (v/v)] and then washed successively with chloroform:methanol [(2:1 (v/v)], chloroform:acetic acid [3:1 (v/v)], and then chloroform:methanol [2:1 (v/v)]. The combined eluates contain dolichol and cholesterol. The DEAE-cellulose column is next washed with chloroform:methanol [2:1 (v/v)] containing 0.1M ammonium acetate. This eluate contains 90–95% of the dolichyl phosphate. After washing with water, the lower phase is removed and dried under nitrogen, and the residue is dissolved in methylene chloride. Dolichyl phosphate is isolated by high power liquid chromatography (5 μm, 22 cm straight phase Brownlee column, Tracor system as described above). The column is pretreated with successive 100-mL vols of hexane:isopropanol [95:5 (v/v)], water, hexane:isopropanol:1.4M H_3PO_4 [95:5:0.5 (v/v/v)], and then equilibrated with hexane:isopropanol:1.4M H_3PO_4 [965:35:0.5 (v/v/v)]. Solvent flow rate is 1.5 mL/min. Dolichyl phosphate elutes at 4.4–4.6 min as a single peak (Fig. 9). Quantitation of dolichyl phosphate is achieved by comparing the peak area with that of a standard curve of authentic (pig liver) dolichyl phosphate. In representative experiments, the purity of dolichyl phosphate can be demonstrated independently by thin layer chromatography [silica gel (Soft-plus) plates developed in chloroform:methanol:ammonium hydroxide (65:35:5)].

6. Conclusions

This chapter has dealt with the principal methods involved in the study of the metabolism of steroids and related isoprenoids. Particular attention is given to synthesis of cholesterol, but discussion is also provided of such other metabolism of steroid compounds as the esterification of cholesterol and the conversion of testosterone to androgenic and estrogenic steroids. Procedures related to the metabolism of the quantitatively minor, but qualitatively critical, nonsteroid isoprenoids ubiquinone and dolichol are reviewed. Throughout, emphasis is placed on analytical techniques, in keeping with the mission of this volume, but the methodology of several critical enzymes is described as well.

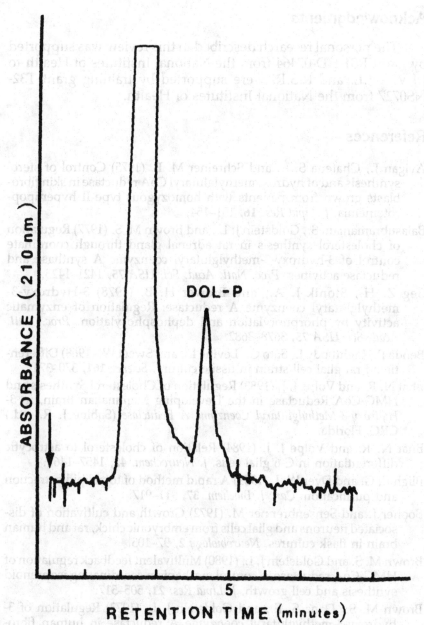

Fig. 9. High-performance liquid chromatography of rat brain dolichyl phosphate, carried out as described in the text.

Acknowledgments

The personal research described in this review was supported by grant R01-HD-07464 from the National Institutes of Health to J.J.V. T.J.L. and R.S.R. were supported by training grant T32-NS0727 from the National Institutes of Health.

References

Avigan J., Ghatena S. J., and Schreiner M. E. (1975) Control of sterol synthesis and of hydroxymethylglutaryl CoA reductase in skin fibroblasts grown from patients with homozygous type II hyperlipoproteinemia. *J. Lipid Res.* **16,** 151–154.

Balasubramaniam S., Goldstein J. L., and Brown M. S. (1977) Regulation of cholesterol synthesis in rat adrenal gland through coordinate control of 3-hydroxy-3-methylglutaryl coenzyme A synthase and reductase activities. *Proc. Natl. Acad. Sci. USA* **75,** 1421–1425.

Beg Z. H., Stonik J. A., and Brewer H. B. (1978) 3-Hydroxy-3-methylglutaryl coenzyme A reductase: Regulation of enzymatic activity by phorphorylation and dephosphosylation. *Proc. Natl. Acad. Sci. USA* **75,** 3678–3682.

Benda P., Lightbody J., Sato G., Levine L., and Sweet W. (1968) Differentiated rat glial cell strain in tissue culture. *Science* **161,** 370–371.

Bhat N. R. and Volpe J. J. (1983) Regulation of Cholesterol Synthesis and HMG-CoA Reductase in the Developing Mammalian Brain, in *3-Hydroxy-3-Methylglutaryl Coenzyme A Reductase* (Sabine J. R., ed.) CRC, Florida.

Bhat N. R. and Volpe J. J. (1984) Relation of cholesterol to astrocytic differentiation in C-6 glial cells. *J. Neurochem.* **42,** 1457–1463.

Bligh E. G. and Dyer W. J. (1959) A rapid method of total lipid extraction and purification. *Can. J. Biochem.* **37,** 911–917.

Booher J. and Sensenbrenner M. (1972) Growth and cultivation of dissociated neurons and glial cells from embryonic chick, rat and human brain in flask cultures. *Neurobiology* **2,** 97–105.

Brown M. S. and Goldstein J. L. (1980) Multivalent feedback regulation of HMG-CoA reductase, a control mechanism coordinating isoprenoid synthesis and cell growth. *J. Lipid Res.* **21,** 505–517.

Brown M. S., Dana S. E., and Goldstein J. L. (1973) Regulation of 3-hydroxy-3-methyglutaryl coenzyme A reductase in human fibroblasts by lipoproteins. *Proc. Natl. Acad. Sci. USA* **70,** 2162–2166.

Brown M. S., Dana S. E., and Goldstein J. L. (1974) Regulation of 3-hydroxy-3-methylglutaryl coenzyme A in human fibroblasts. *J. Biol. Chem.* **249,** 789–796.

Brown M. S., Faust J. R., and Goldstein J. L. (1978) Induction of HMG-CoA reductase in human fibroblasts incubated with compactin, a specific inhibitor of reductase. *J. Biol. Chem.* **253**, 1121–1128.

Carson D. D. and Lennarz W. J. (1981) Relationship of dolichol synthesis to glycoprotein synthesis during embryonic development. *J. Biol. Chem.* **256**, 4679–4686.

Cavenee W. K. and Melnykovych G. (1977) Induction of 3-hydroxy-3-methylglutaryl coenzyme A reductase in Hela cells by glucocorticoids. *J. Biol. Cem.* **252**, 3272–3276.

Cavenee W. K., Johnston D., and Melnykovych G. (1978) Regulation of cholesterol biosynthesis in Hela S3G cells by serum lipoproteins: Dexamethasone-mediated interference with suppression of 3-hydroxy-3-methylglutaryl coenzyme A reductase. *Proc. Natl. Acad. Sci. USA* **75**, 2103–2107.

Chang T.-Y. and Limanek J. S. (1980) Regulation of cytosolic acetoacetyl coenzyme A thiolase, 3-hydroxy-3-methylglutaryl coenzyme A synthase, 3-hydroxy-3-methylglutaryl coenzyme A reductase, and mevalonate kinase by low density lipoprotein and by 25-hydroxycholesterol in Chinese hamster ovary cells. *J. Biol. Chem.* **255**, 7787–7795.

Chen H. W. (1984) Role of cholesterol metabolism in cell growth. *Fed. Proc.* **43**, 126–130.

Chen H. W., Heiniger J. J., and Kandutsch A. A. (1975) Relation between sterol synthesis and DNA synthesis in phytohemagglutinin-stimulated mouse lymphocytes. *Proc. Natl. Acad. Sci. USA* **72**, 1950–1954.

Choi M.-U. and Suzuki K. (1978) A cholesterol-esterifying enzyme in rat central nervous system myelin. *J. Neurochem.* **31**, 879–885.

Clinkenbeard K. D., Sugiyama T., Moss J., Reed W. D., and Lane M. D. (1973) Molecular and catalytic properties of cytosolic acetoacetyl coenzyme A thiolase from avian liver. *J. Biol. Chem.* **248**, 2275–2284.

Cuzner J. L. and Davison A. N. (1968) Lipid composition of rat brain and subcellular fractions during development. *Biochem. J.* **106**, 29–34.

Denef C., Magnus C., and McEwen B. S. (1973) Sex differences and hormonal control of testosterone metabolism in rat pituitary and brain. *J. Endocrinol.* **59**, 605–621.

DeVellis J., Morrison R. S., Deng W. W., and Arenander A. T. (1983) Use of primary cultures in developmental studies of the central nervous system. *Birth Defects* **19**, 67–68.

Dobbing J. and Sands J. (1973) Quantitative growth and development of human brain. *Arch. Dis. Childh.* **48**, 757.

Dobbing J. and Sands J. (1979) Comparative aspects of the brain growth spurt. *Early Hum. Dev.* **3**, 79–83.

Eto Y. and Suzuki K. (1972) Cholesterol esters in devolping rat brain: Concentration and fatty acid composition. *J. Neurochem.* **19,** 109–115.

Fairbanks K. P., Witte L. D., and Goodman D. S. (1984) Relationship between mevalonate and mitogenesis in human fibroblasts stimulated with platelet-derived growth factor. *J. Biol. Chem.* **259,** 1546–1551.

Faust J. F., Goldstein J. L., and Brown M. S. (1979) Synthesis of ubiquinone and cholesterol in human fibroblasts: Regulation of a branched pathway. *Arch. Biochem. Biophys.* **192,** 86–99.

Federoff S. (1977) Primary Cultures, Cell Lines and Cell Strains: Terminology and Characteristics, in *Cell, Tissue and Organ Culture in Neurobiology* (Federoff S. and Hertz L., eds.) Academic, New York.

Federoff S. (1985) Macroglial Cell Lineages, in *Molecular Bases of Neural Development* (Edelman G. M., Fall W. E., and Cowan W. M., eds.) Wiley, New York.

Folch J., Lees M., and Sloane-Stanley G. H. (1957) A simple method for the isolation and purification of total lipids from animal tissues. *J. Biol. Chem.* **226,** 497–509.

Freshney I. (1983) *Culture of Animal Cells: A Manual of Basic Technique* Alan R. Liss, New York.

Goldstein J. L. and Brown M. S. (1984) Progress in understanding the LDL receptor and HMG-CoA reductase, two membrane proteins that regulate the plasma cholesterol. *J. Lipid Res.* **25,** 1450–1459.

Habenicht A. J. R., Glomset J. A., and Ross R. (1980) Relation of cholesterol and mevalonic acid to the cell cycle in smooth muscle and Swiss 3T3 cells stimulated to divide by platelet-derived growth factor. *J. Biol. Chem.* **255,** 5134–5140.

Heftmann E. (1969) *Steroid Biochemistry* Academic, New York.

Hertz L., Juurlink B. H. J., Fosmark H., and Schousboe A. (1982) Methodological Appendix: Astrocytes in Primary Cultures, in *Neuroscience Approached Through Cell Cultures* vol. 1 (Pfeiffer S. E., ed.) CRC, Florida.

Hertz L., Juurlink B. H. J., and Szuchet S. (1985) Cell Cultures, in *Handbook of Neurochemistry* (Lajtha A., ed.) vol. III, Plenum, New York.

Hutchison J. B. and Schumacher M. (1986) Development of testosterone-metabolizing pathways in the avian brain: Enzyme localization and characteristics. *Dev. Brain Res.* **25,** 33–42.

Jagannatha H. M. and Sastry P. S. (1981) Cholesterol-esterifying enzymes in developing rat brain. *J. Neurochem.* **36,** 1352–1360.

James M. J. and Kandutsch A. A. (1980) Elevated dolichol synthesis in mouse testis during spermatogenesis. *J. Biol. Chem.* **255,** 16–19.

Johnson S. M. (1979) A new specific cholesterol assay gives reduced cholesterol/phospholipid molar ratios. *Analyt. Biochem.* **95,** 344–350.

Jones, J. P., Nicholas H. J., and Ramsey R. B. (1975) Rate of sterol formation by rat brain glia and neurons *in vitro* and *in vivo*. *J. Neurochem.* **24**, 123–126.

Kabara J. J. (1973) A critical review of brain cholesterol metabolism. *Prog. Brain. Res.* **40**, 363–382.

Keenan R. W. and Kruczek M. (1975) The preparation of tritiated betulaprenol and dolichol. *Analyt. Biochem.* **69**, 504–509.

Keller R. K., Fuller M. S., Rottler G. D., and Connelly L. W. (1985) Extraction of dolichyl phosphate and its quantitation by straight-phase high performance liquid chromatography. *Analyt. Biochem.* **147**, 166–172.

Kennelly P. J. and Rodwell V. W. (1985) Regulation of 3-hydroxy-3-methylglutaryl coenzyme A reductase by reversible phosphorylation-dephosphorylation. *J. Lipid Res.* **26**, 903–914.

Kishimoto Y., Davies W. E., and Radin N. S. (1965) Developing rat brain-changes in cholesterol, galactolipids and individual fatty acids of gangliosides and glycerophosphatides. *J. Lipid Res.* **6**, 532–536.

Langan T. J. and Volpe J. J. (1987) Oligodendrglial differentiation in glial primary cultures: Requirement for mevalonate. *J. Neurochem.* **48**, 1804–1808.

Langan T. J. and Volpe J. J. (1986) Obligatory relationship between the sterol biosynthetic pathway and DNA synthesis and cellular proliferation in glial primary cultures. *J. Neurochem.* **46**, 1283–1291.

Lennarz W. J. (1983) Glycoprotein synthesis and embryonic development. *CRC Crit. Rev. Biochem.* **14**, 257–282.

Lieberburg I. and McEwen B. S. (1975) Estradiol-17β: A metabolite of testosterone recovered in cell nuclei from limbic areas of adult male rat brains. *Brain Res.* **91**, 171–174.

Maltese W. A. and Aprille J. A. (1985) Relation of mevalonate synthesis to mitochondrial ubiquinone content and respiratory function in cultured neuroblastoma cells. *J. Biol. Chem.* **260**, 11524–11529

Maltese W. A. and Sheridan K. M. (1985) Differentiation of neuroblastoma cells induced by an inhibitor of mevalonate synthesis: Relation of neurite outgrowth and acetylcholinesterase activity to changes in cell proliferation and blocked isoprenoid synthesis. *J. Cell. Physiol.* **125**, 540–558.

Maltese W. A. and Volpe J. J. (1979) Activation of 3-hydroxy-3-methylglutaryl coenzyme A reductase in homogenates of developing rat brain. *Biochem J.* **182**, 367–370.

Maltese W. A., Reitz B. A., and Volpe J. J. (1980) Changes in sterol synthesis accompanying cessation of cell growth in serum-free medium. *Biochem. J.* **192**, 709–717.

Maltese W. A., Reitz B. A., and Volpe J. J. (1981) Effects of prior sterol depletion on neurite outgrowth in neuroblastoma cells. *J. Cell. Physiol.* **108**, 475–482.

Martin J. H. (1985) Development as a Guide to the Regional Anatomy of the Brain, in *Principles of Neural Science* (Kandel E. R. and Schwartz J. H., eds.) Elsevier, New York.

McEwen B. S. (1980) Endocrine Effects on the Brain and Their Relationship to Behavior, in *Basic Neurochemistry* (Siegel G. H., Albers, R. W., Agranoff B., and Katzman R., eds.) Little Brown, Boston.

Mitzen E. J. and Koeppen A. H. (1984) Malonate, malonyl-coenzyme A and acetyl-coenzyme A in developing rat brain. *J. Neurochem.* **43**, 499–506.

Murray M. R. (1971) Nervous Tissues Isolated in Culture, in *Handbook of Neurochemistry* (Lajtha A., ed.) Plenum, New York.

Naftolin F., Ryan K. J., Davies I. J., Reddy V. V., Flores F., Petro Z., Kuhn M., White R. J., Takaoka Y., and Wolin L. (1975) Formation of estrogen by central neuroendocrine tissues. *Rec. Prog. Horm. Res.* **31**, 235–315.

Nambudiri A. M. D., Ranganathan S., and Rudney H. (1980) The role of 3-hydroxy-3-methylglutaryl coenzyme A reductase in the regulation of ubiquinone synthesis in human fibroblasts. *J. Biol. Chem.* **255**, 5894–5899.

Ness G. C. (1983) Regulation of 3-hydroxy-3-methylglutaryl coenzyme A reductase. *Mol. Cell. Biochem.* **53, 54,** 299–306.

Ng Ying Kin N. M. K., Palo J., Haltia M., and Wolfe L. S. (1983) High levels of brain dolichols in neuronal ceroid-lipofuscinosis and senescence. *J. Neurochem.* **40,** 1465–1473.

Nordstrom J. L., Rodwell V. W., and Mitschelen J. J. (1977) Interconversion of active and inactive forms of rat liver hydroxymethylglutaryl coenzyme A reductase. *J. Biol. Chem.* **252,** 8924–8934.

Norton W. T. and Poduslo S. E. (1973) Myelination in rat brain: Changes in myelin composition during maturation. *J. Neurochem.* **21,** 759–773.

Ott P., Binggeli Y., and Brodbeck, U. (1982) A rapid and sensitive assay for determination of cholesterol in membrane lipid extracts. *Biochim. Biophsy. Acta* **685,** 211–213.

Palmer D. N., Anderson M. A., and Jolly R. D. (1984) Separation of some neutral lipids by normal-phase high performance liquid chromatography on a cyanopropyl column: Uniquinone, dolichol and cholesterol levels in sheep liver. *Analyt. Biochem.* **140,** 315–319.

Popjak G. (1969) Enzymes of sterol synthesis in liver and intermediates of sterol biosynthesis. *Meth. Enzymol.* **15,** 393–454.

Potter J. E. R., James M. J., and Kandutsch A. A. (1981) Sequential cycles of cholesterol and dolichol synthesis in mouse spleen during phenyl-hydrazine-induced erythropoiesis. *J. Biol. Chem.* **256**, 2370–2376.

Poulter C. D. and Rilling H. C. (1981) Prenyl Transferases and Isomerase, in *Biosynthesis of Isoprenoid Compounds* (Porter J. W. and Spurgeon S. L., eds.) vol. 1, Wiley, New York.

Pullarkat R. K. and Rehe H. (1982) Accumulation of dolichols in brains of elderly. *J. Biol. Chem.* **257**, 5991–5993.

Quesney-Huneeus V., Galick H., and Siperstein M. (1983) The dual role of mevalonate in the cell cycle. *J. Biol. Chem.* **258**, 378–385.

Raff M. D., Miller R. H., and Noble M. (1983) A glial progenitor cell that develops *in vitro* into an astrocyte or an oligodendrocyte depending on culture medium. *Nature* **303**, 390–396.

Raff M. C., Fields K. F., Hakamori S., Mirsky R., Pruss R. M., and Winter J. (1979) Cell type-specific markers for distinguishing and studying neurons and the major classes of glial cells in culture. *Brain Res.* **174**, 283–308.

Ramsey R. B., Jones J. P., Naqvi S. H. M., and Nicholas H. J. (1971) Biosynthesis of cholesterol and other sterols by brain tissue. 2. Comparison of *in vitro* and *in vivo* methods. *Lipids* **6**, 225–232.

Ranganathan S., Nambudiri A. M. D., and Rudney H. (1981) The regulation of ubiquinone synthesis in fibroblasts: The effect of modulators of β-hydroxy-β-methylglutaryl coenzyme A reductase activity. *Arch. Biochem. Biophys.* **210**, 592–597.

Sakakihara Y. and Volpe J. J. (1984) Dolichol deposition in developing rat brain: Content of free and fatty-acylated dolichol and proportion of specific isoprenologues. *Dev. Brain Res.* **14**, 255–262.

Sakakihara Y. and Volpe J. J. (1985a) Dolichol in human brain: Regional and developmental aspects. *J. Neurochem.* **44**, 1535–1540.

Sakakihara Y. and Volpe J. J. (1985b) Zn^{2+}, not Ca^{2+}, is the most effective cation for activation of dolichol kinase of mammalian brain. *J. Biol. Chem.* **260**, 15413–15419.

Saucier S. E. and Kandutsch A. A. (1979) Inactive 3-hydroxy-3-methylglutaryl coenzyme A reductase in broken cell preparations of various mammalian tissues and cell cultures. *Biochim. Biophys. Acta* **572**, 541–556.

Schein S. H., Britva A., Hess H. H., and Selkoe D. J. (1970) Isolation of hamster brain astroglia by *in vitro* cultivation and subcutaneous growth, and content of cerebroside, ganglioside, RNA and DNA. *Brain Res.* **19**, 497–501.

Schmidt R. A., Glomset J. A., Wight T. N., Habenicht A. J. R., and Ross R. (1982) A study of the influence of mevalonic acid and its metabolites on the morphology of Swiss 3T3 cells. *J. Cell Biol.* **95**, 144–153.

Sexton R. C., Panini S. R., Azran F., and Rudney H. (1983) Effects of 3β-[2-(diethylamino)ethoxy] androst-5-en-17-one on the synthesis of cholesterol and ubiquinone in rat intestinal epithelial cell cultures. *Biochemistry* **22**, 5687–5691.

Shah S. S. (1981) Modulation *in vitro* of 3-hydroxy-3-methylglutaryl coenzyme A reductase in brain microsomes: Evidence for the phosphorylation and dephosphorylation associated with inactivation and activation of the enzyme. *Arch. Biochem. Biophys.* **211**, 439–446.

Volpe J. J. (1979) Microtubules and regulation of 3-hydroxy-3-methylglutaryl coenzyme A reductase. *J. Biol. Chem.* **254**, 2568–2571.

Volpe J. J. and Goldberg R. I. (1983) Effect of tunicamycin on 3-hydroxy-3-methylglutaryl coenzyme A reductase in C-6 glial cells. *J. Biol. Chem.* **258**, 9220–9226.

Volpe J. J. and Hennessy S. W. (1977) Cholesterol biosynthesis and 3-hydroxy-3-methylglutaryl coenzyme A reductase in cultured glial and neuronal cells. *Biochim. Biophys. Acta* **486**, 408–420.

Volpe J. J. and Obert K. A. (1981) Coordinate regulation of cholesterol synthesis and 3-hydroxy-3-methylglutaryl coenzyme A synthase but not 3-hydroxy-3-methylglutaryl coenzyme A reductase in C-6 glia. *Arch. Biochem. Biophys.* **212**, 88–97.

Volpe J. J. and Obert K. A. (1982) Relation of cholesterol to oligodendroglial differentiation in C-6 glial cells. *J. Neurochem.* **40**, 530–537.

Volpe J. J., Hennessy S. W., and Wong T. (1978) Regulation of cholesterol ester synthesis in cultured glial and neuronal cells. Relation to control of cholesterol synthesis. *Biochim. Biophys. Acta* **528**, 424–435.

Volpe J. J., Goldberg R. I., and Bhat N. R. (1985) Cholesterol biosynthesis and its regulation in dissociated cell cultures of fetal rat brain: Developmental changes and the role of 3-hydroxy-3-methylglutaryl coenzyme A reductase. *J. Neurochem.* **45**, 536–543.

Volpe J. J., Iimori Y., Haven G. G., and Goldberg R. I. (1986) Relation of cellular phospholipid composition to oligodendroglial differentiation in C-6 glial cells. *J. Neurochem.* **46**, 475–482.

Volpe J. J., Sakakihara Y., and Rust R. S. (1987) Dolichol kinase and the regulation of dolichyl phosphate levels in developing brain. *Dev. Brain. Res.* **31**, 193–200.

Waechter C. J. and Lennarz W. J. (1976) The role of polyprenol-linked sugars in glycoprotein synthesis. *Annu. Rev. Biochem.* **45**, 95–112.

Waechter C. J. and Scher M. G. (1981) Methods for Studying Lipid-Mediated Glycosyltransferases Involved in Assembly of Glycoproteins in Nervous Tissue, in *Research Methods in Neurochemistry* (Marks N. and Rodnight R., eds.) vol. II., Plenum, New York.

Wells M. A. and Dittmer J. C. (1967) Comprehensive study of the postnatal changes in concentration of lipids in developing rat brain. *Biochemistry* **6**, 3169–3175.

Wernicke J. W. and Volpe J. J. (1986) Glial differentiation in dissociated cell cultures of neonatal rat brain: Noncoordinate and density-dependent regulation of oligodendroglial enzymes. *J. Neurosci. Res.* **15**, 39–47.

Wolfe L. S., Ng Ying Kin N. M. K. Palo J., and Haltia M. (1982) Raised levels of cerebral cortex dolichols in Alzheimer's disease. *Lancet* **ii**, 99–101.

Wolfe L. S., Ng Ying Kin N. M. K., Palo J., and Haltia M. (1983) Dolichols in brain and urinary sediment in neuronal ceroid-lipofuscinosis. *Neurology* **33**, 103–106.

Yachnin S., Toub D. B., and Manickarottu V. (1984) Divergence in cholesterol biosynthetic rates and 3-hydroxy-3-methylglutaryl coenzyme A reductase or a consequence of granulocyte versus monocyte-macrophage differentiation in HL-60 cells. *Proc. Natl. Acad. Sci. USA* **81**, 894–897.

Phospholipids

Yasuhito Nakagawa and Keizo Waku

1. Introduction

Lipids in adult human brain account for 33% of the dry weight of the gray matter and 55% of the white matter (Bowen et al., 1974). The brain contains two major cell types: neurons, which are functionally active cells, and glial cells, which are generally divided into two major subtypes, astrocytes and oligodendrocytes. Investigation of the lipid composition and metabolism in neural tissues is now very important, because it is strongly suspected that lipids, especially phospholipids, play several important physiological roles in neural cells.

For the fractionation of phospholipids in neural tissues, lipids are extracted from the tissues and analyzed by TLC (thin-layer chromatography), column chromatography, HPLC (high-performance liquid chromatography), or GLC (gas-liquid chromatography).

The first part of this chapter describes the analytical methods used for the common phospholipids found in neural tissues. The primary tasks in polar lipid analysis are the separation of the phospholipid classes and that of all the molecular species within each class. Recently, individual molecular species were separated from each other by reverse phase HPLC. The second part describes the enzymological methods for investigating the biosynthesis and catabolism of phosphoglycerides in neural tissues.

2. Analysis of Phospholipids

2.1. Thin-Layer Chromatography (TLC)

The lipids in neural tissues can be extracted almost completely by the method of Folch et al. (1957) or Bligh and Dyer (1959). The latter method is more often used in our laboratory because of its convenience for extraction from the TLC spots.

For the analysis of rather small amounts of phospholipids (10–250 µg), TLC is the most important analytical tool. It is a simple and convenient technique. Silica gel G or H are generally used as the adsorbent, and we use precoated Kieselgel 60, type 5721, 20 × 20 cm Merck plates for all TLC work. Sulfuric acid is a valuable reagent for the quantitative analysis of lipids. The spots can also be visualized with I_2 vapor in a glass chamber. Since iodine reacts with unsaturated fatty chains, it cannot be used for analytical work involving highly unsaturated fatty acids. In this case, iodine is replaced by rhodamine B, 2', 7'-dichlorofluorescein, or primuline.

2.1.1. One-Dimensional TLC

Many one-dimensional TLC systems that have been described are good enough for routine analysis of major lipid components (Table 1). There is no single TLC system that can be used to separate completely all known complex lipids in one dimension. This TLC system is particularly useful, however, for separating the complex lipids from several fractions of DEAE-cellulose column as shown in Table 2.

Table 1: R_f Values of Phospholipids on Silica Gel G

Lipid	Solvent system[a]				
	a	b	c	d	e
Ptd₂Gro (Cardiolipin)	—	0.71	0.38	0.63	—
PtdEtn (Phosphatidylethanolamine)	0.82	0.62	0.41	0.57	—
PtdSer (Phosphatidylserine)	0.56	0.15	0.05	0.27	—
PtdIns (Phosphatidylinositol)	0.46	0.23	0.11	0.18	0.95
PtdCho (Phosphatidylcholine)	0.30	0.33	0.33	0.30	—
CerPCho (Sphingomyelin)	0.17	0.16	0.22	0.18	—
Lyso PtdCho (Lysophosphatidylcholine)	0.09	0.12	0.08	0.14	—
Lyso PtdEtn (Lysophosphatidylethanola-mine)	—	0.28	0.20	0.16	—
PtdOH (Phosphatidic acid)	—	0.74	0.05	0.59	—
PtdGro (Phosphatidylglycerol)	—	0.48	0.37	0.49	—
PtdIns4-P (Phosphatidylinositol-4-phosphate)	—	—	—	—	0.56
PtdIns4,5P_2 (Phosphatidylinositol-4,5-diphosphate)	—	—	—	—	0.24

[a]Solvents:(a) 2% Na_2CO_3-TLC, chloroform:methanol:acetic acid:water (25:15:4:2). (b) Chloroform:methanol:water(65:25:4). (c) Chloroform:methanol: 28% ammonia(65:25:5). (d) Chloroform:methanol:acetone:acetic acid:water (50:10:20:10:5). (e) Potassium oxalate-TLC, chloroform:methanol:4N ammonia(9:7:2).

Table 2
Elution Profiles of Phospholipids on a DEAE-Cellulose Column[a]

Elution solvent[b]	Elution volume[c]	Lipids eluted[d]
C	8–10	Neutral lipids
C-M 9:1	9	LysoPtdCho, PtdCho, CerPCho
C-M 7:3	9	PtdEtn
C-M 1:1	9	LysoPtdEtn
M	10	None
C-HOAc 3:1	10	Free fatty acids
HOAc	10	PtdSer
M	10	None
C-M-NH$_3$-salt	4	Ptd$_2$Gro, PtdOH, PtdGro, PtdIns, and salt

[a]Each sample (100–600 mg) was dissolved in chloroform, then loaded on the column (2.5 × 20 cm) packed with DEAE-cellulose (15 g). The flow rate was 3 mL/min.
[b]Abbreviations for solvents: C, chloroform; M, methanol; HOAc, glacial acetic acid; C-M-NH$_3$-salt, chloroform-methanol (4:1) made to 0.01–0.05M with respect to ammonium acetate, to which was added 20 mL of 28% ammonia/L.
[c]Elution volumes are specified in column volumes.
[d]Abbreviations are the same as in Table 1.

2.1.2. Two-Dimensional TLC

More complex lipid components can be separated on a single TLC plate if two-dimensional systems are used. Many published systems represent only minor adjustments of earlier ones to suit local conditions (e.g., of temperature or humidity). A typical chromatogram from two-dimensional TLC is shown in Fig. 1. In radioactive precursor incorporation experiments on neural tissues, all major phospholipids can be fractionated completely by two-dimensional TLC and stained by iodine vapor; the phosphorous contents and radioactivities can be estimated after extraction of the lipids by the method of Bligh and Dyer (1959).

2.1.3. Separation of Diacyl-, Alkylacyl-, and Alkenylacyl-3-acetyl-glycerols

Since significant amounts of plasmalogens and alkylether phospholipids are present in neural tissues, the TLC technique for the separation of the subclasses of phospholipids (diacyl, alkylacyl,

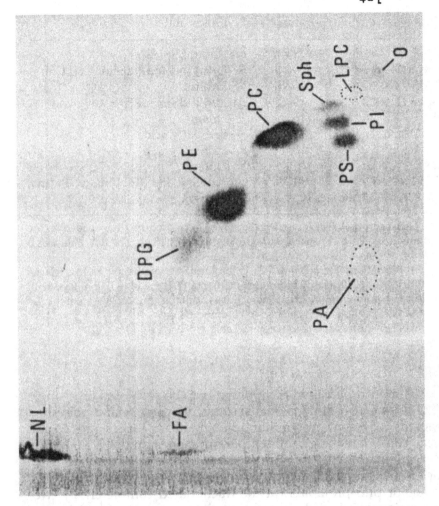

Fig. 1. Two-dimensional thin-layer chromatogram showing the distribution of phospholipids. The solvent system for the first dimension was chloroform:methanol:28% ammonia (65:35:5). The solvent system for the second dimension was chloroform:methanol:acetone:acetic acid:water (50:10:20:13:5). O, origin; LPC, lysophosphatidylcholine; Sph, sphingomyelin; PI, phosphatidylinositol; PS, phosphatidylserine; PC, phosphatidylcholine; PE, phosphatidylethanolamine; PA, phosphatidic acid; DPG, diphosphatidylglycerol; FA, fatty acid; NL, neutral lipid.

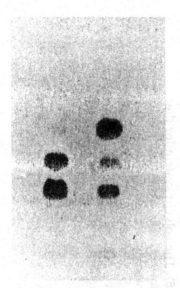

alkenylacyl

alkylacyl

diacyl

CPG   EPG

Fig. 2. TLC separation of 1-alkenyl-2-acyl-3-acetylglycerol, 1-alkyl-2-acyl-3-acetylglycerol, and 1,2-diacyl-3-acetylglycerol derivatives of CPG and EPG from pig lymphocytes. The plate was developed with petroleum ether:ethyl ether:acetic acid (90:10:1), and then with toluene. The spots were visualized with 50% sulfuric acid and charring.

and alkenylacyl) is very important. The preparation of diacyl-, alkylacyl-, and alkenylacyl-3-acetylglycerols from phosphatidyl-ethanolamine is described in the section on HPLC. A typical TLC pattern of these three types of compounds is shown in Fig. 2. The separation of these three types of compounds is also possible with the HPLC method described in the section on HPLC. Each subclass separated by TLC or HPLC is determined by GLC analysis after the preparation of fatty acid methyl esters.

2.2. Column Chromatography

Although rather small amounts of lipid can be fractionated by TLC, it is difficult to fractionate large amounts of lipids (over 5 mg) by two-dimensional TLC. In this case, DEAE-, TEAE-cellulose, and/or silicic acid column chromatography are useful.

2.2.1. DEAE-Cellulose Column Chromatography

DEAE-cellulose column chromatography is generally useful for the separation of the components of complex mixtures of lipids

(Rouser et al., 1976). Lipid classes can be separated through ion-exchange chromatography and differences in polarity and divided into three groups (neutral lipids, nonacidic phospholipids, and acidic phospholipids). A typical elution pattern is shown in Table 2.

2.2.2 Silicic Acid Column Chromatography

The separation of all phospholipids is difficult with a single procedure on a silicic acid column. Silicic acid chromatography is usually used in conjunction with DEAE-cellulose chromatography for complete separation of nonacidic phospholipids from each other. A typical elution pattern is as follows: C:M 20:1, neutral lipids; C:M 5:1, phosphatidylethanolamine; C:M 3:2, phosphatidylcholine; and C:M 1:4, sphingomyelin.

2.3. High-Performance Liquid Chromatography (HPLC)

2.3.1. Separation of Phospholipids

Separation of individual classes of phospholipids by means of HPLC on a silicic acid column has been performed in a number of laboratories (Hanson et al., 1981; Chen and Kou, 1982; Alam et al., 1982; Patton et al., 1982; Yandrasitz et al., 1983; Chen and Chan, 1985; Dugan et al., 1986). Generally, phospholipids eluted from the column are directly detected on the basis of UV absorption. The absorbance of phospholipids around 200 nm is mainly caused by the number of double bonds in the fatty chains attached to the glycerol backbone. Therefore the detection and quantitative analysis of phospholipids by means of UV detection are difficult. The separated phospholipids can be easily collected and independently quantitated by this sensitive method, however. On the other hand, there have been several reports of the formation of derivatives of phospholipids showing specific UV absorbance for the quantitative analysis of phospholipids (Jungalwala et al., 1975; Yandrasitz et al., 1981; Chen et al., 1981). No reagent that reacts quantitatively with all phospholipids in natural tissues is known at the present time, however. Typical chromatograms of commonly occurring intact and derivatized phospholipids are shown in Figs. 3–5.

2.3.2. Separation of Diacyl, Alkylacyl, and Alkenylacyl Analogs of Phospholipids

Although large amounts of ether-linked phospholipids are present in neural tissues, especially plasmenyl ethanolamine (1-*O*-

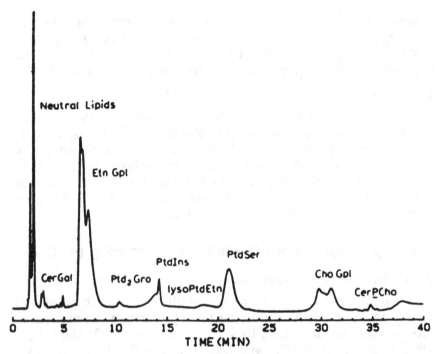

Fig. 3. HPLC chromatogram of lipids from a crude bovine brain plasma membrane fraction (Dugan et al., 1986). A Dupont Zorbax SIL column, 25 cm × 4.6 mm id, 5–6 μm particle size, was maintained at 34°C in a column heater. Peak detection was at 205 nm. The lipid sample was filtered and dissolved in hexane-2-propanol [3:2 (v/v)] before injection on to the column. The flow rate was 1.5 mL/min. The eluting solvent was hexane-2-propanol [3:2 (v/v)] with increasing proportions of water. Water was 2.75% through 9 min, was increased to 4.3% from 9–14 min, then was increased to 5.5% from 23–25 min.

alkenyl-2-acyl-*sn*-glycero-3-phosphoethanolamine), a method for the separation of the diacyl, alkylacyl, and alkenylacyl subclasses of phospholipids without modification of phospholipid structure has not yet been established. All these three subclasses are most easily separated by HPLC after their conversion to 1,2-diradylglycerol derivatives (Nakagawa and Horrocks, 1983; Blank et al., 1983). No satisfactory method that does not involve the removal of the polar head groups of phospholipids has been developed.

2.3.2.1. EXPERIMENTAL PROCEDURES.   Purified ethanolamine glycerophospholipids (EtnGpl) (1–3 mg) of bovine brain was dis-

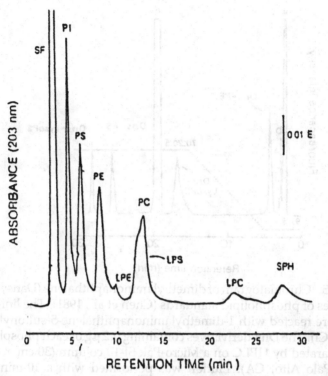

Fig. 4. HPLC chromatogram of phospholipid standards (Chen and Kou, 1982). The mixture of authentic standards, containing 0.5 μg each of phosphatidylserine, phosphatidylethanolamine, and phosphatidylcholine (PS, PE, and PC, respectively), 2.5 μg each of phosphatidylinositol (PI) and sphingomyelin (Sph), and 5 μg each of lysophosphatidylcholine (LPC) and lysophosphatidylethanolamine (LPE), was injected for HPLC on a Micro-pak SI-10 column (30 cm × 4 mm; Varian, Palo Alto, CA). The solvent system was acetonitrile:methanol:85 % phosphoric acid (130:5:1.5). The flow rate was 1 mL/min, and detection was based on the absorption at 203 nm.

solved in 3 mL of diethyl ether and 1 mL of 0.1M Tris-HCl buffer (pH 7.4). The reaction was started by the addition of about 1 mg of phospholipase C (*Bacillus cereus*, Sigma Chemical Co.) and conducted with strong stirring by means of a magnetic bar for 4–6 h at room temperature. The completeness of the reaction (no EtnGpl left at the origin) was checked by TLC using petroleum ether: diethyl ether:acetic acid (60:40:1). The upper phase (diethyl ether) was withdrawn and dried under a N_2 stream. The 1,2-diradylglycerol mixture obtained was acetylated with 0.1 mL of

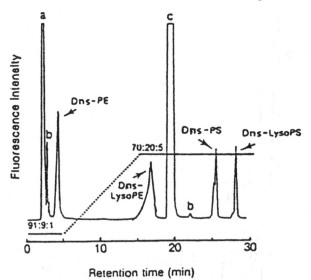

Fig. 5. Chromatogram of dimethylaminonaphthalene (Dansyl, Dns) derivatives of phospholipid standards (Chen et al., 1981). The lipid standards were reacted with 1-dimethylaminonaphthalene-5-sulfonyl chloride (Dns-Cl). The Dns-derivatives containing 0.2 μg of each phospholipid were separated by HPLC on a Micro-Pak SI-10 column (30 cm × 4 mm; Varian, Palo Alto, CA). Elution was performed with a 10-min linear gradient of dichloromethane:methanol:15M NH$_4$OH from 91:9:1 to 70:20:5. The column effluent was monitored by means of fluorescence detection at excitation and emission wavelengths of 342 and 500 nm, respectively. In peak a, unreactive Dns (Dns-Cl and Dns-NH$_4$) was eluted, and peak c was Dns-OH. Peak b comprised unidentified components.

pyridine and 0.5 mL of acetic anhydride at 37°C for 3 h. The solution was evaporated to dryness under a N$_2$ stream at 40°C, and the residue was dissolved in hexane for the separation of diacyl, alkylacyl, and alkenylacyl analogs by HPLC. The HPLC separations were performed with two solvent delivering systems (Model 100, Altex Scientific Co., Berkeley, CA). Elution of lipids was monitored by measurement of the absorbance at 205 nm. The acetylated lipids (1–50 μmol) dissolved in 20 μL of hexane were separated into the three subclasses by HPLC on a 3.9 mm × 30 cm μPorasil column (Waters Associates, Inc., Milford, MA). The solvent system was cyclopentane:hexane:methyl-t-butyl ether:acetic acid (73:24:3:0.03), pumped at a flow rate of 2 mL/min at 36°C. A typical chromatogram of the 1,2-diradyl-3-acetylglycerol mixture

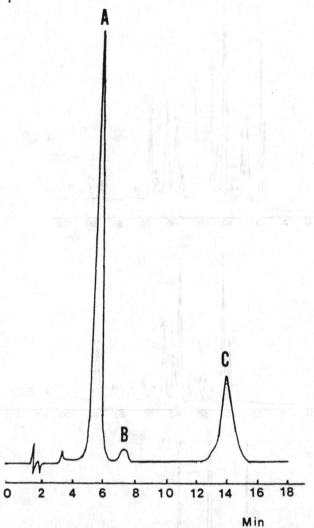

Fig. 6. Separation of alkenylacyl (A), alkylacyl (B), and diacyl (C) analogs derived from ethanolamine glycerophospholipids of bovine brain (0.2 μmol) by normal-phase HPLC (Nakagawa and Horrocks, 1983). The analytical conditions are given in the text.

obtained from EtnGpl of bovine brain is shown in Fig. 6 (Nakagawa and Horrocks, 1983).

2.3.3. Separation of Molecular Species of Phospholipids

Phospholipids exist as a complex mixture of different molecular species, of which the fatty chain at the 1 and 2 position varies

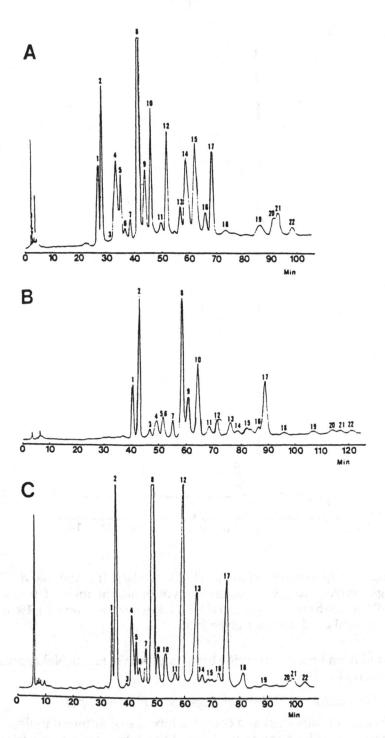

from molecule to molecule. Complete structural analysis of a phospholipid requires that it is separated into molecular species that have a single specific structure. HPLC on a reverse-phase column is more effective for the separation of the molecular species than is AgNO₃ TLC, which is a common technique used for the separation of molecular species because the separation of molecular species by reverse-phase HPLC is based not only on the degree of unsaturation of fatty chains, as in the case of AgNO₃ TLC, but also on the difference in carbon chain length. In recent years, a variety of molecular species of intact (Patton et al., 1982; Jungalwala et al., 1979, 1984) and derivatized phospholipids (Nakagawa and Horrocks, 1983; Hsieh et al., 1981; Blank et al., 1984) has been separated. Most of the methods used, however, are not suitable for the analysis of phospholipids in neural tissues containing large amounts of ether lipids, since no distinction is made between the ether-linked and diacyl subclasses. No previously described HPLC method gives good resolution of all major molecular species in a mixture of ether-linked and diacyl phospholipids. HPLC with a silicic acid column for the separation of ether and diacyl lipids followed by with a reverse-phase column has been successfully used to separate molecular species of phospholipids containing significant amounts of ether phospholipids (Nakagawa and Horrocks, 1983). Typical HPLC chromatograms of molecular species of intact and derivatized phospholipids are shown in Fig. 7 and Table 3.

Fig. 7. HPLC separation of the molecular species of alkenylacyl (A), alkylacyl (B), and diacyl (C) analogs derived from ethanolamine glycerophospholipids of bovine brain (15). Alkenylacyl (1.1 μmol), alkylacyl (0.6 μmol), and diacyl (0.9 μmol) analogs were injected for HPLC on a Zorbax ODS reverse-phase column. Alkylacyl and alkenylacyl analogs were eluted with acetonitrile:2-propanol:methyl-t-butyl ether:water, (63:28:7:2). The separation of the molecular species of diacyl analogs was performed with acetonitrile:2-propanol:methyl-t-butyl ether:water (72:18:8:2). The flow rate was 0.5 mL/min, and the column temperature was maintained at 35°C. Each fraction eluted from the column was collected for the identification and quantitation of each peak by GLC analysis after transmethylation of the 1,2-diradyl-3-acetyl-glycerol species with 0.5N NaOH in methanol. The peak numbers correspond to those in Table 3.

Table 3
Distribution of Molecular Species of Alkenylacyl, Alkylacyl, and Diacyl
Analogs of Bovine Brain

Peak number[a]	Probable molecular species	Alkenyl-acyl	Alkylacyl	Diacyl
1	18:1-22:6(n-3)	2.7 ± 0.4	3.0 ± 0.2	1.7 ± 0.2
2	16:0-22:6(n-3)	4.7 ± 0.3	8.8 ± 0.6	5.5 ± 0.4
3	18:1-22:5(n-3)	0.1 ± 0.7	0.3 ± 0.1	0.1 ± 0.1
4	18:1-20:4(n-6)	4.7 ± 0.4	1.1 ± 0.1	2.9 ± 0.3
	16:0-22:5(n-3)	0.3 ± 0.1	0.7 ± 0.1	0.1 ± 0.1
5	16:0-20:4(n-6)	2.8 ± 0.1	1.6 ± 0.2	1.1 ± 0.2
6	18:1-22:5(n-6)	0.5 ± 0.1	0.6 ± 0.1	0.4 ± 0.1
7	16:0-22:5(n-6)	0.9 ± 0.1	1.4 ± 0.1	0.8 ± 0.1
8	18:0-22:6(n-3)	8.7 ± 1.1	10.2 ± 0.6	25.5 ± 0.5
	18:1-20:3(n-6)	0.3 ± 0.2	0.3 ± 0.1	0.3 ± 0.1
	18:1-18:2(n-6)		0.3 ± 0.1	
9	18:1-22:4(n-6)	4.0 ± 0.9	5.1 ± 0.3	1.3 ± 0.1
	16:0-20:3(n-6)	0.3 ± 0.1	0.3 ± 0.1	0.2 ± 0.1
	16:0-18:2(n-6)		0.5 ± 0.2	
10	16:0-22:4(n-6)	7.3 ± 0.6	10.1 ± 0.7	1.2 ± 0.2
	18:1-20:3(n-9)	0.6 ± 0.3	0.4 ± 0.1	0.6 ± 0.1
11	18:0-22:5(n-3)	0.3 ± 0.1	0.8 ± 0.2	0.3 ± 0.1
	16:0-20:3(n-9)	0.4 ± 0.1	0.6 ± 0.1	0.2 ± 0.1
12	18:0-20:4(n-6)	4.8 ± 0.2	2.2 ± 0.3	15.8 ± 0.5
13	18:0-22:5(n-6)	0.8 ± 0.1	1.3 ± 0.1	3.0 ± 0.3
14	18:1-18:1	21.8 ± 0.2	5.5 ± 0.2	10.1 ± 0.8
15	16:0-18:1	12.9 ± 1.1	6.8 ± 0.2⎱	5.0 ± 0.9
	16:0-16:0	1.9 ± 0.3	3.6 ± 0.2⎰	
	18:1-X		1.6 ± 0.1	0.3 ± 0.1
	18:0-18:2(n-6)		0.4 ± 0.1	
16	18:0-20:3(n-6)	0.3 ± 0.1	0.3 ± 0.1	0.9 ± 0.2
	16:0-X	1.5 ± 0.3	1.7 ± 0.1	0.3 ± 0.1
17	18:0-22:4(n-6)	4.7 ± 0.2	10.3 ± 0.6	8.6 ± 0.9
18	18:0-20:3(n-9)	0.6 ± 0.2	0.3 ± 0.1	1.6 ± 0.2
19	18:1-20:1	3.4 ± 0.2	5.7 ± 0.2	2.2 ± 0.1
20	16:0-20:1	3.0 ± 0.6	8.6 ± 0.3	0.6 ± 0.1
21	18:0-18:1	3.7 ± 0.6	4.5 ± 0.1	8.5 ± 0.5
22	18:0-X	0.7 ± 0.4	1.0 ± 0.1	0.8 ± 0.2
Recovery		98.3 ± 2.1	94.5 ± 3.6	97.6 ± 6.7

[a]Peak numbers correspond to the numbers of the peaks in Fig. 7.

2.4. Gas-Liquid Chromatography (GLC)

2.4.1. Fatty Acid Analysis

Before the fatty acid composition of a lipid can be determined by GLC, it is necessary to prepare the volatile methyl ester derivatives of the fatty acids esterified on the phospholipids. There are two general methods to prepare fatty acid methyl esters from phospholipids. Acid-catalyzed transesterification is accomplished by heating phospholipids with a large excess of anhydrous methanol in the presence of an acidic catalyst (e.g., 5% HCl, 1–2% H_2SO_4, and 12–14% BF_3). Another procedure is base-catalyzed transesterification of phospholipids. Sodium methoxide (0.5M) in anhydrous methanol, which is prepared simply by dissolving sodium in dry methanol, is the most popular reagent. Phospholipids are completely transmethylated in a few minutes at room temperature.

In many column packings, polar polyester liquid phases are much more suited to fatty acid analysis, since they allow clear separations of esters of the same chain length. DEGS (diethyleneglycol succinate), EGSS-X, and EGSS-Y have now become widely accepted as the most useful materials for columns because of their stability of high temperature. Fatty acid methyl esters are usually analyzed at 195°C by GLC with a column (6' × 1/8") packed with chromosorb W. A typical chromatogram is shown in Fig. 8. (*See* the chapter by Sun for a more complete discussion of these techniques.)

2.4.2. Long Chain Fatty Aldehyde Analysis

Plasmalogen-bound fatty aldehydes were obtained in the following manner (Ferrell et al., 1970): a 1-mL aliquot of the phospholipid fraction was evaporated under a N_2 stream to leave a lipid film on the flask. HCl gas was blown for several seconds over the surface of the lipid film that was then allowed to set for 20 min. Nitrogen was then introduced until all of the HCl had been removed. The lipid was dissolved in chloroform and then subjected to TLC with development with hexane:chloroform:methanol (75:25:2). A Pyrex column (5' × 1/8") packed with 15% ethylene glycol succinate-methyl silicone polymer (EGSS-X) coated on 100–120 mesh Gas-chrom P (Applied Science Laboratories, State College, PA) was used. The column temperature for the polar liquid phase was manually programmed to increase from 130–180°C at approximately 3°C/min (Fig. 9).

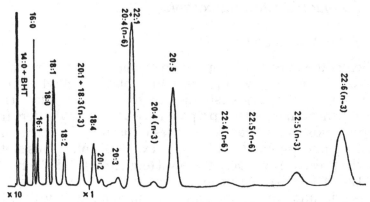

Fig. 8. Separation of a complex mixture of fatty acid methyl ester by GLC. Data obtained with 2 m × 4 mm id glass column packed with 15% EGSS-X on Chromosorb W (100–120 mesh) at 178°C.

Aldehydes are rather reactive. The preparation and analysis of a stable derivative are described in the chapter by Sun.

2.4.3. 1-O-Alkyl Chain Analysis

The alkylacyl analogs obtained by HPLC or by TLC were hydrolyzed with 0.5N NaOH containing 90% methanol at 38°C for 90 min. The resulting alkylglycerols were purified by TLC on silica gel G with development with petroleum ether:diethyl ether:acetic acid, 30:70:1. 1-Alkylglycerol was dissolved in 1 mL of pyridine, and then 0.2 mL of hexamethyldisilazane and 0.1 mL of trimethylchlorosilane were added. The mixture was shaken for 30 s and left standing for 5 min; 5 mL of hexane and water were then added to the mixture, which was shaken vigorously. The water phase was extracted twice with 2 mL of hexane. The hexane extracts were combined and dried over Na_2SO_4. After complete evaporation of the solvent, the 1-O-alkylglycerols were analyzed as their bis-trimethylsilyl derivatives by gas-liquid chromatography on a column packed with 5% SE-30 at 230°C (Fig. 10).

3. Enzymes for Biosynthesis of Phospholipids

3.1. Choline and Ethanolamine Phosphotransferases

The pathways for the biosynthesis of phospholipids in neural tissues are similar to general pathways. The cytidine pathway

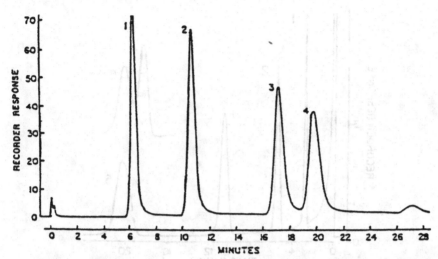

Fig. 9. Representative chromatograms (GLC) obtained on EGSS-X (130–180°C at approximately 3°C/min), showing the resolution of a standard aldehyde mixture (Wood and Harlow, 1969). The numbered peaks are: 1, myristoyl; 2, palmitoyl; 3, stearoyl; and 4, oleoyl aldehydes. The small peak with a retention time of 27 min is a contaminant that was not removed from the synthetic oleoyl aldehyde.

discovered by Kennedy and associates is a key pathway in neural tissues. Binaglia et al. (1973) concluded that the cytidine-dependent enzymic system is concentrated mostly in neuronal cells, rather than in glial cells.

3.1.1. Experimental Procedures

Neuronal and glial fractions were prepared from white rabbits (1.5–1.8 kg body weight) by the methods of Goracci et al. (1973), diluted with large amounts of a 0.32*M* sucrose salt solution, and pelleted by centrifugation for 15 min at 2000*g*. The cell-enriched pellets were homogenized in 0.32*M*-sucrose in 10 m*M* Tris-HCl, pH 7.4, in a glass-Teflon homogenizer at 1000 rev/min with five or six up-and-down strokes. The purity of the neuronal and glial suspensions was assessed by electron microscopy before the final homogenization (Blomstrand, 1970).

3.1.2. Enzyme Assay

The ethanolaminephosphotransferase (EC 2.7.8.1) and cholinephosphotransferase (EC 2.7.8.2) activities were assayed in the following incubation system (final volume, 0.15 mL): 5–6 m*M* soybean diacylglycerol, 50 m*M* Tris-HCl buffer (pH 7.9), 12.3 m*M*

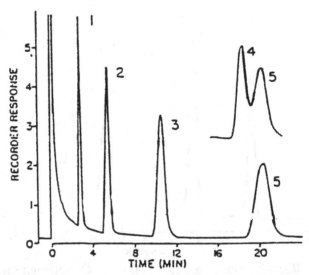

Fig. 10. Chromatograms of glyceryl ether TMS derivatives obtained on a 5' × 1/8" stainless steel column packed with 5% SE-30 operated at 230°C (Wood and Snyder, 1966). Each peak consists of a mixture of 1- and 2-isomers, and the numbered peaks are: 1, 12:0; 2, 14:0; 3, 16:0; 4, 18:1; and 5, 18:0.

cysteine, labeled 1.2 mM CDPcholine (specific activity of about 0.5 Ci/mol), or labeled 1.34 mM CDPethanolamine (specific activity of about 0.5 Ci/mol), neuronal protein (150–200 μg), or glial protein (400–500 μg), and 10 mM MnCl$_2$. The final concentrations of Tween-20 and bovine serum albumin were 0.02 and 0.012%, respectively. Incubation was carried out for 30 min at 39.5°C. Each incubation was performed in a round-bottomed test tube, which was shaken at about 140 strokes/min in a waterbath shaker. At the end of the incubation, the reaction was stopped and the lipids were extracted and analyzed by TLC. Table 4 shows that neuronal components displayed higher incorporation rates of CDPcholine and CDPethanolamine into the corresponding lipid than did glial cells (Binaglia et al., 1973).

3.2. Alkylacylglycerophosphoethanolamine Dehydrogenase

The results of several experiments with intact cells suggested that 1-alkenyl-2-acyl-sn-glycero-3-phosphoethanolamine (ethanolamine plasmalogen, alkenylacyl GPE) is directly synthesized from 1-alkyl-2-acyl-sn-glycero-3-phosphoethanolamine (alkylacyl

Table 4
Comparative Incorporation of CDPCholine and CDPEthanolamine into
Phospholipids of Neuronal and Glial Cell Populations In Vitro[a]

Fraction	Labeled precursor	Lipid labeling[b]
Neurons	CDPcholine	21.8
	CDPethanolamine	17.6
Glia	CDPcholine	4.2
	CDPethanolamine	3.8

[a]Neuronal and glial homogenates were incubated either with CDPcholine or CDPethanolamine. The final concentrations of CDPcholine and CDPethanolamine were 1.1 mM; 5.6 mM diacylglycerol (final concentration) was added for each incubation. The final volume was 0.15 mL, and the incubation was carried out at 39.5°C for 30 min.

[b]Values are expressed in nmol diacyl-Gro$P$Cho or diacyl-Gro$P$Etn/mg protein/30 min.

GPE). The first direct evidence for the existence of an enzyme that catalyzes this conversion was reported by Paltauf (1972) in microsomes from hamster intestinal mucosa.

3.2.1. Assay

The incubation mixture contained 15–60 nmol of 1-[^{14}C]alkyl-2-acyl-GPE, 325–508 μg of homogenate protein, or 498 μg of microsomal protein plus 0.3 mL of the 100,000g supernate, 10 mM ATP, 4 mM MgCl$_2$, 2mM NADH, 6 mM glucose-6-phosphate, 1.64 U of glucose-6-phosphate dehydrogenase per mL, 2 mM glutathione, and 12 mM NaF in a final volume of 1.0 mL that was buffered with 100 mM Tris-HCl, pH 7.1. The lipid substrate was added as a sonicated dispersion in 0.15% Tween-80. The tubes were shaken vigorously for 3 s, then incubated for 60 min at 37°C in a shaking water bath. The reaction was stopped with methanol, and the mixture was extracted with chloroform-methanol.

The substrate was isolated from rat brains after the intracerebral injection of [1-^{14}C]hexadecanol into 14-d-old rats. After 24 h, the rats were killed, and the ethanolamine glycerophospholipid fraction was isolated. Plasmalogens were hydrolyzed with mild acid, and 1,2-diacyl-GPE was removed by hydrolysis with porcine pancreatic lipase.

The dehydrogenation requires a reduced pyridine nucleotide and is strongly inhibited by KCN, but not by CO. Antibodies to NADH-cytochrome b$_5$ reductase inhibit the reaction, and the addi-

Table 5
Alkylacyl-GPE Dehydrogenase Activity of Rat Brain Cells
(Woelk and Peiler-Ichikawa, 1978)

System	Specific activity, nmol/mg protein/h
Neuronal homogenate, 14 d old	2.11
Neuronal homogenate, 21 d old	1.23
Neuronal homogenate, adult	0.91
Astroglial homogenate, 14 d old	2.05
Astroglial homogenate, 21 d old	0.84
Astroglial homogenate, adult	0.36
Oligodendroglial microsomes, 20 d old	2.94
Above plus cytochrome b_5	4.49
(1.2 mg/mg microsomal protein)	
Neuronal microsomes, 20 d old	2.96
Above, 10 d of 80 mg/kg/d piracetam	4.56

tion of cytochrome b_5 stimulates the dehydrogenation of alkyl to alk-1'-enyl groups. Piracetam also stimulates the reaction (Woelk and Peiler-Ichikawa, 1978) (Table 5).

3.3. 1-Acyl-Glycerophosphocholine Acyl-CoA Acyltransferase

Lands (1960) described a CoA and ATP-dependent mechanism in rat liver subcellular particles that actively acylates lysophosphatidylcholine to form phosphatidylcholine. Pieringer and Hokin (1962) reported the formation of phosphatidic acid from lysophosphatidic acid in guinea pig liver and brain. In this section, a study to determine whether or not the brain can also acylate lysophosphatidylcholine (Webster, 1962) is described.

3.3.1. Experimental Procedures

Rat brain tissues were homogenized in ice-cold 0.25M sucrose containing 1 mM EDTA and then centrifuged for 10 min at 1000g to remove nuclei and debris. Mitochondria were then sedimented at 10000g for 20 min and resuspended in a sucrose-EDTA solution so that 1 mL was equivalent to 1 g of fresh tissue. Radioactive lysophosphatidylcholine was prepared through the action of *Agkistrodon p. piscivorus* venom on liver phosphatidylcholine of rats injected intraperitoneally with 100 µCi [methyl-^{14}C]choline chloride 5 h before sacrifice.

3.3.2. Assay

The incubation system contained 1 mL of the rat brain mitochondrial suspension (from 1 g tissue), 100 μmol of phosphate buffer (pH 7.4), 10 μmol of $MgCl_2$, and 3 μmol of radioactive lysophosphatidylcholine (0.131 μCi) in a final volume of 2.3 mL. Incubation was performed at 38°C for 3 h. The reaction was stopped and the lipids were extracted by adding 4 vol of chloroform: methanol [2:1 (v/v)]; the extracted lipids were then separated by silicic acid column chromatography. The radioactivity of the phosphatidylcholine fraction was measured and compared with that of a radioactive phosphatidylcholine standard. The results are shown in Table 6 (Webster, 1962).

4. Enzymes for Catabolism of Phospholipids

4.1. Phospholipases A_1 and A_2

In this section, the distribution of phospholipases A_1 and A_2 in neuronal and glial cells is described (Woelk et al., 1973). Both fractions were incubated with specifically labeled phosphatidyl-choline, and their ability to hydrolyze the compound was evaluated. In brief, it was found that neuronal cells exhibit much higher phospholipase A_1 and A_2 activities per unit of protein than glial cells, and also that the pH optima of the enzymes differ between the two cell types.

4.1.1. Assay

Solutions of phospholipids in chloroform:methanol, 2:1 (v/v), corresponding to 0.50, 0.75, 1.0, 2.0, or 4.0 μmol of substrate, were pipeted into glass-stoppered tubes, and the solvent was removed under a N_2 stream at 30°C. A buffer solution (0.95 mL) (0.1M sodium acetate, pH 4.0–5.4; 0.1M phosphate, pH 6.0–7.2; or 0.1M Tris/HCl, pH 7.5–9.0) containing 3 mg of sodium taurocholate was added to the tubes, and the tubes were shaken to emulsify the substrate. Each mixture was then preincubated for 5 min at 37°C. Incubation was started by adding 0.05 mL of a glial or neuronal suspension. The tubes were agitated vigorously for 3 s, then incubated at 37°C. Control tubes without the enzyme were included in each experiment. The reaction was stopped by adding 2.50 mL of methanol and 1.25 mL of chloroform to the incubation mixture (Figs. 11 and 12) (Woelk et al., 1973; Woelk and Porcellati, 1973).

Table 6
Formation of Labeled Phospholipids by Rat Brain Mitochondria in the
Presence of Labeled Lysophosphatidylcholine[a]

Additions to incubation system	Conversion,[b] %
No additions	6.5
+ 3 μmol of palmitic acid	8.4
+ 3 μmol of oleic acid	8.6
+ 0.2 μmol of CoA and 5 μmol of ATP	11.0
+ 3 μmol palmitic acid, 0.2 μmol of CoA, and 5 μmol of ATP	11.4
+ 3 μmol oleic acid, 0.2 μmol of CoA, and 5 μmol of ATP	13.6
+ 250 μg of palmitoyl-CoA	12.5
+ 1000 μg of palmitoyl-CoA	19.6

[a]The incubation system contained 1 mL of mitochondrial suspension (from 1 g of tissue), 100 μmol of phosphate buffer (pH 7.4), 10 μmol of $MgCl_2$, 3 μmol of radioactive lysophosphatidylcholine (0.131 μCi), and further additions as indicated above, in a final volume of 2.3 mL.

[b]Radioactivity of the phosphatidylcholine fraction after 3 h of aerobic incubation at 38°C expressed as a percentage of the radio-activity added initially as [^{14}C]lysophosphatidylcholine.

4.2. Phospholipase D

The enzyme, obtained from a particulate fraction of brain, liberates phosphatidic acid from phosphatidylcholine (Saito and Kanfer, 1973). The solubilized enzyme from rat brain microsome was stimulated by Ca^{2+} or Mg^{2+}, but neither divalent cation is essential for the hydrolysis of phosphatidylcholine. The pH optimum was 6.0. Lysophospholipase D, which requires Mn^{2+}, but is not stimulated by Ca^{2+}, acting on 1-alkyl- and 1-alkenyl-2-lyso-sn-glycerophosphoethanolamine and -phosphocholine, was also demonstrated in rat brain microsomes (Wykle and Schremmer, 1974).

4.2.1. Experimental Procedures

The partially purified enzyme was prepared from the Sephadex G-25 separation of the solubilized particulate fraction (35000 g) of rat brain (Saito and Kanfer, 1973).

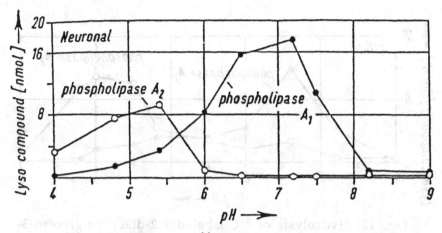

Fig. 11. Hydrolysis of ^{14}C-labeled 1,2-diacyl-*sn*-glycero-3-phosphocholine by neuronal phospholipase A at different pH values. Each incubation mixture contained in a final volume of 1.0 mL: 180 μg of neuronal protein and 1 μmol of 1-acyl-2-[1-^{14}C]oleoyl-*sn*-glycero-3-phosphocholine (for phospholipase A$_1$) or 1 μmol of 2-acyl-1-[1-^{14}C]palmitoyl-*sn*-glycero-3-phosphocholine (for phospholipase A$_2$). Incubation was performed at 37°C for 1 h.

4.2.2. Assay

Reaction mixtures contained 10 μmol of HEPES buffer (pH 7.5), 10 μmol of CaCl$_2$, [1,2-^{14}C]phosphatidylcholine (1.4×10^5 cpm, 0.6 nmol), and enzyme (50–100 μg of protein) in a final volume of 0.2 mL. The samples were incubated for 10 min at 37°C, and the reaction was stopped by the addition of 0.05 mL (500 μg) of BSA and 0.25 mL of 10% TCA. The mixtures were allowed to stand at 0°C for 30 min, and then were centrifuged at 3000 rpm for 10 min. The radioactivity of [^{14}C]choline present in the supernatant was determined.

4.3. Plasmalogenase

Ethanolamine plasmalogens are the major phospholipid components of white matter and myelin of the central nervous system (CNS). CNS plasmalogenase (EC 3.3.2.2, 1-alk-1'-enyl-2-acyl-*sn*-glycero-3-phosphoethanolamine aldehydohydrolase), the enzyme that hydrolyzes the vinyl ether linkage of the plasmalogen molecule, is concentrated in the white matter, and not in myelin, and

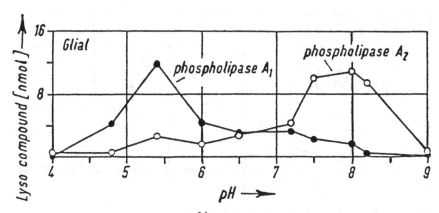

Fig. 12. Hydrolysis of ^{14}C-labeled 1,2-diacyl-sn-glycero-3-phosphocholine by glial phospholipase A at different pH values. Each incubation mixture contained 1180 μg of glial protein. The substrates and enzyme assay conditions were as in Fig. 11.

the plasmalogenase activity parallels myelination. Thus, oligodendroglia may be responsible for the catabolic activity necessary for the turnover of the plasmalogens associated with myelin. In this section, the assay for the plasmalogenase activities in neuronal perikarya, astroglia, and oligodendroglial fractions derived from adult bovine brain is described.

4.3.1. Assay

The source of the added plasmalogens for the assays was myelin, which was isolated and purified from 10–15 g of bovine white matter (Toews et al., 1976). The purified myelin was sonicated in 0.1M glycyl-glycine-HCl buffer, pH 7.4, containing 0.05% Tween-20 to obtain a suspension containing myelin plasmalogens at a concentration of 4.0 mM. A mixture of 1.0 mL of the myelin suspension and 1.0 mL of the dispersed cell fraction was incubated at 37°C. After 0, 20, and 40 min 0.5-mL portions of the incubation mixture were removed and mixed with 2.0 mL of chloroform: methanol [2:1 (v/v)]. The plasmalogens remaining in the lower phase each time were quantitated with the iodine addition method (Gottfried and Rapport, 1962). The oligodendroglia isolated from adult bovine white matter hydrolyzed the plasmalogens of myelin at the rate of 6.72 μmol/mg protein/h. This catabolic activity is 10-fold greater than in the case of neuronal perikarya and six-fold greater than in the case of astroglia (Dorman et al., 1977).

Table 7
Relative Proportions of Arachidonoyl Molecular Species of Alkenylacyl,
Alkylacyl, and Diacyl GroPEtn of Rat Brain[a]

	Alkenylacyl	Alkylacyl	Diacyl
16:0–20:4	27.5 ± 2.5	38.3 ± 2.1	9.0 ± 0.9
18:0–20:4	46.4 ± 1.9	32.3 ± 2.1	79.7 ± 2.3
18:1–20:4	26.0 ± 0.7	28.4 ± 2.1	11.3 ± 1.5

[a]All values are given as mean ± SD for six experiments. The relative proportions of each molecular species were determined by gas-liquid chromatography.

5. Metabolism of Arachidonoyl Molecular Species of Phospholipids of the Brain

The metabolism of individual molecular species of phospholipids is quite different in mammalian tissues (reviewed by Holub and Kuksis, 1978). Studies of metabolism of the arachidonoyl molecular species of phospholipids are interesting since arachidonic acid, which is the precursor of prostaglandins and leukotrienes, is mainly esterified in phospholipids. Little work has been done, however, on the measurement of the biosynthetic rates for individual arachidonoyl molecular species of phospholipids since it is difficult to isolate arachidonoyl molecular species having a single structure.

As described above, a HPLC technique with a reverse-phase column has allowed excellent separation of the molecular species of phospholipids and investigation of the metabolism of individual molecular species, including the arachidonoyl molecular species. The metabolism of the arachidonoyl molecular species (18:1–20:4, 16:0–20:4, and 18:0–20:4 species) of diacyl, alkylacyl, and alkenylacyl glycerophosphoethanolamine (GroPEtn) of rat brain has been examined after an intracerebral injection of [³H]arachidonic acid (Nakagawa and Horrocks, 1986). The procedures for the separation of the molecular species of diacyl, alkylacyl, and alkenylacyl GroPEtn are described in section 2.3. The purified ethanolamine glycerophospholipids (EtnGpl) were converted to 1,2-diradyl-3-acetylglycerols, and the arachidonoyl molecular species were separated from the diacyl, alkylacyl, and alkenylacyl compounds by

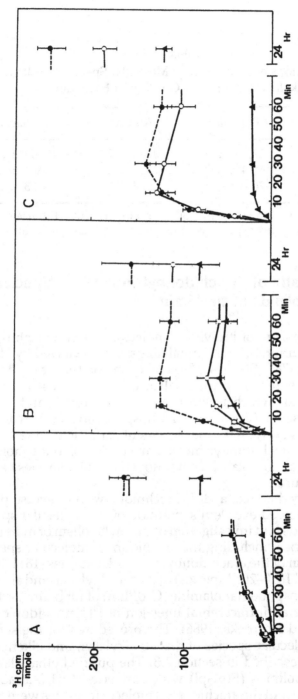

Fig. 13. Specific radioactivity of arachidonoyl molecular species of alkenylacyl (A), alkylacyl (B), and diacyl (C) GroPEtm. The 18:1–20:4 (●), 16:0–20:4 (○), and 18:0–20:4 (▲) species of each EtnGpl subclass were separated by reverse-phase HPLC.

174

reverse-phase HPLC. The alkylacyl subclass contained almost equal proportions of the 18:0–20:4, 16:0–20:4, and 18:1–20:4 species, but alkenylacyl and diacyl subclasses predominantly consisted of the 18:0–20:4 species (Table 7). In all three subclasses, the specific radioactivity of the 18:1–20:4 species was the highest, whereas the 18:0–20:4 species showed the lowest specific radioactivity (Fig. 13). From 60 min to 24 h, the specific radioactivities of all three arachidonoyl molecular species increased equally by five to eight times in the alkenylacyl class and by two times in the alkylacyl class. In diacyl GroPEtn, the specific radioactivities of the 18:1– and 16:0–20:4 species doubled during the same period, but the specific radioactivity of the 18:0–20:4 species at 24 h was more than six-fold greater than that at 60 min. The slower apparent turnover of the 18:0–20:4 species may lead to the greater accumulation of [^{3}H]arachidonic acid in the 18:0–20:4 species than in the 18:1–20:4 and 16:0–20:4 species of diacyl GroPEtn. This may explain the greater abundance of the 18:0–20:4 species of diacyl GroPEtn. The different rates of turnover of the molecular species may be of considerable importance in the maintenance of the characteristic compositions of membrane phospholipids.

References

Alam I., Smith J. B., and Silver M. J. (1982) Novel system for separation of phospholipids by high-performance liquid chromatography. *J. Chromatogr.* **234**, 218–221.

Binaglia L., Goracci G., Porcellati G., Roberti R., and Woelk H. (1973) The synthesis of choline and ethanolamine phosphoglycerides in neuronal and glial cells of rabbit in vitro. *J. Neurochem.* **21**, 1067–1082.

Blank M. L., Gross E. A., Lee T. L., Stephens N., Piantadosi C., and Snyder F. (1983) Quantitative analysis of ether-linked lipids as alkyl- and alk-1-enyl-glycerol benzoates by high-performance liquid chromatography. *Anal. Biochem.* **133**, 430–436.

Blank M. L., Robinson M., Fitzgerald V., and Snyder F. (1984) Novel quantitative method for determination of molecular species of phospholipids and diglycerides. *J. Chromatogr.* **298**, 473–482.

Bligh E. G. and Dyer W. J. (1959) A rapid method of total lipid extraction and purification. *Can. J. Biochem. Physiol.* **37**, 911–917.

Blomstrand C. and Hamberger A. (1970) Amino acid incorporation in vitro into protein of neuronal and glial cell-enriched fractions. *J. Neurochem.* **17**, 1187–1195.

Bowen D. M., Davison A. N., and Ramsey R. B. (1974) The dynamic role of lipids in the nervous system. *Int. Rev. Sci.* (Biochem. Ser. 1) **4,** 141–179.

Chen S. F. and Chan P. H. (1985) One step separation of free fatty acids and phospholipids in brain tissue extracts by high performance liquid chromatography. *J. Chromatogr.* **344,** 297–303.

Chen S. S-H. and Kou A. Y. (1982) Improved procedure for the separation of phospholipids by high-performance liquid chromatography. *J. Chromatogr.* **227,** 25–31.

Chen S. S-H., Kou A. Y., and Chen Y. H-H. (1981) Measurement of ethanolamine- and serine-containing phospholipids by high-performance liquid chromatography with fluorescence detection of their Dns derivatives. *J. Chromatogr.* **208,** 339–346.

Dorman R. V., Toews A. D., and Horrocks L. A. (1977) Plasmalogenase activities in neuronal perikarya, astroglia and oligodendroglia isolated from bovine brain. *J. Lipid Res.* **18,** 115–117.

Dugan L. L., Demediuk P., Pendley II, C. E., and Horrocks L. A. (1986) Separation of phospholipids by HPLC: All major classes, including lysophosphatidylethanolamine. *J. Chromatogr.* **378,** 317–327.

Ferrell W. J., Radloff D. M., and Radloff J. F. (1970) Method for hydrolysis and isolation of plasmalogen-bound fatty aldehydes. *Anal. Chem.* **37,** 227–235.

Folch J., Lees M., and Sloane Stanley G. H. (1957) A simple method for the isolation and purification of total lipides from animal tissues. *J. Biol. Chem.* **226,** 497–507.

Goracci G., Hamberger A., Blomstrand C., and Porcellati G. (1973) Base-exchange enzymic system for the synthesis of phospholipids in neuronal and glial cells and their subfraction. *J. Neurochem.* **20,** 1167–1180.

Gottfried E. and Rapport M. M. (1962) Biochemistry of plasmalogen. *J. Biol. Chem.* **237,** 329–333.

Hanson V. L., Park J. Y., Osborn T. W., and Kiral R. M. (1981) High-performance liquid chromatographic analysis of egg yolk phospholipids. *J. Chromatogr.* **205,** 393–400.

Hattori H. and Kanfer J. (1985) Synaptosomal phospholipase D potential role in providing choline for acetylcholine synthesis. *J. Neurochem.* **45,** 1578–1584.

Holub B. J. and Kuksis A. (1978) Metabolism of Molecular Species of Diacylglycerophospholipids, in *Advances in Lipid Research* (Paoletti R. and Kritchevsky D., eds.) vol. 16, Academic, New York.

Hsieh J. Y-K., Welch D. K., and Turcotte J. G. (1981) General method for the analysis of phosphatidylcholines by high-performance liquid chromatography. *J. Chromatogr.* **208,** 398–403.

Jungalwala F. B., Evans J. E., and McCluer R. H. (1984) Compositional and molecular species analysis of phospholipids by high performance liquid chromatography coupled with chemical ionization mass spectrometry. *J. Lipid Res.* **25**, 738–749.

Jungalwala F. B., Hayssen V., Pasquini J. M., and McCluer R. H. (1979) Separation of molecular species of sphingomyelin by reverse-phase high-performance liquid chromatography. *J. Lipid Res.* **20**, 579–587.

Jungalwala F. B., Turel R. J., Evans J. E., and McCluer R. H. (1975) Sensitive analysis of ethanolamine- and serine-containing phosphoglycerides by high-performance liquid chromatography. *Biochem. J.* **145**, 517–526.

Lands W. E. M. (1960) Metabolism of glycerolipids. *J. Biol. Chem.* **235**, 2233–2237.

Nakagawa Y. and Horrocks L. A. (1983) Separation of alkenylacyl, alkylacyl and diacyl analogues and their molecular species by high performance liquid chromatography. *J. Lipid Res.* **24**, 1268–1275.

Nakagawa Y. and Horrocks L. A. (1986) Different metabolic rates for arachidonyl molecular species of ethanolamine glycerophospholipids in rat brain. *J. Lipid Res.* **27**, 629–636.

Paltauf F. (1972) Plasmalogen biosynthesis in a cell-free system. *FEBS Lett.* **20**, 79–82.

Patton G. M., Fasulo J. M., and Robins S. J. (1982) Separation of phospholipids and individual molecular species of phospholipids by high-performance liquid chromatography. *J. Lipid Res.* **23**, 190–196.

Pieringer R. A. and Hokin L. E. (1962) Biosynthesis of phosphatidic acid from lysophosphatidic acid and palmitoyl coenzyme A. *J. Biol. Chem.* **237**, 659–663.

Rouser G., Kritchevsky A., and Yamamoto A. (1976) Column Chromatographic and Associated Procedures for Separation and Determination of Phosphatides and Glycolipids, in *Lipid Chromatographic Analysis* 2nd ed. (Marinetti G. V., ed.) vol. 3, Marcel Dekker, New York.

Saito M. and Kanfer J. (1973) Solubilization and properties of a membrane-bound enzyme from rat brain catalyzing a base-exchange reaction. *Biochem. Biophys. Res. Commun.* **53**, 391–398.

Toews A. D., Horrocks L. A., and King J. S. (1976) Simultaneous isolation of purified microsomal and myelin fractions from rat spinal cord. *J. Neurochem.* **27**, 25–31.

Webster G. R. (1962) On the acylation of lysolecithin by rat liver and brain mitochondria. *Biochim. Biophys. Acta* **64**, 573–575.

Woelk H. and Peiler-Ichikawa K. (1978) The action of piracetam on the formation of ethanolamine-plasmalogen by neuronal microsomes of the developing rat brain. *Drug Res.* **28**, 1752–1756.

Woelk H. and Porcellati G. (1973) Subcellular distribution and kinetic properties of rat brain phospholipase A_1 and A_2. *Hoppe Seylers Z. Physiol. Chem.* **354,** 90–100.

Woelk H., Goracci G., Gaiti A., and Porcellati G. (1973) Phospholipase A_1 and A_2 activities of neuronal and glial cells of rabbit brain. *Hoppe Seylers Z. Physiol. Chem.* **354,** 729–736.

Wood R. and Harlow R. D. (1969) Gas-liquid chromatographic analysis of free long-chain aldehydes. *J. Lipid Res.* **10,** 463–464.

Wood R. and Snyder F. (1966) Gas-liquid chromatographic analysis of long-chain isomeric glyceryl monoethers. *Lipids* **1,** 62–72.

Wykle R. L. and Schremmer J. M. (1974) A lysophospholipase D pathway in the metabolism of ether-linked lipids in brain microsomes. *J. Biol. Chem.* **249,** 1742–1746.

Yandrasitz J. R., Berry G., and Segal S. (1983) High-performance liquid chromatography of phospholipids. *Anal. Biochem.* **135,** 239–243.

Yandrasitz J. R., Berry G., and Segal S. (1981) High-performance liquid chromatography of phospholipids with UV detection. *J. Chromatogr.* **225,** 319–328.

Determination of Phospholipases, Lipases, and Lysophospholipases

Akhlaq A. Farooqui and Lloyd A. Horrocks

1. Introduction

Phospholipid metabolism in the central nervous system is regulated by a variety of enzymes whose substrate specificities, cellular localizations, and regulatory mechanisms are only beginning to be understood. In brain, different phospholipids turn over at different rates with respect to their cellular structure, location, and the membranes involved (Freysz et al., 1976; Miller et al., 1977; Porcellati, 1983; Porcellati et al., 1983; Farooqui and Horrocks, 1985). The degradation of membrane phospholipids is a stepwise process that is catalyzed by various phospholipases and lysophospholipases (van den Bosch, 1982; Dennis, 1983). Acylglycerols, although quantitatively minor components of mammalian brain tissue (Rowe, 1969; Cook, 1981; Kunze et al., 1984), are very active and important metabolically (Berridge, 1984; Boni and Rando, 1985; Williamson et al., 1985). They are hydrolyzed by lipases called acylglycerol acylhydrolases. Studies on the isolation and characterization of brain lipases, lysophospholipases, and phospholipases are complicated by two major problems. The activities of these enzymes are quite low when compared with other hydrolytic enzymes. Second, rapid and sensitive assay procedures for the determination of their enzymic activities have not been available.

The purpose of this article is to review various procedures available for the determination of activities of lipases, phospholipases, and lysophospholipases. It is hoped that this summary will stimulate further studies on isolation and characterization of various lipolytic enzymes.

The following procedures have been used:

1. Titrimetric procedures
2. Radiochemical procedures
3. Spectrophotometric procedures
4. Fluorometric procedures
5. Immunological procedures

179

2. Titrimetric Procedures

In this method, the released fatty acid is titrated at constant pH, and the activity of the lipolytic enzyme is calculated from the recorded alkali consumption. This procedure can be used for determining initial velocities (deHaas et al., 1971; Wells, 1972). It is restricted to neutral or alkaline pH values, however, because of the pK_a value of the released fatty acid. Thus, at pH 6.0 and 5.0, the titration efficiencies are about 90 and 50%, respectively. Further, this procedure is not very sensitive. The lower limit of detection is about 50–100 nmol/min. Thus it cannot be applied to tissue homogenates and subcellular fractions that usually display activities in the order of 1–5 nmol/min/mg protein. This procedure can be used for the determination of lipolytic enzymes in pancreatic juice and snake venom.

The released fatty acids can be converted to Cu-soaps, which are measured colorimetrically after reaction with diethyl dithiocarbamate (Duncombe, 1963). The original Duncombe procedure (Duncombe, 1963) has been refined and successfully applied to plasma (Hron and Menahan, 1981), milk (Shipe et al., 1983), and single oat grains (Sahasrabudhe, 1982). In spite of these refinements, the procedure is insensitive and the lower limit of detection is 50–100 nmol/min.

3. Radiochemical Procedures

3.1. Substrates for Phospholipases

Radioactive glycerophospholipids can be prepared either chemically (Baer and Buchnea, 1959) or biochemically (van den Bosch and Aarsman, 1979; Waite et al., 1981). The starting material for chemical synthesis is *sn*-glycero-3-phosphocholine (GroPCho). The acylation of this compound results in formation of phosphatidycholine (PtdCho) (Baer and Buchnea, 1959; Pugh and Kates, 1975; Gupta et al., 1977). The acyl group at the *sn*-2 position of diacyl-GroPCho is removed by phospholipase A$_2$. Subsequently, a labeled fatty acid is introduced at the *sn*-2 position to yield either single- or dual-labeled PtdCho (van den Bosch and van Deenen, 1965).

In the biochemical procedure, the labeled fatty acids are introduced by incubation with a lyosphospholipid. This reaction is catalyzed by rat liver microsomes. Phospholipase A_2 substrate is prepared by incubating 1-acyl lysophospholipid with radioactive unsaturated fatty acid (arachidonic acid) (Robertson and Lands, 1962). The incubation mixture consists of 2 μmol of 1-acyl GroP-Cho, 20 μCi of [1-^{14}C]arachidonic or oleic acid (56 μCi/μmol), and 1.6 μmol of unlabeled fatty acid in 0.1*M* sodium phosphate, pH 7.4, containing 600 μmol of $MgCl_2$, 6 μmol of CoA, 600 μmol of ATP, and 60 mg of rat liver microsomes in a total volume of 4 mL. After incubating for 2 h, the reaction is stopped, and labeled phosphatidylcholine is isolated by silicic acid chromatography (Prescott and Majerus, 1983).

It is somewhat difficult to prepare radiolabeled substrate for phospholipase A_1. The 2-acyl lysophospholipid is prepared by I_2 cleavage of plasmalogen (choline or ethanolamine) by the method of Lands and Merkl (1963). The 2-acyl lysophospholipid is incubated with rat liver microsomes in the presence of radiolabeled fatty acid, ATP, CoA, and $MgCl_2$ at pH 7.4. After 1 h, the reaction is terminated by the addition of chloroform:methanol [2:1 (v/v)]. The product can be separated either by TLC or HPLC. An alternative method to label glycerophospholipids at position 1 is by the exchange of the fatty acid at position 1 with labeled fatty acid in the presence of *Rhizopus delemar* lipase (Brockerhoff et al., 1976). Briefly, the incubation mixture consists of 100 m*M* sodium acetate buffer, pH 3.4, 1.72 m*M* phosphatidylethanolamine (PtdEtn), 0.45 m*M* butylated hydroxytoluene (to prevent peroxidation), and [^{14}C]oleic acid in 1.8 mL. The mixture is sonified for 20 s and mixed with 2 mg of *Rhizopus delemar* lipase (1200 U) in 0.2 mL of water. After 4 h shaking at 4°C under N_2, the lipids are extracted with a mixture of 2 mL of chloroform and 1.6 mL of methanol and separated by TLC on silica gel with chloroform:methanol:water:acetic acid [64:32:2:2 (v/v)]. Eight percent of the radioactivity can be found in PtdEtn, the rest in the free fatty acid, and none in lyso-PtdEtn. On hydrolysis of this PtdEtn with snake venom phospholipase A_2, 100% of the radioactivity can be found in the lyso-PtdEtn.

The substrate for phospholipase C, 1-acyl-2-[^{3}H]oleoyl-GroPIns (GroPIns = *sn*-glycero-3-phosphoinositol) or 1-acyl-2-[^{14}C]arachidonoyl-GroPIns, can be prepared from 1-acyl-GroPIns and [9,10-^{3}H]oleic acid or [1-^{14}C]arachidonic acid using rat liver microsomes. The radiolabeled PtdIns can be separated to greater than 96% purity either by TLC on silica gel GF or HPLC (Woelk et al., 1973; Rittenhouse, 1982).

Taki and Kanfer (1979) used uniformly labeled [U-^{14}C]PtdCho for the determination of phospholipase D activity. The reaction product, phosphatidic acid (PtdoH), is extracted by the addition of chloroform:methanol [2:1 (v/v)]. PtdoH is separated by a two-dimensional TLC system (Rouser et al., 1969). The area corresponding to PtdoH standard is visualized autoradiographically and is scraped into a scintillation vial and counted. Choline or ethanolamine labeled phospholipids can also be used as substrate. Here the reaction is terminated by heating the mixture at 100°C for 1 min. An aliquot of the mixture is applied directly to a thin layer chromatogram with added carrier base and developed with a solvent system of chloroform:methanol:water [65:25:4 (v/v)]. Radioactivity on the TLC plate is visualized autoradiographically. Labeled PtdoH can also be separated by HPLC using the procedure of Dugan et al. (1986). The radiochemical procedures are sensitive, but they are time consuming and expensive. Enzymic methods for the determination of released choline, using choline oxidase from *Arthrobacter globiformis*, have also been reported (Imamura and Horiuti, 1979), but the sensitivity of this procedure is less than that of radiochemical procedures. The optimal conditions for the hydrolysis of labeled phospholipids by various phospholipases are given in Table 1.

3.2 Substrates for Mono- and Diacylglycerol Lipases

1,2-Diacylglycerol is prepared from labeled PtdCho using either *Bacillus cereus* phospholipase C (Little et al., 1975) or *Clostridium welchii* phospholipase C (Mauco et al., 1984). Briefly, 1.5 μmol of labeled PtdCho dissolved in 5 mL of diethyl ether are mixed with 5 mL of 50 mM Tris·HCl buffer, pH 7.4, containing 5 mM CaCl$_2$, 5 mM MgCl$_2$, and 3 mg *Bacillus* or *Clostridium* phospholipase C. Incubation is performed for 2 h at 37°C with vigorous mixing in glass-stoppered tubes. Labeled diacylglycerols are extracted four times with diethyl ether and after evaporation under reduced pressure at 20°C, dissolved in petroleum ether, and kept under nitrogen at –20°C. Storage in petroleum ether under N$_2$ avoids conversion into 1,3-diacylglycerol. This procedure gave a specific fatty acid (e.g., arachidonic acid) at the *sn*-2 position. Labeled diacylglycerols can also be prepared from labeled triacylglycerols using pancreatic lipase (Fauvel et al., 1984). This procedure gives diacyglycerols labeled at both 1 and 2 positions.

The monoacylglycerol lipase substrates containing either [^{14}C]arachidonate or [^{3}H]oleate in the *sn*-1 or *sn*-2 position are prepared by incubating the corresponding diacylglycerols with purified pancreatic lipase. Labeled diacylglycerols are suspended in 1.0 mL of 0.18*M* Tris-HCl buffer, pH 6.5, containing 2.27 mg of bovine serum albumin, 18 m*M* CaCl$_2$, and 0.145 mg of Triton X-100. The mixture is sonicated for 4 min at 4°C in a Bransonic 220 bath sonicator. Purified pancreatic lipase (5 mg) is added, and the sample is incubated for 1 h at 37°C. The reaction is terminated by the addition of 8.0 mL of chloroform:methanol [2:1 (v/v)]. The lower phase is retained, and the aqueous upper phase is washed twice with chloroform. The chloroform extracts are combined, taken to dryness rapidly under N$_2$, and resuspended in 3 mL of petroleum ether. The monoacylglycerol is purified by chromatography on a 10% borate Unisil silicic acid column (Okazaki et al., 1981). This procedure preferentially gives 2-acylglycerol. Boric acid is added to the silicic acid in order to retard acyl migration. Some leakage of boric acid from the column, which would contaminate the purified sample, may occur during elution. Care should be taken to wash the final organic phase with water to remove the boric acid before drying. The optimal conditions for the hydrolysis of labeled mono-, di-, and triacylglycerols are given in Table 1.

4. Spectrophotometric Procedures

Spectrophotometric procedures have an advantage over radiochemical procedures in that the spectrophotometric substrates can be synthesized in larger amounts than their radioactive counterparts (analogs). Unfortunately, the reaction products of phospholipase reactions, i.e., lysophospholipids and fatty acids, are not detectable with a spectrophotometer. Modifications have been introduced in phospholipid molecules. Aarsman et al. (1976) and Aarsman and van den Bosch (1977) were the first to develop a spectrophotometric assay procedure for phospholipases using thioester substrate analogs. They synthesized a glycol analog of PtdCho with a thioester linkage in place of an oxyester linkage. Hydrolysis of this substrate by phospholipase resulted in exposure of the thiol group, which was determined using 5,5'-dithiobis (2-nitrobenzoic acid) (DTNB). The color produced (yellow) during

Table 1

Optimal Conditions for the Determination of Brain Phospholipases, Lysophospholipases, and Lipases with Radiochemical Substrates[a]

Enzyme	Buffer	pH Value	Substrate	Optimal conc (μM)	Cofactor	Reference
Phospholipase A_1						
Neutral	Sodium phosphate	7.2	1-Acyl-2-[1-^{14}C]oleoyl-GroPCho	1000	—	Woelk et al., 1973
Alkaline	Glycylglycine/glycine	9.25	1-[^{3}H]stearoyl-2-acyl-GroPEtn	200	Ca^{2+}	Rooke and Webster, 1976
Phospholipase A_2						
Acidic	Sodium acetate	5.4	2-Acyl-1-[1-^{14}C]palmitoyl-GroPCho	1000	Ca^{2+}	Woelk et al., 1973
Neutral	Tris-maleate-acetate	7.5	1-acyl-2-[1-^{14}C]oleoyl-GroPIns	1000	Ca^{2+}	Gray and Strickland, 1982
Phospholipase C	3-(N-Morphilino) propane sulfonic acid	7.0	1-[^{3}H]Palmitoyl-2-acyl GroPCho	1000	Ca^{2+}	Edgar and Freysz, 1982

Enzyme	pH	Buffer	Substrate		Reference
Phospholipase D	6.0	Hepes	[U-^{14}C]Phosphatidylcholine	1300 Ca^{2+}	Taki and Kanfer, 1979
Lysophospholipase	8.0	Tris-HCl	[1-^{14}C]Palmitoyl-2-lyso-Gro$P$Cho	1000	Fujikura and Baistel, 1985
Monoacylglycerol lipase Acidic	4.8	Sodium acetate	1-[1-^{14}C]oleoyl-glycerol	400	Cabot and Gatt, 1978
Alkaline	8.0	Tris-HCl	1-[1-^{14}C]oleoyl-glycerol	400	Cabot and Gatt, 1978
Diacylglycerol lipase Acidic	4.8	Sodium acetate	1,2-[1-^{14}C]oleoyl-glycerol	400	Cabot and Gatt, 1978
Alkaline	8.0	Tris-HCl	1,2-[1-^{14}C]oleoyl-glycerol	400	Cabot and Gatt, 1978
Triacylglycerol lipase	7.4	Tris-HCl	1,2,3-[1-^{14}C]oleoylglycerol	50	Mizobuchi et al., 1981

[a]Abbreviations: Gro$P$Cho, sn-glycero-3-phosphocholine; Gro$P$Etn, sn-glycero-3-phosphoethanolamine.

this reaction was monitored continuously at 412 nm. The thioester substrate analog assay procedure has several advantages over classical titrimetric and radiochemical procedures. Assays with thioesters substrates are fast, specific, continuous, and convenient. The relatively short acyl chains make substrate suspension easier and obviate the problem of unsaturated side chain oxidation. It is should also be recognized, however, that the relatively short chain lengths are not necessarily physiological and may not represent exact physiologic K_m and V_{max} values. The thioester substrate assay procedure has resulted in complete purification of pancreatic and brain lysophospholipases (Aarsman and van den Bosch, 1979, 1981; Farooqui et al., 1985; Pendley et al., 1981). Two diacylglycerol lipases (Farooqui et al., 1986b) have also been purified to near homogeneity.

The most serious drawback of the thioester substrate assay is the necessary use of thiol capturing reagents. These compounds may inhibit enzymic activity (Aarsman et al., 1977) and also exclude measurement with samples containing mercaptoethanol or dithiothreitol. In the following section we discuss the synthesis of thioester substrate analogs of phospholipids, lysophospholipids, and mono- and diacylglycerols.

4.1. Spectrophotometric Determination of Phospholipases Using Thioester Substrate Analogs

4.1.1. Determination of Phospholipase A_1 Activity

Cox and Horrocks (1981) reported the preparation of thioester substrate analogs of PtdCho and PtdEtn for the determination of phospholipase A_1 activity (Fig. 1).

4.1.1.1. RAC-1,2-S,O-DIDECANOYL-3-PHOSPHOCHOLINE-1-MERCAPTO-2,3-PROPANEDIOL (III) SYNTHESIS.   The didecanoylthioglycerol (II) (3.45 mmol) is dissolved in 15 mL of CHCl$_3$ and added dropwise to a rapidly stirred solution of bromoethylphosphorusoxydichloride (13.8 mmol) and 28 mmol triethylamine, at 0–5°C. After the addition is complete, the mixture is constantly stirred at room temperature for 12 h; this is followed by hydrolysis, trimethylamination, and purification, as described by Aarsman et al. (1976). Following treatment with the ion exchange resin, the crude products are taken up in CHCl$_3$ and eluted from a 100-g silicic acid column with CHCl$_3$:MeOH [4:1, 1:1, 2:3, and 1:3 (v/v)]. The thiophosphatidylcholine (III) elutes in the last two fractions in 40% yield. On TLC, it migrates identically to

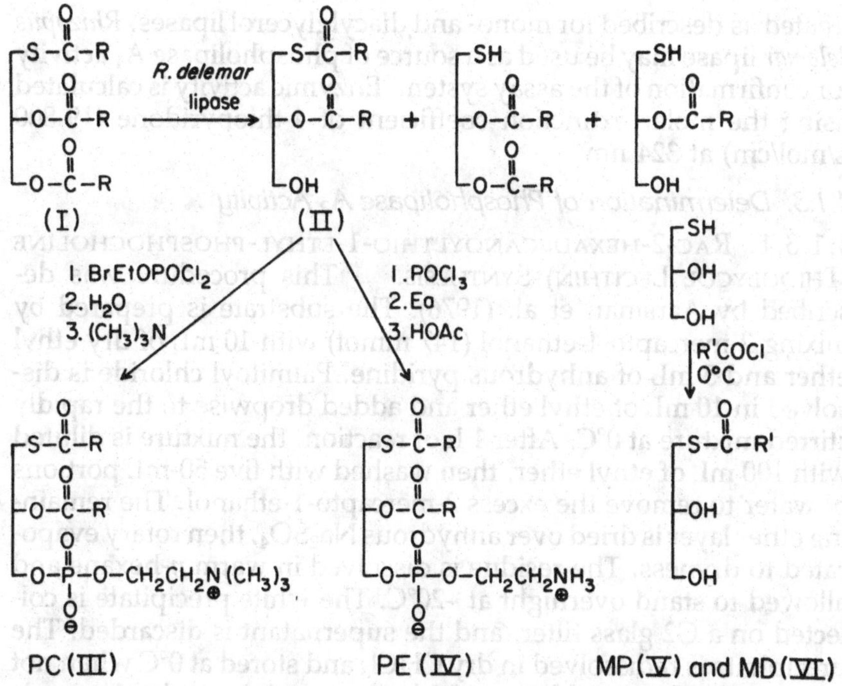

Fig. 1. Synthetic scheme of thioester substrates for phospholipase A_1 and monoacylglycerol lipase. R, $CH_3(CH_2)_8$; R', $CH_3(CH_2)_{14}$ for (V) and $CH_3(CH_2)_8$ for (VI).

didecanoylphosphatidylcholine and stains positively for both choline and phosphate ester.

4.1.1.2. RAC-1,2-S,O-DIDECANOYL-3-PHOSPHOETHANOLAMINE-1-MERCAPTO-2,3-PROPANEDIOL (IV) SYNTHESIS.   The didecanoyl-thioglycerol (11) (3.45 mmol) is converted to the thiophosphatidy-lethanolamine by the method of Eibl (1978) and purified on a 100-g silicic acid column eluted with $CHCl_3$:MeOH [(9:1, 6:1, 5:1, and 4:1 (v/v)]. Thiophosphatidylethanolamine (IV) is eluted in the last fraction in 30% yield. On TLC, it migrates just below bovine brain ethanolamine glycerophospholipids and stains positively for both phosphate ester and primary amine.

4.1.2. Assay Conditions for Phospholipase A_1 Determination

Substrate suspensions are prepared in 50 mM MOPS buffer (pH 7.5). The assay mixture contains 100 μmol of MOPS buffer (pH 7.5), 0.05 μmol of thioester or oxyester substrate, and 0.6 μmol of 4,4'-dithiodipyridine, in a total volume of 1.0 mL. Samples are treated as described for mono- and diacylglycerol lipases. *Rhizopus delemar* lipase may be used as a source of phospholipase A_1 activity for confirmation of the assay system. Enzymic activity is calculated using the molar extinction coefficient of 4-thiopyridone (19,800 L/mol/cm) at 324 nm.

4.1.3. Determination of Phospholipase A_2 Activity

4.1.3.1. RAC-2-HEXADECANOYLTHIO-1-ETHYL-PHOSPHOCHOLINE (THIOGLYCOL LECITHIN) SYNTHESIS. This procedure was described by Aarsman et al. (1976). The substrate is prepared by mixing 2-mercapto-1-ethanol (147 mmol) with 10 mL of dry ethyl ether and 5 mL of anhydrous pyridine. Palmitoyl chloride is dissolved in 10 mL of ethyl ether and added dropwise to the rapidly stirred mixture at 0°C. After 1 h of reaction, the mixture is diluted with 100 mL of ethyl ether, then washed with five 50-mL portions of water to remove the excess 2-mercapto-1-ethanol. The remaining ether layer is dried over anhydrous Na_2SO_4, then rotary evaporated to dryness. The residue is dissolved in warm *n*-hexane and allowed to stand overnight at –20°C. The white precipitate is collected on a G2 glass filter, and the supernatant is discarded. The precipitate is redissolved in dry $CHCl_3$ and stored at 0°C when not in use. To the hexadecanoylthiomonoacylglycerol mixture is added 20 mmol of 2-bromoethyl-phosphorusoxydichloride and 20 mmol of triethylamine in $CHCl_3$. This reaction is allowed to continue at room temperature for 20 h. Hydrolysis of the 2-hexadecanoyl-thio-1-ethyl-phosphomonochloride-ethyl-bromoester to 2-hexa-decanoylthio-1-ethyl-phosphoethyl-bromoester and the introduction of the trimethylamine group is achieved by the method of Eibl et al. (1967). Purified 2-hexadecanoylthio-1-ethyl-phosphocholine (Fig. 2) is isolated by silicic acid chromatography in yields of 33%. The isolated compound has no free thiol groups, and on a weight basis the phosphorus analysis revealed a purity of 98%. The assay mixture for the estimation of phospholipase A_2 activity contains 200 µmol of Tris-maleic acid buffer (pH 7.5), 100 nmol of 2-hexadecanoylthio-1-ethyl-phosphocholine, 1 µmol of DTNB, and 0.6 µmol of EDTA. The addition of hog pancreatic phospholipase A_2 in the presence of EDTA has no effect on the absorbance because this enzyme has an absolute requirement for Ca^{2+} (van den Bosch et al., 1973). After the addition of Ca^{2+}, a sharp linear increase in the absorbance at 412 nm is seen, which can be stopped by the addition of excess EDTA. Volwerk et al. (1979) synthesized a short chain dithioester analog of PtdCho, 2,3-bis(hexanoylthio)-

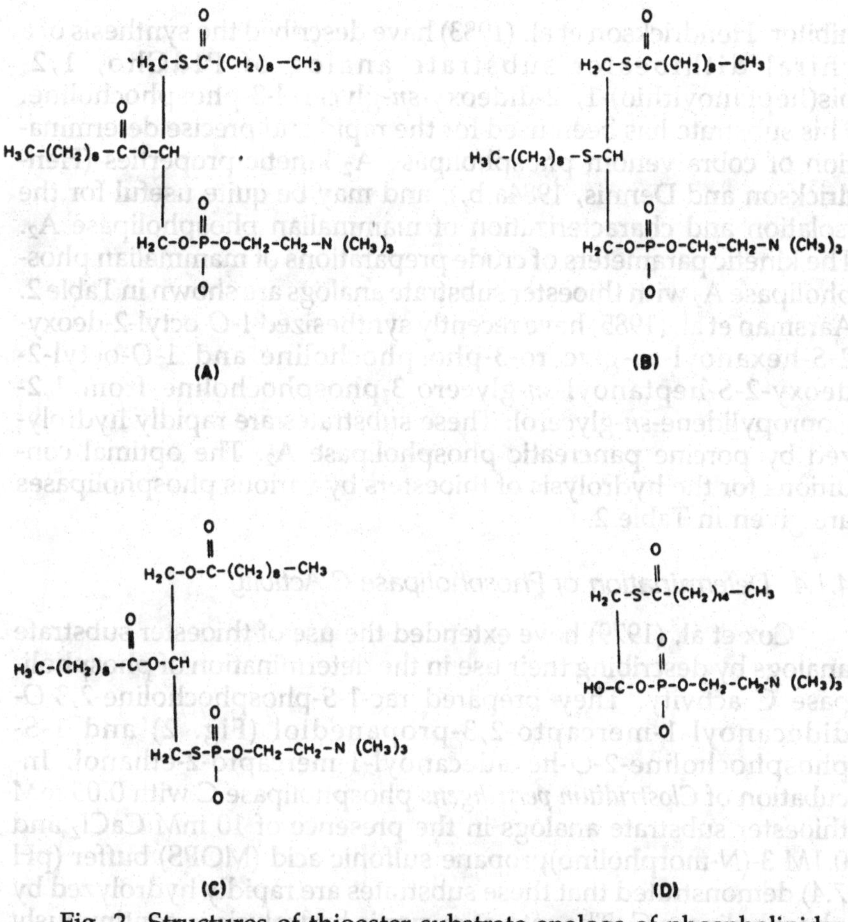

Fig. 2. Structures of thioester substrate analogs of phospholipids. (A) rac-1,2-S,O-didecanoyl-3-phosphocholine-1-mercapto-2,3-propanediol (phospholipase A₁), (B) rac-1, 2-S,S-didecanoyl-3-phosphocholine-1,2-dimercapto-2,3-propanediol (phospholipase A₂), (C) rac-1,2-O,O-didecanoyl-3-phosphocholine-3-mercapto-2,3-propanediol (phospholipase C), and (D) 2-hexadecanoylthio-1-ethyl-phosphocholine (lysophospholipase).

propylphosphocholine, in which both fatty acids were bound to glycerol by a thioester linkage. This thioester analog has an advantage over long chain diacylphospholipids in that it produces a clear suspension. It is a racemic mixture, however, and therefore may not be suitable for studying enzyme kinetic properties, since the enzymically inactive enantiomer may act as a competitive inhibitor. Hendrickson et al. (1983) have described the synthesis of a chiral dithioester substrate analog of PtdCho, 1,2-bis(heptanoylthio)-1, 2-dideoxy-*sn*-glycerol-3-phosphocholine. This substrate has been used for the rapid and precise determination of cobra venom phospholipase A$_2$ kinetic properties (Hendrickson and Dennis, 1984a,b,), and may be quite useful for the isolation and characterization of mammalian phospholipase A$_2$. The kinetic parameters of crude preparations of mammalian phospholipase A$_2$ with thioester substrate analogs are shown in Table 2. Aarsman et al. (1985) have recently synthesized 1-*O*-octyl-2-deoxy-2-*S*-hexanoyl-*sn*-glycero-3-phosphocholine and 1-*O*-octyl-2-deoxy-2-*S*-heptanoyl-*sn*-glycero-3-phosphocholine from 1,2-isopropylidene-*sn*-glycerol. These substrates are rapidly hydrolyzed by porcine pancreatic phospholipase A$_2$. The optimal conditions for the hydrolysis of thioesters by various phospholipases are given in Table 2.

4.1.4. Determination of Phospholipase C Activity

Cox et al. (1979) have extended the use of thioester substrate analogs by describing their use in the determination of phospholipase C activity. They prepared rac-1-*S*-phosphocholine-2,3-*O*-didecanoyl-1-mercapto-2,3-propanediol (Fig. 2) and 1-*S*-phosphocholine-2-*O*-hexadecanoyl-1-mercapto-2-ethanol. Incubation of *Clostridium perfringens* phospholipase C with 0.05 mM thioester substrate analogs in the presence of 10 mM CaCl$_2$ and 0.1M 3-(*N*-morpholino)propane sulfonic acid (MOPS) buffer (pH 7.4) demonstrated that these substrates are rapidly hydrolyzed by phospholipase C. The rate of enzymic hydrolysis is continuously monitored with 4,4'-dithiodipyridine at 324 nm.

4.2. Spectrophotometric Determination of Lysophospholipase Activity

Van den Bosch's group (Aarsman et al., 1976, 1977) prepared thiodeoxylysolecithin (3-hexadecanoylthio-1-propyl-phosphocholine) (Fig. 2) by a procedure similar to that described for the phospholipase A$_2$ substrate, except that 3-mercapto-1-propanol is substituted for 2-mercapto-1-ethanol. This substrate is rapidly hydrolyzed by beef liver lysophospholipases and has been successfully used in the isolation and characterization of these enzymes (Aarsman et al., 1977).

Table 2: Optimal Conditions for the Determination of Phospholipases, Lysophospholipases, and Lipases

Enzyme	Buffer	pH Value	Substrate	Optimal substrate conc (μM)	Enzyme source	Reference
Phospholipase A₁	3-(N-Morphilino)propane sulfonic acid	7.4	Rac-1-2-S,O-didecanoyl-3-phosphocholine-1-mercapto-2,3-propanediol	10000	Rhizopus delemar	Cox and Horrocks, 1981
Phospholipase A₂	Potassium phosphate	7.4	Rac-2-hexadecanoylthio-1-ethyl-phosphocholine	1000	Hog pancreatic	Aarsman et al., 1976
Phospholipase C	3-(N-Morphilino)propane sulfonic acid	7.4	Rac-1-S-phosphochocholine-2,3-O-didecanoyl-1-mercapto-2,3-propanediol	500	Clostridium perfringens	Cox and Horrocks, 1981
Lysophospholipase	3-(N-Morphilino)propane sulfonic acid	7.4	2-Hexadecanoylthio-1-ethyl-phosphocholine	300	Ox brain	Farooqui et al., 1985; Pendley and Horrocks (unpublished)
Monoacylglycerol lipase	3-(N-Morphilino)propane sulfonic acid	7.4	Rac-1-S-decanoyl-1-mercapto-2,3-propanediol	250	Ox brain	Strosznajder et al., 1984
Diacylglycerol lipase	3-(N-Morphilino)propane sulfonic acid	7.4	Rac-1,2-S,O-didecanoyl-1-mercapto-2,3-propanediol	250	Ox brain	Farooqui et al., 1985, 1986a,b
Triacylglycerol lipase	Tris-HCl	8.5	2,3-Dimercaptopropan-1-ol-tributyrate	20000	Human plasma	Kurioka et al., 1977

Our laboratory has also obtained homogeneous preparations of two forms of bovine brain lysophospholipases (Farooqui et al., 1985; Pendley, 1982) using 2-hexadecanoylthio-1-ethyl-phosphocholine. The purified enzyme has a molecular weight of 36,000 and is strongly inhibited by EDTA, Triton X-100, and Tween-20. The kinetic parameters of bovine brain lysophospholipase with thioester substrate analogs are shown in Table 3. Our results (Farooqui et al. 1985), as well as that of Aarsman et al. (1976, 1977), strongly support the view that the rates of hydrolysis of 2-hexadecanoylthio-1-ethyl-phosphocholine by lysophospholipases are proportional to their true activities.

Dithioester substrate analogs of PtdCho with hexanoyl groups have been used successfully for determining the activity of bovine milk lipoprotein lipase (Shinomiya et al., 1984). The preparation of this substrate is essentially that described by Cox and Horrocks (1981). The apparent V_{max} value for the dihexanoyldithio analog of PtdCho is 0.12 μmol/min/mg protein, which is considerably lower than that for the dihexanoyloxy form of phosphatidylcholine of 5.0 μmol/min/mg protein. The apparent K_m values are also considerably lower at 1900 μM vs. 4000 μM, respectively (Table 2). A precise explanation of the lower V_{max} value for the thioester than for the oxyester substrate is not known. Shinomiya et al. (1984) suggest, however, that the lower V_{max} value may be caused by differences in the nature of the chemical bond (C-S-C vs. C-O-C) and to steric factors.

4.3. Spectrophotometric Determination of Mono- and Diacylglycerol Lipases

Thioester substrate analogs of mono- and diacylglycerol are prepared according to the method of Cox and Horrocks (1981).

4.3.1. Rac-1-S-acyl-1-mercapto-2,3-propanediol Synthesis

Monoacyl thioester substrate analogs are synthesized by the method of Aarsman et al. (1976) for the monoacylation of 2 mercapto-1-ethanol. The sulfhydryl group of mercaptoglycerol is selectively acylated with decanoyl or hexadecanoyl chloride, under controlled conditions, to form the monodecanoylthioglycerol (V) and monohexadecanoylthioglycerol (VI), respectively (Figs. 1 and 2). Mercaptoglycerol (100 mmol) is dissolved in 100 mL of dry diethyl ether with 60 mmol pyridine. Acyl chloride (25 mmol), in 30 mL of dry diethyl ether, is added dropwise with rapid stirring at

Table 3
Kinetic Parameters for Phospholipases, Lysophospholipases, and Lipases

Enzyme	pH Optimum	V_{max} nmol/min/mg protein	K_m μM	Enzyme source	Reference
Monoacylglycerol lipase	7.4	3.7	56	Bovine brain homogenate	Strosznajder et al. (1984)
Diacylglycerol lipase	7.4	1.5	30	Bovine brain homogenate	Farooqui et al. (1985,1986a,b)
Phospholipase A_2	8.0	2.2	700	Porcine pancreas	Volwerk et al. (1979)
Phospholipase A_2	8.5	440,000	180	Purified cobra venom enzyme	Hendrickson and Dennis (1984a,b)
Lysophospholipase	7.4	0.55	18	Bovine brain homogenate	Farooqui et al. (1985)
Lysophospholipase	7.4	13	N.D.[a]	Bovine liver homogenate	Aarsman and van den Bosh (1977)
Lysophospholipase	7.4	140	N.D.[a]	Purified Acholeplasma laidlawii strain B	Aarsman and van den Bosh (1977)
Lipoprotein lipase	8.0	120	1900	Purified from bovine milk	Shinomiya et al. (1984)

[a]N.D., Not determined.

0°C. After the addition is complete, the reaction mixture is allowed to stir for 2 h at room temperature. The mixture is then washed with four 50 mL portions of water. The organic phase is dried over Na$_2$SO$_4$ and rotary evaporated to yield white crystals. The crystals are redissolved in warm *n*-hexane and allowed to stand overnight at –20°C. The precipitate is collected on a Buchner funnel. Recrystallization is repeated until the precipitate is >95% pure by TLC with chloroform:methanol [96:4 (v/v)]. If the purity of the product after several recrystallizations is not acceptable, further purification on a neutralized silicic acid column (100 g) eluted with CHCl$_3$:MeOH [98:2, 96:4 (v/v)] can be achieved. The monodeca-noyl-thioglycerol elutes in the last fraction. The greater the number of recrystallizations, the lower the yield. Yields of about 15% are common with this method, however.

4.3.2. Rac-1,2-S,O-didecanoyl-1-mercapto-2,3-propanediol(II) Synthesis

Mercaptoglycerol is triacylated with decanoyl chloride, under controlled conditions, to form tridecanoylthioglycerol (I), which is then partially hydrolyzed with *Rhizopus delemar* lipase (6000 U/mg) to obtain the desired didecanoylthioglycerol (II) (Cox and Hor-rocks, 1981) (Fig. 1). Mercaptoglycerol (35 mmol) is dissolved in 100 mL of chloroform containing 210 mmol pyridine. Decanoyl chloride (175 mmol) is added dropwise with rapid stirring at 0°C. After cooling, the mixture is allowed to stand in the dark for 2 d. It is then extracted with 500 mL of hexane and washed with three 100-mL portions of water. Rotary evaporation of the organic phase yields 30 g of brown oil. The crude product (5 g) is purified on a 200 g silicic acid column eluted with hexane:diethyl ether [(98:2, 95:5, and 85:15 (v/v)].

Tridecanoylthioglycerol (I) is obtained in 95% yield as a light yellow oil that migrates just ahead of trimyristin on TLC with hexane:diethyl ether:acetic acid (90:10:1, by vol) with an R_f of 0.3 and is then partially hydrolyzed with *R. delemar* lipase using conditions described by Fujumato et al. (1964).

I (10 mmol) is incubated with 5 mg of lipase in 60 mL of 0.2M sodium acetate buffer (pH 5.6) containing 10 mM CaCl$_2$. The mixture is stirred rapidly, and the reaction progress is followed by TLC with hexane:diethyl ether:acetic acid (60:40:1, by vol). After 1 h of incubation at room temperature, five components of the reaction mixture are detected. They are: unreacted I (R_f = 0.98); 2,3-didecanoyl-1-mercapto-2,3-propanediol (R_f = 0.95); decanoic acid

(R_f = 0.68); 1,2-S,O-didecanoyl-1-mercapto-2,3-propanediol (II) (R_f = 0.47), and 2-O-decanoyl-1-mercapto-2,3-propanediol (R_f = 0.26). The concentration of didecanoylthioglycerol (II) reaches a maximum after approximately 3 h of incubation. The reaction mixture is then extracted with 4 vol of $CHCl_3$:MeOH [2:1 (v/v)]. After rotary evaporation at 30°C, the residual oil is fractionated on a neutralized silicic acid column (200 g) by elution with hexane: diethyl ether [90:10, 80:20, and 70:30 (v/v)]. II is eluted in the last fraction in an approximate yield of 26%.

4.3.3. Optimal Conditions for Determining Mono- and Diacylglycerol Lipase Activities

4.3.3.1. MONOACYLGLYCEROL LIPASE.   Strosznajder et al. (1984) have assayed monoacylglycerol lipase activity using rac-1-S-decanoyl-1-mercapto-2,3-propanediol (Fig. 3) and its corresponding oxyester substrate. Semistable suspensions (1.5 mM) are prepared by separately vortexing the thioester and oxyester substrates with 2.5 mM lysoPtdCho in 0.1M Tris-HCl buffer (pH 7.4), 0.75 μmol of 5,5'-dithiobis (2-nitrobenzoic acid) (DTNB), 0.625 μmol of lysoPtdCho, and 0.375 μmol of the appropriate substrate, in a total volume of 1.0 mL. The enzyme protein (150 μg) is first reacted with DTNB, in 0.5 mL total volume, in both the reference and the sample compartments of a spectrophotometer. When the slopes of the recorder tracings reach zero, the reaction is started by adding monodecanoylthioglycerol and monodecanoylglycerol to the sample and reference cuvettes, respectively. The formation of free thiol groups and their reaction with DTNB is monitored continuously at 412 nm. Enzymic activity is calculated from the slope of the recorder tracing using the extinction coefficient for DTNB of 12,800 L/mol/cm. Nonenzymic background activity caused by substrate acyl migration is monitored at 412 nm and subtracted from the observed reaction rate to yield the actual enzymic activity.

4.3.3.2. DIACYLGLYCEROL LIPASE.   Enzymic activity is determined by the method of Farooqui et al. (1985, 1986a,b), with 1,2-S,O-didecanoyl-1-mercapto-2,3-propanediol (Fig. 2) as the substrate. Suspensions of didecanoylthioglycerol and dicaprin (1 mM) are prepared by separately sonicating each with 1 mM lysoPtdCho in 0.05M MOPS buffer (pH 7.4). Sonicated suspensions are stable for 4–5 h at room temperature. The assay mixture contains 100 μmol of MOPS buffer (pH 7.4), 0.75 μmol of DTNB, 0.25 μmol of thioester or oxyester didecanoylglycerol, and up to 300 μg of enzyme protein, in a total volume of 1.0 mL. Enzymic activity is

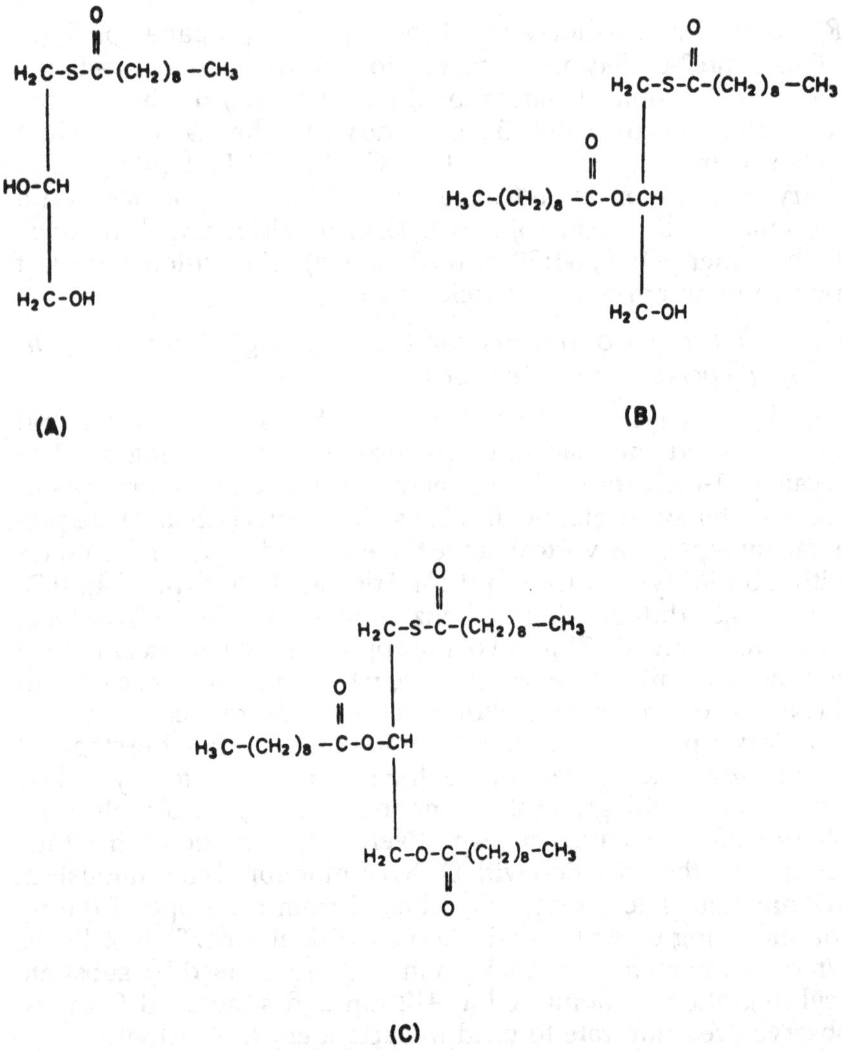

(A)

(B)

(C)

Fig. 3. Structures of thioester substrate analogs of mono-, di- and triacylglycerols. (A) 1-S-decanoyl-1-mercapto-2,3-propanediol, (monoacylglycerol lipase), (B) rac-1,2-S,O-didecanoyl-2,3-propanediol (diacylglycerol lipase), and (C) rac-1,2,3-S,O,O-tridecanoyl-1-mercapto-2,3-propanediol (triacylglycerol lipase).

determined as described for monoacylglycerol lipase. The action of diacylglycerol lipase on the thioester substrate analog and the reaction of the thiol group with DTNB is shown in Fig. 4. DTNB is usable only in the pH range of from 6 to 10, but other thiol reagents,

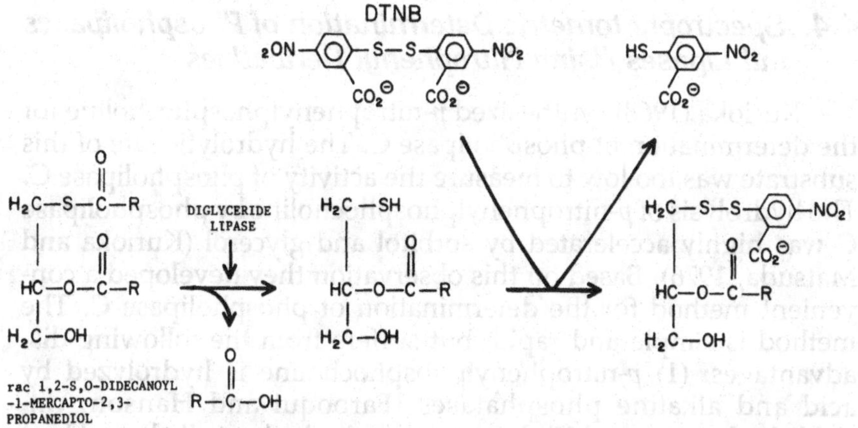

Fig. 4. Action of diacylglycerol lipase on the thioester substrate analog and the reaction of DTNB with sulfhydryl group.

such as 4,4'-dithiodipyridine (Grassetti and Murray, 1967), can also be used for the pH range of from 3 to 7. With the use of these two thiol capture reagents, the complete pH range necessary for the assay of lipases can be covered.

Kinetic parameters of bovine brain mono- and diacylglycerol lipases using thioester substrate analogs are given in Table 3. These substrates follow simple Michaelis-Menten kinetics and show saturation at 250 μM.

Development of continuous spectrophotometric procedures using thioester substrate analogs has aided the study of various lipases. Using rac-1,2-S,O-didecanoyl-1-mercapto-2,3-propanediol, Farooqui et al. (1985, 1986a,b) have shown that bovine brain contains two diacylglycerol lipases. One is localized in the microsomal and the other in the crude plasma membrane fraction. The microsomal enzyme is markedly stimulated by Triton X-100 and Ca^{2+}. The crude plasma membrane enzyme is strongly inhibited by Triton X-100, however, whereas Ca^{2+} has no effect on its enzymic activity. We have also separated mono- and diacylglycerol lipases (Farooqui et al., 1984) using heparin-Sepharose affinity chromatography (Farooqui and Horrocks, 1984). Microsomal diacylglycerol lipase is completely retained on a heparin-Sepharose column and can be eluted with 0.5M NaCl, whereas monoacylglycerol lipase is not retained and is therefore washed from the column. Further studies on the isolation and characterization of mono- and diacylglycerol lipases are currently in progress in this laboratory.

4.4. Spectrophotometric Determination of Phospholipases and Lipases Using Nitrophenyl Derivatives

Kurioka (1968) synthesized p-nitrophenylphosphocholine for the determination of phospholipase C. The hydrolytic rate of this substrate was too low to measure the activity of phospholipase C. The hydrolysis of p-nitrophenylphosphocholine by phospholipase C was highly accelerated by sorbitol and glycerol (Kurioka and Matsuda, 1976). Based on this observation they developed a convenient method for the determination of phospholipase C. The method is simple and rapid, but suffers from the following disadvantages: (1) p-nitrophenylphosphocholine is hydrolyzed by acid and alkaline phosphatases (Farooqui and Hanson, unpublished work), and (2) the nonenzymic blanks at alkaline pH are quite high.

Gupta and Wold (1980) used p-nitrophenylphosphocholine for determination of phospholipase D. They claimed that phospholipase D hydrolyzed p-nitrophenylphosphocholine to p-nitrophenylphosphate and choline. The subsequent addition of acid phosphatase liberated p-nitrophenol, which can be read at 400 nm. In our opinion, p-nitrophenylphosphocholine cannot be used for the assay of phospholipases C and D because of the occurrence of high activities of acid and alkaline phosphatases in various tissues. Acid and alkaline phosphatases hydrolyze p-nitrophenylphosphocholine seven to ten times faster than phospholipases C and D (Farooqui and Hanson, unpublished).

4.5. Spectrophotometric Determination of Phospholipases and Lipases Using ω-Nitrophenylaminolaurylglycerol Derivatives

Goldberg et al. (1978) synthesized ω-trinitrophenylaminolauric acid (TNPAL) by the method of Satake et al. (1960). Glycerol esters of TNPAL were prepared by the method of Bhattacharya and Hilditch (1930). Briefly, TNPAL (11 nmol) is mixed with dry naphthalenesulfonic acid (24 mg) and double vacuum-distilled glycerol (285 mg). The mixture is reacted at 150°C for 6 h at 35–55 mm Hg. The products are extracted with $CHCl_3$, and the solvent is evaporated under a stream of nitrogen. Equal volumes of benzene and $0.1M$ $NaHCO_3$: ethylene glycol [(1:1 (v/v)] are added, and the mixture is shaken to separate most of the residual unreacted fatty acid. The benzene phase is washed twice with

NaHCO$_3$:ethylene glycol and concentrated in vacuo. Toluene is much less toxic and probably can be substituted for benzene. Mono-, di-, and tri-esters of TNPAL are separated by preparative TLC on 1 mm thick silica gel H plates developed in CHCl$_3$:MeOH:acetic acid [(98:2:1 (v/v)]. The bands that migrate in parallel with authentic mono-, di-, and triacylglycerol standards are scraped, placed in small columns, and eluted with the above solvent system.

The assay mixture contains 25 μmol of Tris-HCl buffer (pH 8.0), 0.5 μmol of CaCl$_2$, 0.25 μmol of mono-TNPAL-glycerol, and 100–200 μg of rat brain microsomal protein, in a total volume of 0.5 mL. Incubation is carried out for 5 min at 37°C. The reaction is terminated with 0.5 mL of ethylene glycol/30 nmol of NaOH, followed by the addition of 2 mL of benzene. The mixture is vortexed for 30 s and centrifuged for 5 min at 2000 rpm, and the upper benzene phase is removed. The lower phase is again washed with benzene, and 0.1 mL of conc HCl is added to the aqeuous ethylene glycol phase. The yellow fatty acid is extracted into 2 mL of benzene, and the absorbance of the benzene phase is measured at 330 nm.

With 1,3-di-TNPAL-glycerol, the apparent K_m and V_{max} values for *Rhizopus delemar* lipase are 16 μM and 200 μmol/min/mg protein, respectively. The hydrolysis rates of mono-, di-, and tri-TNPAL-glycerol by rat brain microsomal lipase, hog pacreatic lipase, and *R. delemar* lipase are of similar orders of magnitude as those determined with labeled oleoylglycerol derivatives. This method has two serious disadvantages. The trinitrophenyl derivatives have solubility properties that are different from the normal triacylglycerols. Therefore, an exact comparison of the hydrolytic rates of triolein and the trinitroaminolauryltriester cannot be made. The other disadvantage is that these substrates provide a fixed-time discontinuous assay procedure and thus require several samples to be performed in order to assure linearity and reproducibility.

5. Fluorometric Procedures

Hendrickson and Rauk (1981) developed a fluorometric assay for phospholipase A$_2$ using an excimer emitting pyrene-labeled PtdCho (1,2-bis[4-(1-pyreno) butanoyl]-*sn*-glycero-3-phosphocholine), which on hydrolysis gives monomer fluorescence that can be measured at 382 nm. Pyrene-labeled phosphatidylcholine was synthesized by the method of Sunamoto et al. (1980). Briefly, carbonyldiimidazole was reacted with 4-(1-pyreno)butanoic acid in dry benzene for 48 h at room temperature under nitrogen. The benzene was evaporated *in vacuo*, and *sn*-glycero-3-phosphocholine-cadmium chloride complex in dry dimethylsulfoxide was added (Hendrickson and Rauk, 1981). The mixture was stirred at 42°C under nitrogen for 4 d. Water (1 mL) was added, and stirring was continued for 30 min. The mixture was passed through a Bio-Rad AG501-X8 column packed in chloroform:methanol:water [5:4:1 (v/v)]. The solvent was evaporated *in vacuo*, and the water was removed by azeotropic distillation with *n*-propanol. The crude product was purified by preparative HPLC using a Polygosil 60-1525 column. Thuren et al. (1983, 1985) synthesized 1-triacontanyl-2-(pyren-1-yl)-hexanoyl-*sn*-glycero-3-phosphocholine and used it for the determination of human pancreatic phospholipase A_2. They reported a lower limit of detection of 6 pmol/min/mL, which is 100 and 1000 more sensitive than the spectrophotometric and titrimetric procedures, respectively. With this substrate, human pancreatic phospholipase A_2 showed optimal activity at pH 8.0 The K_m and V_{max} values of this enzyme were 5 μM and 1.8 μmol/min/mg serum protein. This procedure was successfully used for determining phospholipase A_2 activity in pancreatitis. Phospholipase A_2 activity in serum of healthy volunteers and subjects with pancreatitis averaged 69 and 1092 pmol/min/mL, respectively (Thuren et al., 1985).

Wittenauer et al. (1984) used a fluorescent phospholipid analog, 1-acyl-2-[6-[(7-nitro-2,1,3-benzoxadiazol-4-yl)amino]-caproyl]-*sn*-glycero-3-phosphocholine, to assay porcine pancreatic phospholipase A_2 as well as bovine milk lipoprotein lipase. Hydrolysis of this substrate by these enzymes resulted in a more than 50-fold fluorescence enhancement, with no shift in the emission maximum at 540 nm.

Fugman et al. (1984) entrapped 6-carboxyfluorescein into liposomes prepared from egg yolk PtdCho. Porcine pancreatic phospholipase A_2 and bovine milk lipoprotein lipase catalyze the hydrolysis of PtdCho, which causes the release of the entrapped dye. This procedure provides a simple, accurate, and convenient method for determining the rate of phospholipase A_2 and lipoprotein lipase catalyzed hydrolysis of PtdCho vesicle substrates.

6. Determination of Lipases and Phospholipases by Treatment with Enzymes

The release of fatty acids and glycerol has been successfully assayed using bioluminescence procedures. Ulitzur and Heller (1981) have isolated mutants of a luminous organism, *Benecke harveyi*, that via a luciferase respond to as little as 1 pmol of 14:0 and 100–200 pmol of 16:0 or 18:1 with the production of light. The luminescence can be measured in a photomultiplier photometer. The activity of porcine pancreatic lipase phospholipases A_2 and C has been successfully determined. Similarly, glycerol released during lipolysis has been determined by bioluminescence (Bjorkhem et al., 1981). This glycerol-dependent ATP utilization can be monitored with a kit containing luciferon and luciferase (Sigma) and 0.5 μmol/L of glycerol are successfully determined.

7. Immunological Procedures for the Determination of Lipases and Phospholipases

Quantities of chicken adipose tissue and human pancreatic lipases (Cheung et al., 1979; Grenner et al., 1982) have been determined using radioimmunoassay procedures. Cheung et al. (1979) incubated antilipoprotein lipase immunoglobulins coupled to hydrophilic beads with the lipase and immunoglobulins labeled with ^{125}I-immunoglobulin and reacted with the antigen (lipase) associated with the immunoabsorbent. The quantity of lipase in the sample was proportional to the amount of ^{125}I bound to the immunoabsorbent. Grenner et al. (1982) developed an enzyme-linked immunoabsorbent assay (ELISA) for human pancreatic lipase. Antibodies to human pancreatic lipase were prepared in sheep and dogs. Canine antibodies were conjugated with peroxidase. Serum, containing pancreatic lipase, was incubated with the sheep antibodies, and the peroxidase-bound antibodies were then added to form the sandwich. The quantity of bound lipase was determined by the colorimetric determination of peroxidase with H_2O_2 and O-phenylenediamine. The procedure of Grenner et al. (1982) has an advantage over the method of Cheung et al. (1979) in that one does not have to work with ^{125}I. Both immunological procedures require purified preparations of lipases. The fluoroimmunoassay of human pancreatic phospholipase A_2 has been recently described (Eskola et al., 1983). This procedure is very sensitive for phospholipase A_2 in sera of patients suffering from pancreatic diseases such as acute and chronic pancreatitis and pancreatic cancer.

8. Summary and Conclusion

Methods for determination of lipolytic enzymes have been reviewed. Titrimetric procedures are insensitive and can only be used for extracellular lipolytic enzymes such as pancreatic and snake venom phospholipases. Radiochemical procedures are sensitive, but suffer from the disadvantages of being discontinuous, time-consuming, and expensive. Furthermore, handling of radioactivity is very undesirable. The determination of lipolytic enzymes using nitrophenyl derivatives also suffers from several drawbacks. The nitrophenyl derivatives have different solubility properties compared to the native phospholipids and provide a fixed-time discontinuous assay procedure. Further, p-nitrophenyl derivatives of phosphocholine are rapidly hydrolyzed by acid and alkaline phosphatases. Fluorometric procedures are sensitive and rapid, but limited to extracellular phospholipases (serum and pancreatic). Further, the presence of albumin and some other unidentified serum protein produces a monomeric pyrene fluorescence signal indistinguishable from phospholipase A_2. Immunological procedures are sensitive, but can only be used for the determination of the quantities of lipolytic enzymes.

In our opinion, thioester substrate assay procedures provide a relatively rapid, sensitive, specific, continuous, and convenient method for the determination of lipolytic enzymes. A relatively short acyl chain makes substrate suspension easier and obviates the problem of unsaturated side chain oxidation. The most serious disadvantage of the thioester substrate assay is that it cannot be used if the sample contains thiol group-containing compounds. Use of this procedure has already resulted in complete purification of lysophospholipases and mono- and diacylglycerol lipases of bovine brain. It is hoped that thioester substrate analogs of neutral lipids and phospholipids will become available commercially for more extensive characterization of intracellular lipolytic enzymes.

Acknowledgment

Preparation of this review was supported in part by NIH research grants NS-08291 and NS-10165.

References

Aarsman A. J., van Deenen L. L. M., and van den Bosch H. (1976) Studies on lysophospholipases. *Bioorg. Chem.* **5**, 241–253.

Aarsman A. J. and van den Bosch H. (1977) A continuous spectrophotometric assay for membrane-bound lysophospholipase using thioester substrate analog. *FEBS Lett.* **79**, 317–320.

Aarsman A. J. and van den Bosch H. (1979) A comparison of acyl-oxyester and acyl thioester substrate for some lipolytic enzymes. *Biochim. Biophys. Acta* **572**, 519–530.

Aarsman A. J. and van den Bosch H. (1981) Comparative action of lysophospholipases on acyloxyester and acylthioester substrates in micellar and membrane bound form. *Chem. Phys. Lipids* **29**, 267–275.

Aarsman A. J., Hille J. D. R., and van den Bosch H. (1977) Application of a continuous spectrophotometric assay using thioester substrate analogs. *Biochim. Biophys. Acta* **489**, 242–246.

Aarsman A. J., Roosenboom C. F. P., Van der Marel G. A., Shadid B., van Boom J. H., and van den Bosch H. (1985) Synthesis of acylthioester phospholipids and their hydrolysis by phospholipase A_2. *Chem.Phys. Lipids* **36**, 229–242.

Baer E. and Buchnea D. (1959) Synthesis of saturated and unsaturated L-α-lecithins. Acylation of the cadmium chloride compound of L-α-glycerylphosphorylcholine. *Can. J. Biochem. Physiol.* **37**, 953–959.

Berridge M. J. (1984) Inositol trisphosphate and diacylglycerol as second messengers. *Biochem. J.* **220**, 345–360.

Bhattacharya R. and Hilditch T. P. (1930) The structure of synthetic mixed triglycerides. *Proc. Roy. Soc. (Lond.) A.* **129**, 468–476.

Bjorkhem I., Arner P., Thore A., and Ostman J. (1981) Sensitive kinetic bioluminescent assay of glycerol release from human fat cells. *J. Lipid Res.* **22**, 1142–1147.

Boni L. T. and Rando R. R. (1985) The nature of protein kinase C activation by physically defined phospholipid vesicles and diacylglycerols. *J. Biol. Chem.* **260**, 10819–10825.

Brockerhoff H., Schmidt P. C., Fong J. W., and Tirri L. J. (1976) Introduction of labeled fatty acid in position 1 of phosphoglycerides. *Lipids* **11**, 421–422.

Cabot M. C. and Gatt S. (1978) The hydrolysis of triacylglycerol and diacylglycerol by rat brain microsomal lipase with acidic pH optimum. *Biochim. Biophys. Acta* **530,** 508–512.

Cheung A. H., Bensadoun A., and Chen C. F. (1979) Direct solid phase radio-immunoassay for chicken lipoprotein lipase. *Anal. Biochem.* **94,** 346–357.

Cook H. W. (1981) Metabolism of triacylglycerol in developing rat brain. *Neurochem. Res.* **6,** 1217–1229.

Cox J. W., Snyder W. R., and Horrocks L. A. (1979) Synthesis of choline and ethanolamine phospholipids with thiophosphoester bonds as substrate for phospholipase C. *Chem. Phys. Lipids* **25,** 369–380.

Cox J. W. and Horrocks L. A. (1981) Preparation of thioester substrates and development of continuous spectrophotometric assays for phospholipase A_1 and monoacylglycerol lipase. *J. Lipid Res.* **22,** 496–505.

deHaas G. H., Bonsen P. P. M., Pieterson W. A., and van Deenen L. L. M. (1971) Studies on phospholipase A and its zymogen from porcine pancreas. *Biochim. Biophys. Acta* **239,** 252–266.

Dennis E. A. (1983) Phospholipases. *Enzymes* **14,** 307–353.

Dugan L. L., Demediuk P., Pendley II C. E., and Horrocks L. A. (1986) Separation of phospholipids by HPLC: All major classes, including ethanolamine and choline plasmalogens, and most minor classes, including lysophosphatidylethanolamine. *J. Chromatog.* **378,** 317–327.

Duncombe W. G. (1963) The colorimetric micro-determination of long-chain fatty acids. *Biochem J.* **88,** 7–12.

Edgar A. D. and Freysz L. (1982) Phospholipase activities of rat brain cytosol: Occurrence of phospholipase C activity with phosphatidylcholine. *Biochim. Biophys. Acta* **711,** 224–228.

Eibl H., Arnold D., Weltzien U., and Westphal O. (1967) Zur Synthese von α- und β-Lecithinen und ihren Atheranaloga. *Liebig's Ann. Chem.* **709,** 226–230.

Eibl H. (1978) Phospholipid synthesis: Oxazophospholanes and dioxaphospholanes as intermediates. *Proc. Natl. Acad. Sci. USA* **75,** 4074–4077.

Eskola J. U., Nevalainen T. J., and Lovgren N. E. (1983) Time-resolved fluoroimmunoassay of human pancreatic phospolipase A_2. *Clin. Chem.* **29,** 1777–1780.

Farooqui A. A., Taylor W. A., and Horrocks L. A. (1984) Separation of bovine brain mono and diacyl glycerol lipases by heparin Sepharose affinity chromatography. *Biochem. Biophys. Res. Commun.* **129,** 1241–1246.

Farooqui A. A. and Horrocks L. A. (1984) Heparin-Sepharose affinity chromatography. *Adv. Chromatogr.* **23,** 127–148.

Farooqui A. A., Pendley II C. E., Taylor W. A., and Horrocks L. A. (1985) Studies on Diacylglycerol Lipases and Lysophospholipases of Bovine Brain, in *Phospholipids in Nervous System* vol. 2 (Horrocks L. A., Kanfer J. N., and Porcellati G., eds.) Raven, New York.

Farooqui A. A. and Horrocks L. A. (1985) Metabolic and Functional Aspects of Neural Membrane Phospholipids, in *Phospholipids in the Nervous System* vol. 2 (Horrocks L. A., Kanfer J. N., and Porcellati G., eds.) Raven, New York.

Farooqui A. A., Taylor W. A., and Horrocks L. A. (1986a) Characterization and solubilization of membrane bound diacylglycerol lipases from bovine brain. *Int. J. Biochem.* **18,** 991–997.

Farooqui A. A., Taylor W. A., and Horrocks L. A. (1986b) Membrane Bound Diacylglycerol Lipases in Bovine Brain: Purification and Characterization, in *Phospholipids in the Nervous System: Pharmacological Aspects* (Horrocks L. A., Toffano G., and Freysz L., eds.) Liviana, Padova, Italy.

Fauvel J., Chap M., Roques V., Sarda L., and Douste-Blazy L. (1984) Substrate specificity of two cationic lipases with high phospholipase A_1 activity purified from guinea-pig pancreas. *Biochim. Biophys. Acta* **792,** 65–71.

Freysz L., Lastennet A., and Mandel P. (1976) Metabolism of brain sphingomyelins: Half-lives of sphingosine, fatty acids and phosphate from two types of rat brain sphingomyelin. *J. Neurochem.* **27,** 355–359.

Fugman D. A., Shirai K., Jackson R. L., and Johnson J. D. (1984) Lipoprotein lipase and phospholipase A_2 catalyzed hydrolysis of phospholipid vesicles with an encapsulated fluorescent dye. *Biochim. Biophys. Acta* **795,** 191–195.

Fujikura Y. and Baistel D. (1985) Purification and characterization of a basic lysophospholipase in germinating barley. *Arch. Biochem. Biophys.* **243,** 570–578.

Fujumato J., Iwai M., and Tsujisaka Y. (1964) Studies on lipase IV. Purification and properties of a lipase secreted by *Rhizopus delemar. J. Gen. App. Microbiol.* **10,** 257–265.

Goldberg R., Barenholz Y., and Gatt S. (1978) Synthesis of trinitrophenylaminolauric acid and the use of its glyceryl esters for assaying lipase by a spectrophotometric procedure. *Biochim. Biophys. Acta* **531,** 237–241.

Grassetti D. R. and Murray J. F. (1967) Determination of sulfhydryl groups with 2,2 or 4,4' dithiodipyridine. *Arch. Biochem. Biophys.* **119,** 41–49.

Gray N. C. C. and Strickland K. P. (1982) The purification and characterization of a phospholipase A_2 activity from the 106,000 × g pellet (microsomal fraction) of bovine brain acting on phosphatidylinositol. *Can. J. Biochem.* **60,** 108–117.

Grenner G. V., Deutsch G., Schmidtberger R., and Dati F. (1982) A highly sensitive enzyme immunoassay for the determination of pancreatic lipase. *J. Clin. Chem. Clin. Biochem.* **20,** 515–519.

Gupta C. M., Radkakrishnan R., and Khorana H. G. (1977) Glycerophospholipid synthesis: Improved general method and new analogs containing photoactivable groups. *Proc. Natl. Acad. Sci. USA* **74,** 4315–4319.

Gupta M. N. and Wold F. (1980) A convenient spectrophotometric assay for phospholipase D using p-nitrophenyl phosphocholine as substrate. *Lipids* **15,** 594–596.

Hendrickson H. S. and Rauk R. N. (1981) Continuous fluorometric assay for phospholipase A_2 with pyrene labeled lecithin as a substrate. *Anal. Biochem.* **116,** 553–558.

Hendrickson H. S., Hendrickson E. K., and Dybvig R. H. (1983) Chiral synthesis of a dithioester analog of phosphatidylcholine as a substrate for the assay of phospholipase A_2. *J. Lipid Res.* **24,** 1532–1537.

Hendrickson H. S. and Dennis E. A. (1984a) Kinetic analysis of the dual phospholipid model for phospholipase A_2 action. *J. Biol. Chem.* **259,** 5734–5739.

Hendrickson H. S. and Dennis E. A. (1984b) Analysis of the kinetics of phospholipid activation of cobra venom phospholipase A_2. *J. Biol. Chem.* **254,** 5740–5744.

Hron W. T. and Menahan L. A. (1981) A sensitive method for the determination of free fatty acids in plasma. *J. Lipid Res.* **22,** 377–381.

Imamura S. and Horiuti Y. (1979) Purification of *Streptomyces chromofucus* phospholipase D by hydrophobic affinity chromatography on palmitoyl cellulose. *J. Biochem.* (Tokyo) **85,** 79–95.

Kunze D., Rustow B., Rabe H., and Ullrich K. P. (1984) Diacylglycerol species in microsomes and mitochondria of normal and dystrophic human muscle. *Clin. Chim. Acta* **140,** 215–222.

Kurioka S. (1968) Synthesis of p-nitrophenylphosphorylcholine and the hydrolysis with phospholipase C. *J. Biochem.* (Tokyo) **63,** 678–680.

Kurioka S. and Matsuda M. (1976) Phospholipase C assay using p-nitrophenylphosphocholine together with sorbitol and its application to studying the metal detergent requirement of the enzyme. *Anal. Biochem.* **75,** 281–289.

Kurioka S., Okamoto S., and Hashimoto M. (1977) A novel and simple colorimetric assay for human serum lipase. *J. Biochem.* (Tokyo) **81,** 361–369.

Lands W. E. and Merkl I. (1963) Metabolism of glycerolipids III. Reactivity of various acyl esters of coenzyme A with α-acylglycerophosphorylcholine, and positional specificities in lecithin synthesis. *J. Biol. Chem.* **238,** 898–903.

Little C., Aurebekk B., and Otnaess A. B. (1975) Purification by affinity chromatography of phospholipase C from *Bacillus cereus*. *FEBS Lett.* **52,** 175–179.

Mauco G., Fauvel J., Chap H., and Douste-Blazy L. (1984) Studies on enzymes related to diacylglycerol production in activated platelet. *Biochim. Biophys. Acta* **796,** 169–177.

Miller S. L., Benjamins J. A., and Morell P. (1977) Metabolism of glycerophospholipids in myelin and microsomes in rat brain. *J. Biol. Chem.* **252,** 4025–4031.

Mizobuchi M., Shirai K., Matsuoka N., Saito Y., and Kumagai A. (1981) Studies on lipase in brain. *J. Neurochem.* **36,** 301–303.

Okazaki T., Sagawa N., Okita J. R., Bleasdale J. E., MacDonald P. C., and Johnston J. M. (1981) Diacylglycerol metabolism and arachidonic acid release in human fetal membranes and decidua vera. *J. Biol. Chem.* **256,** 7316–7321.

Pendley II C. E., Singh H., and Cox J. (1981) Bovine brain diglyceride lipase: Assay and copurification with a lysophospholipase. *Fed. Proc.* **40,** 1708.

Pendley II C. E. (1982) Purification and partial characterization of bovine brain lysophospholipase. PhD Dissertation, The Ohio State University, Columbus, Ohio.

Porcellati G., Goracci G., and Arienti G. (1983) Lipid Turnover, in *Handbook of Neurochemistry* 2nd edn., vol. 5 (Lajtha A., ed.) Plenum New York.

Porcellati G. (1983) Phospholipid Metabolism in Neural Membranes, in *Neural Membranes* (Sun G. Y., Bazan N., Wu J. Y., Porcellati G., and Sun A. Y., eds.) Humana, New Jersey.

Prescott S. M. and Majerus P. W. (1983) Characterization of diacylglycerol hydrolysis in human platelets. *J. Biol. Chem.* **256,** 7316–7321.

Pugh E. L. and Kates M. (1975) A simplified procedure for synthesis of di-[^{14}C]acyl-labeled lecithins. *J. Lipid Res.* **16,** 392–397.

Rittenhouse S. E. (1982) Preparation of selectively labeled phosphatidylinositol and assay of phosphatidylinositol-specific phospholipase C. *Meth. Enzymol.* **86,** 3–11.

Robertson A. F. and Lands W. E. (1962) Positional specificities in phospholipid hydrolysis. *Biochemistry* **1,** 804–810.

Rooke J. A. and Webster G. R. (1976) Phospholipase A in human brain: A_1-type at alkaline pH. *J. Neurochem.* **27,** 613–620.

Rouser G., Kritchevsky G., Yamamoto A., Simon G., Galli C., and Bauman A. J. (1969) Diethylaminoethyl and triethylaminoethyl cellulose column chromatographic procedures for phospholipids, glycolipids and pigments. *Meth. Enzymol.* **14,** 272–317.

Rowe C. E. (1969) The measurement of triglyceride in brain and the metabolism of brain triglyceride *in vitro*. *J. Neurochem.* **16,** 205–214.

Sahasrabudhe M. R. (1982) Measurement of lipase activity in single grains of oat (Avena sativa L). J. Am. Oil Chem. Soc. **59**, 354–355.

Satake K., Okuyama T., Ohasi M., and Shinoda T. (1960) The spectrophotometric determination of amine, amino acid and peptide with 2,4,6-trinitrobenzene 1-sulfonic acid. J. Biochem. (Tokyo) **47**, 654–660.

Shinomiya M., Epps D. E., and Jackson R. L. (1984) Comparison of acyl-oxyester and acyl-thioester lipids as substrate for bovine milk lipoprotein lipase. Biochim. Biophys. Acta **795**, 212–220.

Strosznajder J., Singh H., and Horrocks, L. A. (1984) Monoacylglycerol lipase: Regulation and increased activity during hypoxia and ischemia. Neurochem. Path. **2**, 139–147.

Sunamoto J., Kondo H., Nomura T., and Okamoto M. (1980) Liposomal membranes. 2. Synthesis of a novel pyrenelabeled lecithin and structural studies on liposomal bilayers. J. Am. Chem. Soc. **102**, 1146–1152.

Taki T. and Kanfer J. N. (1979) Partial purification and properties of a rat brain phospholipase D. J. Biol. Chem. **254**, 9761–9765.

Thuren T., Vainio P., Virtanen J. A., and Kinnunen P. K. J. (1983) Hydrolysis of 1-triacontanyl-2-(pyren-1-yl)-hexanoyl-sn-glycero-3-phosphocholine by human pancreatic phospholipase A_2. Chem. Phys. Lipids **33**, 283–292.

Thuren T., Virtanen J. A., Lalla M., and Kinnunen P. K. J. (1985) Fluorometric assay for phospholipase A_2 in serum. Clin. Chem. **31**, 719–717.

Ulitzur S. and Heller M. (1981) Bioluminescent assay for lipase, phospholipase A_2 and phospholipase C. Meth. Enzymol. **72**, 338–346.

van den Bosch H. (1982) Phospholipases, in Phospholipids (Hawthorne J. N., and Ansell G. B., eds.) Elsevier Biomedical, Amsterdam.

van den Bosch H., Aarsman A. J., Dejong J. G. N., and van Deenen L. L. M. (1973) Studies on lysophospholipases. Biochim. Biophys. Acta **296**, 94–104.

van den Bosch H. and Aarsman A. J. (1979) A review on methods of phospholipase A determination. Agents Actions **9**, 382–389.

van den Bosch H. and van Deenen L. L. M. (1965) Chemical structure and biochemical significance of lysolecithins from rat liver. Biochim. Biophys. Acta **106**, 326–337.

Volwerk J. J., Dedieu A. G. R., Verheij H. M., Dijkman R., and de Haas G. H. (1979) Hydrolysis of monomeric substrates by porcine pancreatic (pro) phospholipase A_2. Recl. Trav. Chim. Pays-Bas **98**, 214–220.

Waite M., Rao R. H., Griffin H., Franson R., Miller C., Sisson P., and Frye J. (1981) Phospholipases A_1 from lysosomes and plasma membranes of rat liver. Meth. Enzymol. **71**, 674–682.

Wells M. A. (1972) A kinetic study of the phospholipase A_2 catalyzed hydrolysis of 1,2-dibutyryl-*sn*-glycero-3-phosphorylcholine. *Biochemistry* **11**, 1030–1041.

Williamson J. R., Cooper R. H., Joseph S. K., and Thomas A. P. (1985) Inositol trisphosphate and diacylglycerol as intracellular second messengers in liver. *Am. J. Physiol.* **248**, C203–C216.

Wittenauer L. A., Shirai K., Jackson R. L., and Johnson J. D. (1984) Hydrolysis of a fluorescent phospholipid substrate by phospholipase A_2 and lipoprotein lipase. *Biochem. Biophys. Res. Commun.* **118**, 894–901.

Woelk H.,Goracci G., Gaiti A., and Porcellati G. (1973) Phospholipase A_1 and A_2 activities of neuronal and glial cells of the rabbit brain. *Hoppe Seylers Z. Physiol. Chem.* **354**, 729–736.

Isolation, Separation, and Analysis of Phosphoinositides from Biological Sources

Amiya K. Hajra, Stephen K. Fisher, and Bernard W. Agranoff

1. Introduction

The phosphoinositides, namely phosphatidylinositol (PI), phosphatidylinositol 4-phosphate (PIP), and phosphatidylinositol 4,5-bisphosphate (PIP_2), are acidic lipids that are present in all animal tissues. The markedly polar character of these lipids makes conventional lipid extraction and assay procedures unsuitable for their analysis. Hence, specific analytical procedures have been devised in a number of laboratories for the quantitative extraction, separation, and assay of this class of lipids. These methods are reviewed, and details of analytical procedures used in our laboratory are given below.

2. Extraction of Phosphoinositides from Tissues

The early work of Folch (1942, 1949) showed that typical organic solvent mixtures, such as $CHCl_3$: methanol at neutral pH, do not extract PIP or PIP_2 from brain (even though it contains relatively large amounts of these two lipids—collectively termed the polyphosphoinositides). Folch and collaborators, however, made the interesting discovery that these lipids can be extracted from an acetone powder of brain by highly nonpolar organic solvents such as petroleum ether. Apparently in the acetone powder, these lipids exist as Ca^{2+} and Mg^{2+} salts, which are soluble in organic solvents, whereas the protein-bound (electrostatic-linkage) alkali metal salts of these acidic lipids, as they exist in tissues, are not. Based on these findings, Folch and coworkers developed a procedure for the preparative isolation of inositides from brain acetone powder. The dried residue of brain following homogenization in acetone is extracted with petroleum ether and then fractionated with ethanol to yield a "cephalin" fraction. This fraction is dissolved in chloroform and further fractionated by the addition of methanol to yield a fraction rich in phosphoinositides (brain fraction II). The details of the procedure have been described by Lees (1957). Though the extraction method is not quantitative and the fractions are not pure, this is an excellent way to obtain gram quantities of crude polyphosphoinositides, which can be further purified to yield the individual inositol lipids.

Folch and LeBaron also showed that acidified solvents quantitatively extract these polar lipids from whole brain (Folch and LeBaron, 1953; LeBaron and Folch, 1956). Apparently the uncharged (protonated) phosphoinositides are soluble in organic solvents, especially in a mixture of $CHCl_3$ and methanol. Based on this finding, Dittmer and Dawson (1961) developed an acid extraction procedure for the isolation of phosphoinositides from brain. Bovine brain is first extracted with $CHCl_3$:methanol (1:1) to remove most other lipids. The residue is then repeatedly extracted with $CHCl_3$:methanol (2:1) containing conc. HCl (0.25%). After partitioning the lipid extract by adding 0.2 vol of 0.9% NaCl, most of the phosphoinositides, especially PIP_2, are recovered from the interfacial precipitate. This precipitate is reextracted with acidic solvent and, after further fractionation, the extract is neutralized with alkali or NH_3. Almost pure PIP_2 ("triphosphoinositide" Na^+, K^+, or NH_4^+ salt) is precipitated. This procedure is useful for the preparative isolation of PIP_2 from brain. Dawson and Eichberg (1965) also developed a similar two-step procedure to isolate phosphoinositides from different tissues. A major drawback to both of these methods is a partial loss of phosphoinositide to the initial neutral solvent extract. Hajra et al. (1968) developed a one-step acidic extraction procedure for the extraction of labeled PIP from incubation mixtures. This procedure, a modification of the Bligh and Dyer (1959) extraction, has been used in one form or another for the isolation of trace quantities of phosphoinositides from tissue homogenates and incubation mixtures. It uses an acidified solvent for extraction and washing to keep the phosphoinositides in the $CHCl_3$-rich organic phase.

2.1. Method

To 1.2 mL of tissue homogenate (up to 5 mg protein) or incubation mixture, 4.5 mL of $CHCl_3$:methanol (1:2) and 0.1 mL of

$6N$ HCl are added. After mixing vigorously (Vortex mixer), an additional 1.5 mL of $CHCl_3$ and 1.5 mL of $2M$ KCl containing $0.5M$ H_3PO_4 are added with additional mixing, followed by centrifugation at low speed (1000 g for 5 min) to separate the phases. The lipids partition to the lower phase, and the proteins precipitate out as a white disc at the interface. The upper layer is suctioned off, and the lower layer is transferred to another tube and washed once or twice with 5 mL of acidified upper phase [1 drop of $6N$ HCl added to a mixture of $CHCl_3$:methanol:water (1:12:12) to remove non-lipid contaminants]. The lipids are recovered from the washed lower phase by evaporating off the solvents under a stream of nitrogen at 40°C. Acid-labile lipids, especially plasmalogens, are decomposed during the evaporation because the acidity of the extract increases as the volume is being reduced. This is not important in the case of inositol lipids, since they contain little if any plasmalogen. This acid extraction quantitatively extracts all acidic lipids including PIP and PIP_2. To preserve most of the acid-labile lipids, the extracts can be washed once with neutral upper phase to remove most of the acid, after which a drop of $6N$ methanolic ammonia solution is added to the lower phase to neutralize the remaining acidity prior to solvent removal under N_2.

3. Separation of Inositides

3.1. Paper Chromatography

Hörhammer and coworkers (1959, 1960) developed a chromatographic process using formaldehyde-treated paper to separate the phosphoinositides from each other as well as from other lipids. Historically, this method first established that brain "diphosphoinositide" is composed of the three inositol-containing lipids (Hörhammer et al., 1960). The method is actually a reverse-phase chromatography, the filter paper being converted to a hydrophobic stationary phase by prior treatment with formaldehyde. The formaldehyde treatment is accomplished by autoclaving the papers with formalin solution, then washing off the excess formaldehyde by water. The paper is developed with the upper phase of a butanol-acetic acid-water mixture. The inositides migrate more slowly than other lipids (R_f = 0.23, 0.35, and 0.6 for PIP_2, PIP, and PI, respectively). Although the method works well, subsequently developed thin-layer chromatography (TLC) methods are faster and more convenient.

Silica-impregnated glass filter paper was employed for separation of inositol lipids many years earlier, antedating TLC (Agranoff et al., 1958). Silica gel-impregnated paper or glass fiber paper has also been used for the separation of polyphosphoinositides (Santiago-Calvo et al., 1964; Hokin-Neaverson, 1980). These methods are similar to the thin-layer chromatographic methods described in a later section of this article.

3.2. Column Chromatography

Ion exchange chromatography has been successfully used to separate the phosphoinositides. Hendrickson and Ballou (1964) used a DEAE-cellulose column and a gradient elution with ammonium acetate in $CHCl_3$:methanol:water (0–0.6M ammonium acetate) for the preparative isolation of each inositide from Folch fraction II of brain. In this elution scheme, PI is eluted first from the column, followed by the contaminating phosphatidylserine; pure PIP and PIP_2 are eluted at higher concentrations of ammonium acetate. This chromatographic procedure is excellent for the preparative isolation of inositol-containing lipids. The Ca^{2+}-Mg^{2+} salts of the phosphoinositides do not bind to the DEAE-cellulose and the lipids must be converted to alkali metal salts before chromatography. These authors used an EDTA wash and Chelex column to convert the alkaline earth salts of these acidic lipids to alkali metal salts; it is evident from the published chromatography profile that the conversion was not complete, however. Probably an acid wash followed by neutralization with NaOH or KOH would be a more efficient method for the preparation of Na^+ or K^+ salts of these lipids.

Immobilized neomycin has been used for the chromatographic isolation of polyphosphoinositides. Neomycin is an aminoglycoside antibiotic that is postulated to have special affinity for polyphosphoinositides (Lodhi et al., 1979). Schacht (1978) employed neomycin immobilized on glass beads as an "affinity column" for the preparative isolation of PIP and PIP_2. The neomycin was attached to the periodate-treated, glycol-coated glass beads through Schiff's base formation, and the linkage stabilized by subsequent reduction with borohydride. The acidic solvent extract of brain was applied to the column, and the lipids were eluted with a mixture of $CHCl_3$ and methanol containing ammonium acetate. Under the chromatographic conditions employed, most of the lipids (including PI, but not PIP or PIP_2) did not bind to the column.

PIP is eluted with 0.35M ammonium acetate, and PIP$_2$, with 1M ammonium acetate or 1N NH$_4$OH (in CHCl$_3$-methanol) solution. Further details of this procedure have been recently described (Schacht, 1981). Palmer (1981) showed that because neomycin is basic, the immobilized neomycin may act as a conventional ion-exchanger so that all acidic lipids can be chromatographically separated. He described a modified procedure, using formic acid and ammonium formate step gradient in CHCl$_3$-methanol for the isolation of various acidic lipids, in addition to PIP and PIP$_2$ using this column. Because of the high capacity and high flow rate of the neomycin column rather than its affinity properties, it is claimed to be superior to DEAE-cellulose chromatography (Palmer, 1981).

3.3. Thin-Layer Chromatography

Thin-layer chromatography (TLC) is currently the most widely used method for the analysis of lipids. Because of the high polarity of the phosphoinositides, standard TLC systems that are employed for the separation of most phospholipids are not particularly useful. In the systems commonly used for lipid separations, the phosphoinositides barely migrate from the origin so that a number of TLC systems using acidic, basic, or even neutral polar solvents have been devised for their separation. A list of several chromatographic systems described from different laboratories is found in Table 1. In many of these systems, it is observed that the phosphoinositides sometimes streak from the origin instead of giving discrete TLC spots. This is probably because of the partial separation of their multiple salt forms. This problem was solved by Gonzales-Sastre and Folch (1968) by using TLC plates coated with oxalate, which sequestered Ca^{2+} and Mg^{2+}, thus eliminating the streaking. This TLC method has been widely used for the analysis of trace quantities of phosphoinositides present in different tissues. A number of modifications of this method have been described. One is to impregnate commercially available silica gel-coated plates with oxalate solutions (Jolles et al., 1981) instead of adding oxalate to a silica gel slurry used in the production of the plates. In the system originally described by Gonzales-Sastre and Folch (1968), PIP and PIP$_2$ are well separated from other lipids, but other polar lipids, especially PI and PA, are not separated from each other. These lipids are highly labeled in tissues incubated with ^{32}P$_i$ (as are PIP and PIP$_2$), and it is desirable to have a single system that will separate all four labeled lipids. It is also more

Table 1
Thin-Layer Chromatographic Systems for Phosphoinositides

Stationary phase	Mobile phase	Reference
Silica gel-coated paper	Phenol:ammonia (99:1)	Santiago-Calvo et al. (1964)
Silica gel-coated glass fiber paper (ITLC)	Chloroform:methanol:acetic acid:water 50:25:8:0.4	Hokin-Neaverson (1980)
Silica gel H containing K-oxalate	Chloroform:methanol:$4N$ NH$_4$OH (9:7:2)	Gonzales-Sastre and Folch (1968)
Silica gel 60 (with (organic binder, E. Merck)	Chloroform:methanol:$17N$ NH$_4$OH:H$_2$O (45:45:4:11)	Schacht et al. (1974)
Silica gel with magnesium acetate	Chloroform:methanol:water (90:65:20)	Farese et al. (1979)
Silica gel H	Chloroform:methanol:$4.3M$ NH$_4$OH (90:65:20) followed by propanol:$4.3M$ NH$_4$OH containing 10 mM cyclohexanediamine tetracetate	Hauser et al. (1971)
Silica gel HPTLC impregnated with K-oxalate	Chloroform:acetone:-methanol:acetic acid:water (40:15:13:12:8)	Jolles et al. (1981)
Silica gel 60 (E. Merck (impregnated with K-oxalate	Chloroform:acetone:-methanol:acetic acid:water (40:15:13:12:8)	Van Rooijen et al. (1983)

convenient to modify commercially produced TLC plates than to produce them in one's laboratory. Systems using acidic developing solvent on oxalate-treated silica gel plates have been described (Palmer and Verapoorte, 1971; Jolles et al., 1981; Van Rooijen et al., 1983). The details of a procedure that has been successfully used in our laboratory (Van Rooijen et al., 1983) follow.

3.4. Method

E. Merck silica gel 60 plates are coated with oxalate by running them in a solution containing 1.2 g of potassium oxalate in 40 mL of methanol and 60 mL of water. The solvent is allowed to migrate to the top of the plate. The plates are removed, air-dried, and activated at 110°C for 30 min. An aliquot of dried lipid extract is dissolved in 50–100 μL of $CHCl_3$: methanol : water [75 : 25 : 2 (v/v/v)] and spotted on the plate. The plate is developed with a mixture of $CHCl_3$: acetone : methanol : glacial acetic acid : water (40 : 15 : 13 : 12 : 7 by vol). In this system, PIP_2 has an R_f of 0.25, PIP has an R_f of 0.35, and PI has an R_f of 0.45. Most of the other lipids migrate above the PI in this solvent system. The PIP and PIP_2, along with other lipids, are visualized with common lipid-locating agents such as I_2 or molybdenum blue (Dittmer and Lester, 1964) spray for phosphate. In many organs, however, the amount of PIP and PIP_2 isolated from tissues is small and difficult to visualize with common TLC lipid or phospholipid detection sprays. The copper acetate charring method is shown to be very sensitive, and less than a nanomole of lipid is reported to be visualized by this spray (see next section). Radioactive lipids are readily detected by radioautography (Van Rooijen et al., 1985).

4. Deacylation and Separation of Water-Soluble Products

These polar lipids, although difficult to handle in the intact state, can be deacylated to highly acidic water-soluble products, i.e., the glycerophosphoryl backbone, which can be analyzed either by ion-exchange chromatography or by electrophoresis. Dawson and coworkers deacylated the lipid with alkali, and then used a two-dimensional chromatographic and electrophoretic procedure to separate the water-soluble deacylated products (Dittmer and Dawson, 1961; Dawson and Dittmer, 1961). Brockerhoff and Ballou (1961) separated these products by chromatography on a Dowex-1, Cl⁻ column using a LiCl gradient elution. The products were eluted according to their charges, i.e., glycerophosphorylinositol, followed by the monophosphate and bisphosphate derivatives. This method provided a better separation of the components than the ion-exchange chromatography of the corresponding lipids. The purified products also proved useful for the identification and elucidation of the structure of the parent lipids.

High-voltage paper electrophoresis has been successfully

used to separate these acidic water-soluble deacylated products. Seiffert and Agranoff (1965) developed a high-voltage paper electrophoresis method at pH 1.5 for the resolution of phosphate esters, especially the inositol phosphates. This method can also be used to separate the glycerophosphorylinositol phosphates (Hajra et al., 1968). The detailed procedure for the deacylation (alkaline methanolysis) and electrophoresis are given below (Hajra et al., 1968; Hajra and Agranoff, 1968).

4.1. Method 1: High-Voltage Electrophoresis

Acidic lipid extracts from tissues containing up to 10 μmol of phospholipid phosphates are dissolved in 2 mL of $CHCl_3$ to which 1 mL of $0.3N$ NaOH in methanol is added. After mixing, the reaction mixture is allowed to stand at room temperature for 15 min. The reaction is stopped by adding 0.75 mL of water, mixed, and centrifuged. The upper layer containing the deacylated products is passed through a small column that is packed at the bottom with Dowex-50 Na^+ (0.5 × 0.6 cm) and on the top with Dowex-50 H^+ (2.5 × 0.6 cm). A Pasteur pipet plugged with glass wool can be used as the column. This ion-exchange column effectively removes the excess alkali with the Dowex 50 H^+ resin and then reconverts the deacylated products to the Na^+ salt with the Na^+ form of the resin. The lower layer of the alkaline methanolysate is washed with 2 mL of $CHCl_3$:methanol:water (3:47:48), the washings are passed through the same Dowex column, and the eluates are combined. The eluates are evaporated to dryness (Savant Speedvac), dissolved in a small volume of water, and used for electrophoresis.

The high-voltage electrophoresis is performed on Whatman No. 1 paper in Na^+ oxalate buffer, pH 1.5, for 20–30 min at 4000 V (Seiffert and Agranoff, 1965; Agranoff et al., 1983). Up to 0.1 μmol of phosphate esters (10–30 μL) can be spotted on each lane. Phosphate-containing compounds are detected by a molybdate spray (Bandurski and Axelrod, 1951), and the radioactive compounds are detected by radioautography. In this sytem, the migration rate of glycerophosphorylinositol (GPI) relative to that of inorganic phosphate is 1.7. The corresponding value for GPIP is 2.4, and for $GPIP_2$ is 2.8.

As mentioned above, this electrophoretic method has been used to separate different inositol phosphates from each other. Separation of inositol phosphates has received much recent attention, since their release from phosphoinositides via phosphodiesteratic cleavage is implicated in second messenger generation (Berridge and Irvine, 1984). Thus, PI, PIP, and PIP_2 give rise to *myo*-inositol-D-1-phosphate (IP_1), 1,4-bisphosphate (IP_2), and 1,4,5-trisphosphate (IP_3), respectively. Labeled inositol phosphates can be separated by high-voltage paper electrophoresis (Seiffert and Agranoff, 1965; Agranoff et al., 1983). Ion-exchange chromatography using a Dowex-50, formate column, and ammonium formate step gradient has been conveniently used to separate and isolate the inositol phosphates (Berridge et al., 1982). A method presently used in our laboratory is given below.

4.2. Method 2: Ion Exchange

Reaction mixtures (0.5 mL final vol) are terminated by the addition of 1.5 mL of $CHCl_3$: methanol (1 : 2, by vol). Following the addition of 1 mL of chloroform and 0.5 mL of water, tubes are vortexed and phases are separated by centrifugation. The upper aqueous phase is removed and warmed at 50°C to remove the last traces of $CHCl_3$. The volume of the aqueous phase is then adjusted to 3.0 mL by the addition of 1.7 of mL of water (to reduce the ionic strength), and tubes are rewarmed if necessary. For batch isolation of a total inositol phosphate fraction, 0.5 mL of a 50% (w/v) slurry of Dowex-1 resin (formate form) is added. Tubes are then vigorously mixed and centrifuged. The supernatant is aspirated, and the resin washed four times with 2.5 mL of 5 m*M* *myo*-inositol. A total inositol phosphate fraction containing glycerophosphorylinositol, inositol mono-, bis-, tris-, and tetrakis-phosphates is eluted from the resin with 1 mL of 1*M* ammonium formate/0.1*M* formic acid, and a 0.8-mL aliquot is removed for determination of radioactivity after the addition of 10 mL of ACS scintillation fluid (Berridge et al, 1982). For separation of the individual inositol phosphates, the water-soluble components are first applied to a Dowex-1 resin (formate form) anion exchange column [1.0 mL of a 50% (w/v) slurry of the resin in a Pasteur pipet with a glass fiber plug]. Free *myo*-inositol is eluted with 30 mL of water. Glycerophosphoinositol, IP, IP_2, and IP_3 fractions are eluted with 3 × 6 mL vol each of 5 m*M* sodium tetraborate/60 m*M* sodium formate, 0.1*M* formic acid/ 0.2*M* ammonium formate, 0.1*M* formic acid/0.4*M* ammonium formate, and 0.1*M* formic acid/1.0*M* ammonium formate, respectively (Berridge et al., 1983). Elution of the Dowex-formate columns with 0.1*M* formic acid/0.8*M* ammonium formate, followed by 0.1*M* formic acid/1.0M ammonium formate, permits the separation of IP_3 and IP_4 (Batty et al., 1985).

Ion-exchange HPLC has been used to separate inositol phosphates and even to effect a partial separation of the different positional isomers of IP_3 (Irvine et al., 1984). An elaborate automated system for quantitation of the inositol phosphates based on phosphate content employs an HPLC ion exchange column followed by degradation on an immobilized alkaline phosphatase column and spectrophotometric determination of phosphate (Meek and Nicoletti, 1986). Recently, using this method, nmol quantities of the inositol phosphate isomers present in different tissues have been analyzed (Meek, 1986).

Inositol phosphates and their isomers can also be separated by paper chromatography (Tomlinson and Ballou, 1961), which is time-consuming and cumbersome, but remains in many instances definitive. For separation of higher inositol phosphates, i.e., IP_4, IP_5, and IP_6, high-voltage paper electrophoresis (Seiffert and Agranoff, 1965) is at present the most convenient technique.

5. Quantitative Analysis

In most tissues, the phosphoinositides are present only in trace quantities and have a very high turnover rate. Dawson and coworkers have demonstrated that in brain, where these lipids are present at relatively high concentrations, there is a rapid postmortem decrease in their amounts (Dawson and Eichberg, 1965; Eichberg and Dawson, 1965). This may also be true for other organs. Therefore, for the quantification of these lipids, care should be taken to avoid postmortem decreases in their amounts by quickly freezing the tissues *in situ* (Eichberg and Dawson, 1965). Rapid heating of the tissues, especially brain, by microwave irradiation to inactivate degradative enzymes has also been used to prevent postmortem hydrolysis of inositides in brain (Soukup et al., 1978; Van Dongen et al., 1985). Nishihara and Keenan (1985) compared three different methods, i.e., freezing the brain in liquid N_2, microwave irradiation, and freeze-blowing, to arrest the postmortem changes in the amounts of inositol lipids in rat brain. These authors found that both microwave irradiation and freeze-blowing prevent the rapid postmortem hydrolysis of polyphosphoinositides, freeze-blowing being somewhat more effective than microwave irradiation. Microwave irradiation has also been shown to prevent the rapid postmortem changes in the concentrations of inositol phosphates in different tissues (Meek, 1986).

The common method of analysis of these lipids is to extract and separate them by thin-layer or column chromatographic methods, and then measure the phosphorus content of the purified lipids. The spots from TLC can be scraped off and phosphorus directly measured by digesting the silica gel powder with perchloric acid (Rouser et al., 1969) or H_2SO_4-H_2O_2 (Bartlett, 1959) to convert the organic phosphates to inorganic phosphate, which is then estimated spectrophotometrically as the molybdate complex (Rouser et al., 1969; Bartlett, 1959). These methods are sensitive at the nanomolar level, especially for PIP and PIP_2, which have multiple P atoms.

However, quantitation by phosphate assay may not be sensitive enough for small amounts of sample (<1 nmol) such as is present in in vitro incubation mixtures. For such purposes, Van Dongen et al. (1985) recently described a densitometric micromethod. After separation by TLC, the lipid spots are charred with copper acetate reagent (Fewster et al, 1969), and the brown spots formed are densitometrically scanned at 450 nm. These authors showed that as little as 0.1 nmol of inositide can be measured by employing this method. However, it should be pointed out that because of variations in thickness of the stationary phase, migration rate of the lipids, and charring procedure, it is difficult to replicate the densitometric measurements from one TLC plate to another. Therefore, standard lipid samples should be run side by side with each unknown to minimize variation. A possible sensitive method to measure these lipids would be to analyze the fatty acid content by gas-liquid chromatography (GLC). A number of workers have shown that the phosphoinositides contain predominantly (70–80%) one diacylglycerol species, the 1-stearoyl-2-arachidonoyl derivative (Hendrickson and Ballou, 1964; Keough et al., 1972). Since capillary GLC is extremely sensitive, picomole quantities of these lipids should be measurable from the amounts of stearic and arachidonic acid recovered following degradation of TLC-purified lipid spots.

References

Agranoff B. W., Bradley R. M., and Brady R. O. (1985) The enzymatic synthesis of inositol phosphatide. *J. Biol. Chem.* **233,** 1077–1083.

Agranoff B. W., Murthy P., and Seguin E. B. (1983) Thrombin-induced phosphodiester cleavage of phosphatidylinositol bisphosphate in human platelets. *J. Biol. Chem.* **258**, 2076–2078.

Bandurski R. S. and Axelrod B. (1951) The chromatographic identification of some biologically important phosphate esters. *J. Biol. Chem.* **193**, 405–410.

Bartlett G. R. (1959) Phosphorus assay in column chromatography. *J. Biol. Chem.* **234**, 466–468.

Batty I. R., Nahorski S. R., and Irvine R. F. (1985) Rapid formation of inositol 1,3,4,5-tetrakisphosphate following muscarinic receptor stimulation of rat cerebral cortical slices. *Biochem. J.* **232**, 211–215.

Bell M. E., Peterson R. G., and Eichberg J. (1982) Metabolism of phospholipids in peripheral nerve from rats with chronic streptozotocin-induced diabetes: Increased turnover of phosphatidylinositol-4,5-bisphosphate. *J. Neurochem.* **39**, 192–200.

Berridge M. J., Dawson R. M. C., Downes C. P., Heslop J. P, and Irvine R. F. (1983) Changes in the levels of inositol phosphates after agonist-dependent hydroysis of membrane phosphoinositides. *Biochem. J.* **212**, 473–482.

Berridge M. J. and Irvine R. F. (1984) Inositol trisphosphate, a novel second messenger in cellular signal tranduction. *Nature* **312**, 315–321.

Berridge M. J., Downes C. P., and Hanley M. R. (1982) Lithium amplifies agonist-dependent phosphatidylinositol responses in brain and salivary glands. *Biochem. J.* **206**, 587–595.

Bligh E. G. and Dyer W. J. (1959) A rapid method of total lipid extraction and purification. *Can. J. Biochem.* **37**, 911–917.

Brockerhoff H. and Ballou C. E. (1961) The structure of phosphoinositide complex of beef brain. *J. Biol. Chem.* **236**, 1907–1911.

Dawson R. M. C. and Dittmer J. C. (1961) Evidence for the structure of brain triphosphoinositide from hydrolysis of degradation products. *Biochem. J.* **81**, 540–545.

Dawson R. M. C. and Eichberg J. (1965) Diphosphoinositide and triphosphoinositide in animal tissues. *Biochem. J.* **96**, 634–643.

Dittmer J. C. and Dawson R. M. C. (1961) The isolation of a new lipid, triphosphoinositide, and monophosphoinositide from ox brain. *Biochem. J.* **81**, 535–540.

Dittmer J. C. and Lester R. L. (1964) A simple, specific spray for the detection of phospholipids on thin-layer chromatograms. *J. Lipid Res.* **5**, 126–127.

Eichberg J. and Dawson R. M. C. (1965) Polyphosphoinositides in myelin. *Biochem. J.* **96**, 644–650.

Farese R. V., Sabir A. M., and Vandor S. L. (1979) Adrenocorticotropin actually increases adrenal phosphoinositides. *J. Biol. Chem.* **254,** 6842–6844.

Fewster M. E., Burns B. J., and Mead J. F. (1969) Quantitative densitometric thin-layer chromatography of lipids using copper acetate reagent. *J. Chromatogr.* **43,** 120–126.

Folch J. (1942) Brain cephalin, a mixture of phosphatides. *J. Biol. Chem.* **146,** 35–44.

Folch J. (1949) Complete fractionation of brain cephalin: Isolation from its phosphatidyl serine, phosphatidyl ethanolamine and diphosphoinositide. *J. Biol. Chem.* **177,** 497–504.

Folch J. and LeBaron F. N. (1953) Isolation of phosphatido-peptides, a new group of brain phosphatides. *Fed. Proc.* **12,** 203.

Gonzales-Sastre F. and Folch J. (1968) Thin-layer chromatography of the phosphoinositides. *J. Lipid Res.* **9,** 532–533.

Hajra A. K. and Agranoff B. W. (1968) Acyl dihydroxyacetone phosphate: Characterization of a ^{32}P-labeled lipid from guinea pit liver mitochondria. J. Biol. Chem. **243,** 1617–1622.

Hajra A. K. and Agranoff B. W. (1968) Rapid labeling of mitochondrial lipids by labeled orthophosphate and adenosine triphosphate. *J. Biol. Chem.* **243,** 1609–1616.

Hauser G., Eichberg J., and Gonzales-Sastre, F. (1971) Regional distribution of polyphosphoinositides in rat brain. *Biochim. Biophys. Acta* **248,** 87–95.

Hendrickson H. S. and Ballou C. M. (1964) Ion exchange chromatography of intact brain phosphoinositides or DEAE-cellulose by gradient salt elution in a mixed solvent system. *J. Biol. Chem.* **239,** 1369–1373.

Hokin-Neaverson M. (1980) Actions of chlorpromazine, haloperidol and pimozide on lipid metabolism in guinea pig brain slices. *Biochem. Pharmacol.* **29,** 2697–2700.

Horhammer L., Wagner H., and Richter G. (1959) Zur Papierchromatographischen auftrennung von Phosphatiden. *Biochem. Z.* **331,** 155–161.

Horhammer L., Wagner H., and Holzl J. (1960) Uber die Inositphosphatide des Rinderhirns. *Biochem. Z.* **332,** 269–276.

Irvine R. F., Letcher A. J., Lander D. J., and Downes C. P. (1984) Inositol trisphosphates in carbachol-stimulated rat parotid glands. *Biochem J.* **223,** 237–243.

Jolles J., Zwiers H., Dekker A., Wirtz K. W. A., and Gispen W. H. (1981) Corticotropin-(1-24)-tetracosapeptide affects protein phosphorylation and polyphosphoinositide metabolism in rat brain. *Biochem. J.* **194,** 283–291.

Keough K. M. W., Macdonald G., and Thompson W. (1972) A possible relation between phosphoinositides and the diglyceride pool in rat brain. *Biochim. Biophys. Acta* **270**, 337–347.

LeBaron F. H. and Folch J. (1956) The isolation from brain tissue of a trypsin-resistant protein fraction containing combined inositol and its relation to neurokeratin. *J. Neurochem.* **1**, 101–108.

Lees M. (1957) Preparation and analysis of phosphatides. *Meth. Enzymol.* **3**, 328–345.

Lodhi S., Weiner N. D., and Schacht, J. (1979) Interactions of neomycin with monomolecular films of polyphosphoinositides and ether lipids. *Biochim. Biophys. Acta* **557**, 1–8.

Meek J. L. (1986) Inositol bis-, tris-, and tetrakis(phosphate)s: Analysis in tissues by HPLC. *Proc. Natl. Acad. Sci. USA* **83**, 4162–4166.

Meek J. L. and Nicoletti F. (1986) Detection of inositol trisphosphate and other organic phosphates by high-performance liquid chromatography using an enzyme-loaded post-column reactor. *J. Chromat.* **351**, 303–311.

Nishihara M. and Keenan R. W. (1985) Inositol phospholipid levels of rat forebrain obtained by freeze-blowing method. *Biochim. Biophys. Acta* **835**, 415–418.

Palmer F. B. S. C. (1981) Chromatography of acidic phospholipids on immobilized neomycin. *J. Lipid Res.* **22**, 1296–1300.

Palmer F.B.S.C. and Verapoorte J. A. (1971) The phosphorus components of solubilized erythrocyte membrane protein. *Can. J. Biochem.* **49**, 337–347.

Rouser G., Fleischer S., and Yamamoto A. (1969) Two dimensional thin-layer chromatographic separation of polar lipids and determination of phospholipids by phosphorus analysis of spots. *Lipids* **5**, 494–496.

Santiago-Calvo E., Mule S., Redman C. M., Hokin M. R., and Hokin L. E. (1964) The chromatographic separation of polyphosphoinositides and studies on their turnover in various tissues. *Biochim. Biophys. Acta* **84**, 550–562.

Schacht J. (1978) Purification of polyphosphoinositides by chromatography on immobilized neomycin. *J. Lipid Res.* **19**, 1063–1067.

Schacht J. (1981) Extraction and purification of polyphosphoinositides. *Meth. Enzymol.* **72**, 626–631.

Schacht J., Neale E., and Agranoff B. W. (1974) Cholinergic stimulation of phospholipid labeling for [^{32}P]orthophosphate in guinea pig cortex synaptosomes *in vitro*: Subsynaptosomal localization. *J. Neurochem.* **23**, 211–218.

Seiffert U. B. and Agranoff B. W. (1965) Isolation and separation of inositol phosphates from hydrolysate of rat tissues. *Biochim. Biophys. Acta* **98**, 574–581.

Soukup J. F., Friedel R. O., and Schanberg S. M. (1978) Microwave irradiation fixation for studies of polyphosphoinositide metabolism in brain. *J. Neurochem.* **27**, 1273–1276.

Tomlinson R. V. and Ballou C. E. (1961) Complete characterization of the myoinositol polyphosphates from beef brain phosphoinositide. *J. Biol. Chem.* **236**, 1902–1906.

Van Dongen C. J., Zwiers H., and Gispen W. H. (1985) Microdetermination of phosphoinositides in a single extract. *Anal. Biochem.* **144**, 104–109.

Van Rooijen L. A. A., Seguin E. B., and Agranoff B. W. (1983) Phosphodiesteratic breakdown of endogenous polyphosphoinositides in nerve ending membranes. *Biochem. Biophys. Res. Commun.* **112**, 919–926.

Van Rooijen L. A. A., Hajra A. K., and Agranoff B. W. (1985) Tetraenoic species are conserved in muscarinically enhanced inositide turnover. *J. Neurochem.* **44**, 540–543.

Analysis of Prostaglandins, Leukotrienes, and Related Compounds in Retina and Brain

Dale L. Birkle, Haydee E. P. Bazan,
and Nicolas G. Bazan

1. Introduction

The eicosanoids are oxygenated products of arachidonic acid. Prostaglandins (PGs) are synthesized by cyclooxygenase; hydroxyeicosatetraenoic acids (HETEs) and leukotrienes (LTs) are synthesized by lipoxygenase (for reviews, see Hammarstrom, 1983; Lands, 1979). Prostaglandins are variations of the 20-carbon fatty acid prostanoic acid, and their nomenclature is based on the oxygen-derived substitutions in the five-membered ring and the number of double bonds in the two side chains. The carboxylic acid function is present in the α side chain, and the terminal methyl group is present in the β side chain. Thromboxanes and prostacyclin have a slightly modified ring structure. Lipoxygenase reaction products can occur as various isomers, depending upon where addition of oxygen takes place. The immediate precursors of HETEs are hydroperoxy derivatives (HPETEs). Further metabolism of HPETEs by dehydration to epoxides and addition of glutathione results in the synthesis of leukotrienes. Eicosanoids can also be synthesized from other fatty acids, such as eicosatrienoic acid (20:3, n-6) and eicosapentaenoic acid (20:5, n-6). Prostaglandins from arachidonic acid are the 2-series, e.g., PGE_2. The 1-series are products of 20:3 (n-6), and the 3-series are products of 20:5 (n-6).

Prostaglandins, HETEs, and leukotrienes have potent actions in many, if not all, cells and tissus (for reviews, see Wolfe, 1982; Hammarstrom, 1983). They act as mediators and modulators of inflammation by influencing chemotaxis, phagocytosis, and lysosomal enzyme release. They have potent vascular effects, causing vasodilation, vasoconstriction, changes in capillary permeability, and platelet aggregation. Prostaglandins are important mediators of thermoregulation, gastric acid secretion, and various aspects of the reproductive cycle. Leukotrienes are extremely potent bronchoconstrictors and are mediators of the asthma of anaphylaxis. There is increasing evidence that eicosanoids are modulators of central and peripheral neurotransmission.

These metabolites are produced in extremely small quantities in any given tissue; therefore, quantitative and qualitative analyses present an important challenge. The major problems are: (1) extraction of the compounds of interest, with appropriate removal of water-soluble contaminants and interfering lipids, and (2) separation of the various classes of eicosanoids and the chemically similar isomers and derivatives. This chapter will review some of the commonly employed procedures for extraction and separation of eicosanoids in the central nervous system, where the high concentrations of lipids present special problems.

2. Extraction Procedures

Eicosanoids are acidic lipids, relatively polar compared to triacylglycerols and fatty acids. There are many methods in the literature (for review, *see* Hensby, 1977) describing the extraction of eicosanoids, most of them based on extraction with organic solvents. These procedures allow removal of proteins and other water-soluble compounds that can interfere with the subsequent analysis of eicosanoids. More recently, solid phase extraction procedures have been described (Luderer et al, 1983; Metz et al., 1982; Schulz and Seeger, 1986; Varhagen et al., 1984). The type of extraction depends upon three major factors determined by the investigation: (1) what kind of analytical procedure will be used [i.e., thin-layer chromatography (TLC), gas chromatography-mass spectrometry (GC-MS), high-performance liquid chromatography (HPLC), radioimmunoassay (RIA)]; (2) what components are present in the sample initially (i.e., large amounts of proteins, neutral lipids, water); and (3) what is to be recovered (i.e., all lipids, prostaglandins, leukotrienes). In many cases, extraction also serves to stop enzymatic reactions by denaturation of proteins. The major concern regarding any extraction procedure is adequate recovery of the compounds of interest (ideally greater than 90%). Two phenomena appear to offer the main problems for eicosanoid extraction, protein binding and degradation by acidic conditions. Protein binding can be minimized by allowing the tissue, homogenized in organic solvent, to be extracted overnight under nitrogen at −20°C. When this is not feasible, several re-extractions of the tissue usually increases recovery. Care must be taken to avoid strongly acidic conditions, i.e., pH < 3, during volume reduction by using volatile organic acids whenever possible and by constantly diluting samples with acetone or methanol. When removal of water-soluble components and recovery of glycerolipids are not major concerns, tissue can be homogenized in ethanol or methanol. This procedure is best suited for "extraction" of cell-free media, where the major function of the alcohol is to provide an azeotropic mixture for rapid evaporation. It has been applied to brain, but the resulting extract contains high concentrations of proteins and proteolipids that must be removed by another method (*see below*).

When tissues have been incubated ex vivo for metabolic studies (e.g., radioactive precursors such as arachidonic acid are added to the incubation medium), the reaction can be stopped with ethanol, methanol, or acetone, and the precipitated protein separated by centrifugation. Formic acid, citric acid, or acetic acid can be used to acidify the medium to pH 3.5. The extraction is followed by a partition with organic solvents such as chloroform, ethyl acetate, or diethyl ether. The organic extraction should be repeated twice and the combined organic phases dried under nitrogen or vacuum.

If other lipids such as glycerolipids are to be analyzed in the samples, preparation of a total lipid extract by homogenizing the tissue in hexane:2-propanol (3:2, by vol) yields good recoveries (Saunders and Horrocks, 1984). If the tissue has been incubated and large amounts of water are present, the reaction can be stopped with chloroform:methanol (1:2, by vol), centrifuged, the supernatant removed, and the pellet rehomogenized in hexane:2-propanol (3:2). This mixture is centrifuged, the supernatant is combined with the chloroform:methanol extraction, and the solvents are removed under nitrogen.

3. Bulk Separation Procedures

3.1. Fractionation by Silica Gel Column Chromatography

Lipid extracts can be fractionated by silica gel chromatography into four fractions using stepwise elution with solvents of increasing polarity (Claeys et al., 1985). Small columns of 0.5 mg silica gel are prepared in Pasteur pipets with a slurry of silica gel in diethyl ether:acetic acid (100:0.5, by vol; solvent system C) and washing with 6 mL of the same system. After preconditioning the columns with 6 mL of hexane:diethyl ether:acetic acid (90:10:0.5, by vol; system A), the sample is applied onto the columns in 150 μL of system C while the column is temporarily plugged off (the solubility of the sample in system A is incomplete). After evaporation of the diethyl ether on the top of the column with a stream of N_2, elution is started with 6 mL of system A to elute triacylglycerols, cholesterol esters, and free fatty acids. A fraction containing monohydroxy acids, free fatty acids, and diacylglycerols is obtained by subsequent elution with 6 mL of a mixture of hexane-:diethyl ether:acetic acid (60:40:0.5, by vol; system B). Following elution of the monohydroxy acids, elution of dihydroxy acids is performed with 6 mL of system C. Subsequently, elution of trihydroxy acids and PGs is done with 6 mL of a mixture of diethyl ether:methanol:acetic acid (90:10:0.5, by vol; system D). The solvents are evaporated under N_2, and the residues are redissolved into an appropriate volume of methanol and stored at −20°C for further analysis.

3.2. Solid Phase Extractions (SPEs)

Solid phase extraction is a relatively new procedure particularly suited to the extraction of organic components from aqueous mixtures (Van Horne, 1985). Samples, typically in large volume, are applied in H_2O to a cartridge packed with bonded phase silica gel. These packings are commercially available in a wide variety of polarities. The cartridge is washed to remove water-soluble contaminents, and the compounds of interest are eluted with an organic solvent such as methanol.

We have attempted various modifications of published SPE procedures to obtain a purified preparation of eicosanoids from brain, with moderate success. The best procedure utilizes a NH_2-bonded phase cartridge (Analytichem, Van Horne, 1985), which is a fairly polar bed. Lipid extracts, prepared as described above, are resuspended in chloroform:2-propanol (2:1), in the ratio of 1 g tissue:10 mL solvent, and applied to a NH_2 column that has been prewashed with 10 mL of hexane. The column is washed with 3 mL of chloroform:2-propanol (2:1), which elutes triacylglycerols, diacylglycerols, cholesterol, and cholesterol esters. The column is then washed with 5 mL of diethyl ether, which elutes HETEs and free fatty acids. Finally the phospholipids and PGs are eluted with

5 mL of methanol. This procedure is useful for the removal of glycerolipids prior to analysis by HPLC. An intermediate wash with ethyl acetate, prior to methanol, will elute PGs separately from phospholipids. Unfortunately this method is not useful for peptide LTs, which are not retained and elute with the neutral lipid fraction. There are several published methods for SPE of LTs . (Luderer et al., 1983; Metz et al., 1982; Schulz and Seeger, 1986; Varhagen et al., 1984), but we have had limited success when applying these methods to brain. The extraction and isolation of LTs from tissues containing large amounts of lipid continue to present a challenge.

4. Analytical Procedures

4.1. TLC Methods

Thin-layer chromatography has been used extensively to isolate eicosanoids from other lipids and each other. The procedure is simple, economical, and reliable. Commercially available plates give the best results, and we found that activation of these plates at 110°C for 30 min produces better resolution of components.

Several of these methods have been published (Salmon and Flower, 1982). In our laboratory, we have successfully used high-performance silica gel plates (LHP-K plates, Whatman, Inc., NY) with solvent system of ethyl acetate:acetone:acetic acid (90:5:1, by vol) (Smith et al., 1983) for separation of prostaglandins. The samples are applied to the plates with a clean microsyringe; before spotting the samples, a standard mixture containing 10 µg each of PGE_2, $PGF_{2\alpha}$, 6-keto-$PGF_{1\alpha}$, PGD_2, and TXB_2 is applied to allow visualization by iodine vapors. The tanks, lined with paper saturation pads, are equilibrated with the solvent system for 2 h. The plates are developed twice. Two 10 × 10 cm plates with six samples per plate are run simultaneously. Table 1 shows the R_f of eicosanoids isolated by this technique. Blanks (incubations of boiled tissue) are run in parallel to provide a subtractable background for each plate. The smaller particle size of high performance plates makes tailing of the samples less of a problem.

For lipoxygenase products (HETEs), 20 × 20 cm silica gel plates (LK-5, Whatman, Inc., NY) are developed in petroleum ether:diethyl ether:acetic acid (50:50:1, by vol). This is a reliable method for separating diHETE, 5-HETE, 12-HETE, and arachidon-

Table 1
Separation of Prostaglandins by Thin-Layer Chromatography[a]

Compound	R_f
Arachidonic acid	0.85
PGD$_2$	0.70
TXB$_2$	0.54
PGE$_2$	0.49
PGF$_{2\alpha}$	0.40
6-Keto-PGF$_{1\alpha}$	0.31

[a]The solvent system consists of ethyl acetate:acetone:acetic acid (90:50:1).

ic acid. Methylated HETEs can be separated on silica gel G plates developed in benzene:ethyl acetate (20:1, by vol).

Another suitable solvent for separating HETEs is the organic phase of ethyl acetate:2,2,4-trimethylpentane:acetic acid:water (100:60:20:100, by vol). One advantage of this method is that when other lipids such as diacylglycerols are present in the samples, they run with an R_f higher than arachidonic acid, and there is no contamination in the HETEs (Table 2).

After the plates are developed, the spots can be detected by iodine vapors, autoradiography, or spraying a solution of 10% phosphomolybdic acid in ethanol and heating the plates for several min at 110°C. The amount of radioactive eicosanoids can be determined by scraping off zones in the plates and determining radioactivity by liquid scintillation counting.

4.2. HPLC Methods

High-performance liquid chromatography offers several advantages over analysis of eicosanoids by the alternative technique, TLC. HPLC is a nondestructive technique, and therefore provides a convenient method for the purification and collection of eicosanoids. Because of the greater number of theoretical plates in a HPLC column, HPLC has better resolving power than TLC has. Also, recovery of compounds by collection of HPLC eluent is simpler and more efficient than recovery from a TLC plate, which requires scraping the silica gel and eluting the compounds. Quantitation of labeled compounds can be done most conveniently

Table 2
Separation of Hydroxyeicosanoids by Thin-Layer Chromatography

Compound	System A[a]	System B[b]
	R_f	
15-HETE	0.42	0.52
12-HETE	0.42	0.52
5-HETE	0.29	0.45
Arachidonic acid	0.59	0.67
diHETEs	0.08	0.38
Diacylglycerols	0.50	0.73

[a]Petroleum ether:diethyl ether:acetic acid (50:50:1)
[b]Organic phase of ethyl acetate:isoctane:acetic acid:water (100:60:20:100).

by on-line flow scintillation detection, thus avoiding the tedious process of collection of fractions, taking aliquots, and counting in a liquid scintillation counter. Because in many cases interferences and cross-reactivity produce uncertainty in the quantitation of eicosanoids by RIA, HPLC is a very suitable method for separation of the components in a sample prior to RIA. Similarly, purification of the sample by HPLC is necessary prior to quantitative or qualitative analysis of eicosanoids by GC-MS.

Eicosanoids can be separated by a variety of methods using reverse phase or straight phase HPLC (for review, see Hamilton and Karol, 1982). In our laboratory, a reverse phase system using a C18 (Waters Associates) column and gradient elution with acetonitrile and buffered water (pH 3.5) has yielded the best results. Reverse phase HPLC offers several advantages over straight phase HPLC. First, the separation of PGE_2 and PGD_2 is very difficult, if not impossible, with a straight phase system, but these two PGs separate by several minutes on a reverse phase system. Second, in straight phase the unconverted arachidonic acid, which can represent a substantial amount of radioactivity, elutes before the eicosanoids and can contaminate the eicosanoid peaks by tailing. Third, in straight phase we have found that radiolabeled diacylglycerols can cause a significant contamination problem because they are partially extracted with acetone or alcohols and co-elute with the monohydroxy derivatives of polyunsaturated fatty acids (Fig. 1; Claeys et al., 1986).

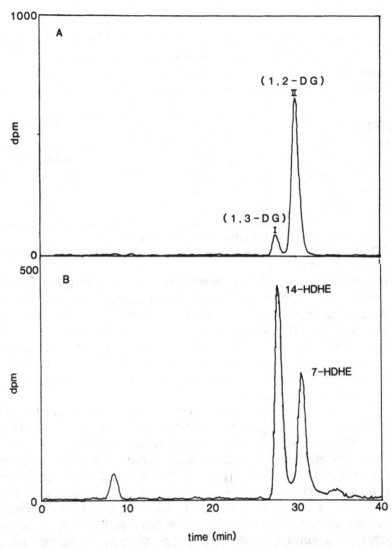

time (min)

Fig. 1. Coelution of radiolabeled diacylglycerols (DG, panel A) and monohydroxy derivatives of docosahexaenoic acid (HDHE, panel B) from unfractionated retinal lipid extracts by straight phase HPLC. Arachidonic acid metabolites elute in this system with retention times similar to the corresponding docosanoids (e.g., 5-HETE co-elutes with 7-HDHE). Samples were injected onto a Waters μPorasil Radial-PAK column and eluted with a gradient of increasing 2-propanol in acidic hexane.

We have used a modification of the method of Eling et al. (1986) to separate radiolabeled prostaglandins and hydroxy fatty acids (Fig. 2). Radioactivity can be monitored with a flow scintillation detector (Flo-One β, Radiomatic) using ReadySolv EP scintillation cocktail (Beckman) at a ratio of 2:1 (v:v), cocktail to HPLC solvent (Birkle et al., 1988). Counting efficiency for ^{3}H is 30% and 70% for ^{14}C. Alternatively fractions can be collected and aliquots counted by liquid scintillation counting. The solvent system is also amenable to UV detection at 198 nm for PGs and 235 nm for HETEs. Samples are dissolved in 200 μL initial conditions solvent (26% acetonitrile, 74% buffered water). The concentration of acetonitrile in the injection solvent can be increased to 100% if solubility is a problem, without affecting the separation. Buffered water is prepared by adding 1 mL of glacial acetic acid to 1 L of water and adjusting the pH to 3.5 with ammonium hydroxide (8 drops of 5M solution). Initial conditions are maintained for 24 min for the elution of PGs. The acetonitrile concentration is increased linearly to 40% over 3 min and maintained at this level for 28 min for the elution of dihydroxy derivatives (e.g., LTB$_4$). Acetonitrile is then increased to 50% over 4 min, and maintained for 11 min for elution of monohydroxy derivatives. The acetonitrile concentration is increased to 80% over 4 min and maintained for 10 min to elute free fatty acids, monoacylglycerols, diacylglycerols, and triacylglycerols. Increasing acetonitrile to 100% for 10 min elutes the remaining glycerolipids. The column can be re-equilibrated at initial conditions for 15 min before injection of the next sample. This system provides an excellent separation of eicosanoids. Since the neutral lipids elute as a single peak, however, an alternative method, such as TLC, is needed to assess the quantity of free arachidonic acid. Also, because of the low pH (3.5), peptide LTs do not elute until the final condition of 100% acetonitrile.

As mentioned previously, preparation of the sample for injection is very important, particularly in light of the concentration of glycerolipids in the sample. For most HPLC analyses, samples are dissolved in a solvent that contains around 75% water; large amounts of glycerolipids are insoluble in this solvent. This is a major problem in the analysis of eicosanoids in a fatty tissue such as brain, in which the levels of glycerolipids far exceed the levels of eicosanoids. Not only do glycerolipids present a solubility problem, but extremely nonpolar glycerolipids (mainly triacylglycerols) can be difficult to elute from C18 columns, causing carryover into subsequent analyses. This phenomenon can be ameliorated by

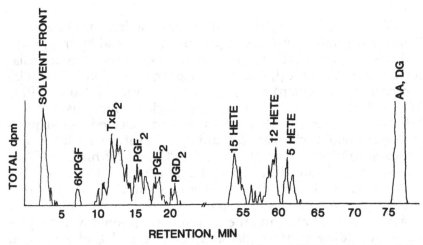

Fig. 2. Reverse phase separation of PGs and HETEs. Details of the method are provided in the text (reprinted with permission from Birkle et al., 1988).

using guard columns that bind the nonpolar material and by removing glycerolipids by bulk separation procedures described above. An additional problem with lipid samples is the presence of protein or proteolipid that is insoluble in the injection solvent. This can be removed by filtering the sample through a 0.4 μm filter fitted onto a syringe.

For the analysis of eicosanoids by GC-MS, we have found that straight phase HPLC is useful for further purification of compounds after reverse phase HPLC. Peaks from the reverse phase system are collected, converted to methyl esters, and rechromatographed on straight phase HPLC with a solvent system of hexane:2-propanol:acetic acid (Fig. 3) prior to further derivatization for GC-MS (Careaga and Sprecher, 1984).

Leukotrienes can be separated on a μBondapak C18 column (10 cm × 8 mm, 10 μm) with an isocratic system (Abe et al., 1985) of acetonitrile:methanol:water:acetic acid (403:65:532:10) adjusted to pH 5.6 with 5 M ammonium hydroxide (solvent A). The samples are injected in 200 μL of solvent A, and the flow rate is 2 mL/min. By UV and liquid scintillation detection of known standards, we have determined that LTC_4 has a retention time of 5 min, and LTD_4 elutes at 7 min. At 30 min the solvent is changed to 75% acetonitrile in water (0.1% acetic acid, pH 3.5), and hydroxy derivatives (HETEs) are eluted: 12-HETE elutes at 42 min, and 5-HETE at 44.5 min. This system can be modified slightly to afford separation of

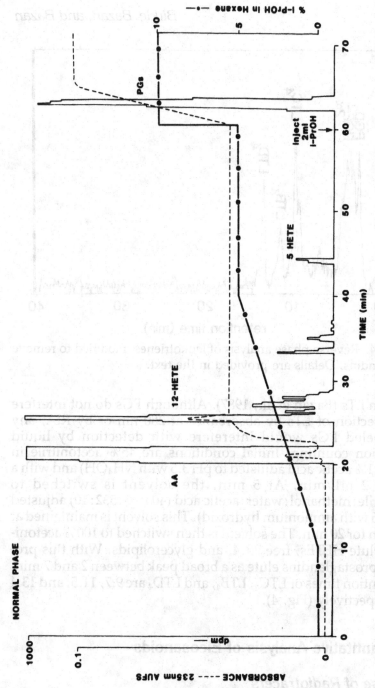

Fig. 3. Straight phase HPLC analysis of HETEs. Samples of bovine retina extracts were applied to a μPorasil Radial-PAK column and eluted with a hexane:2-propanol:acetic acid gradient (——). Absorbance at 235 nm (– – –) and radioactivity detection (——) by flow scintillation detection is also shown (reprinted with permission from Birkle and Bazan, 1984b).

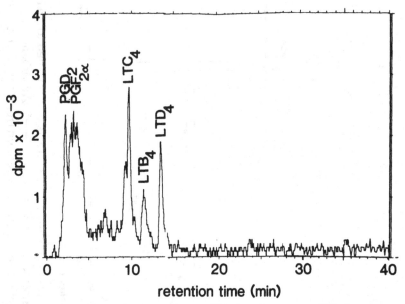

Fig. 4. Reverse phase analysis of leukotrienes, modified to remove prostaglandins. Details are provided in the text.

PGs from LTs (Bazan et al., 1987). Although PGs do not interfere with detection of LTs by absorbance at 280 nm or by RIA, any radiolabeled PGs would interefere with detection by liquid scintillation counting. Initial conditions are 45% acetonitrile in water (0.1% acetic acid adjusted to pH 3.5 with NH_4OH) and with a flow of 2 mL/min. At 5 min, the solvent is switched to acetonitrile:methanol:water:acetic acid (403:65:532:10) adjusted to pH 5.6 with ammonium hydroxid). This solvent is maintained at 2 mL/min for 20 min. The solvent is then switched to 100% acetonitrile to elute HETEs, free 20:4, and glycerolipids. With this procedure, prostaglandins elute as a broad peak between 2 and 7 min. The retention times of LTC_4, LTB_4, and LTD_4 are 9.7, 11.5, and 13.4 min, respectively (Fig. 4).

5. Quantitative Analysis of Eicosanoids

5.1. Use of Radiotracers

One of the major obstacles in the study of eicosanoids is the quantitation of various components. The use of radiolabeled precursors allows estimation of relative amounts of eicosanoids and relative changes in synthesis. Introduction of high specific activity radiotracers to a tissue results in labeling of various metabolic pools. These techniques require several assumptions: (1) adequate mixing of radiolabel with endogenous pools; (2) no physiological effect of addition of radiolabel; and (3) equal recovery of radiotracer and endogenous components.

We have used radiotracer techniques in several experimental protocols to study the metabolism of eicosanoids in the central nervous system (Pediconi et al., 1982, 1983; Reddy and Bazan, 1983; Birkle and Bazan, 1984a, b). A useful method is the in vivo prelabeling technique. Radiolabeled arachidonic acid is introduced in vivo into brain by intraventricular injection or into the eye by intravitreal injection. Preliminary time course data are obtained to determine maximum, steady-state incorporation of the precursor into the lipid pools of interest. Because of loss of radiolabeled precursor to the blood and peripheral tissues (Birkle and Bazan, 1984a), relatively large amounts must be injected; about 1 μCi per eye and 2 μCi per rat brain for arachidonic acid. High specific activity radiotracers represent very little mass, however, so effects are minimal. This point is very important in the case of fatty acids, which have detergent effects on cell membranes. After the appropriate prelabeling time, stimuli can be applied in vivo (e.g., electroconvulsive shock, flicker light, cryogenic injury, carotid occlusion), and the effects on various eicosanoids can be assessed. Alternatively or additionally, prelabeled tissues can be removed and subjected to experimental manipulation in vitro.

The prelabeling paradigm is advantageous because it allows mixing and incorporation of the radiotracer in a more physiological manner. This method is not always practical, however; for example, when the tissue source is a large animal like cow or when postmortem tissues are used. In these situations, in vitro incubation of the tissues with radiotracer is the method of choice. A pulse-chase design can approximate the prelabeling paradigm, in which tracer is added, incubated, and the excess is washed out, and the stimulus is applied.

A third use of radiotracer is in the measurement of endogenous eicosanoids. The tissue sample is obtained, and known quantities of very high specific activity standards are added. The presence of the standards allows detection of the compounds of interest by scintillation counting and also serves as a control for the recovery of compounds through the extraction and isolation procedures. Addition of a combined deuterated/tritiated internal standard allows quantitation of endogenous levels by GC-MS (Green et al., 1973; Nicosia and Galli, 1976).

5.2. Radioimmunoassay

One of the most common methods for quantitation of eicosanoids is RIA, and the availability of commercial RIA kits has made this technique popular. The basic procedure is very straightforward. Known quantities of [^{3}H]ligand and antiserum are added to the sample prepared in phosphate-buffered saline. Samples are incubated for 16 h at 4°C to allow binding to reach equilibrium. Charcoal-dextran solution is added to bind unbound radioligand, and the tubes are centrifuged to pellet the charcoal fraction. Radioactivity in the supernatant, reflecting the antibody–antigen complex, is measured by liquid scintillation counting in scintillation cocktail designed for use with aqueous samples. Quantitation is by interpolation from a standard curve.

Although the procedure for RIA is very simple, there are several important controls that must be used to ensure that the binding one measures truly reflects the amounts of compounds of interest. One such control is to eliminate interference by the sample matrix. This is accomplished by adding known concentrations of the compound to the samples. A parallel with the standard curve should be obtained. It is highly recommended that lipid extracts are purified by HPLC prior to RIA, because cross-reactivity of the antibody can produce erroneous results (Fig. 5). RIA should be done on fractions collected from a blank HPLC injection to eliminate any interference caused by the HPLC solvents or bleed from the column.

6. Conclusions

The difficulties inherent in the quantitative analysis of eicosanoids in brain have hindered progress concerning the definition of the physiological and pathological roles of these compounds. Aside from problems of quantitative extraction and isolation, the rapid postmortem stimulation of eicosanoid synthesis has confounded the results of many studies. The technique of head-focused microwave irradiation has ameliorated the complications of postmortem effects, but also limits the types of experiments that

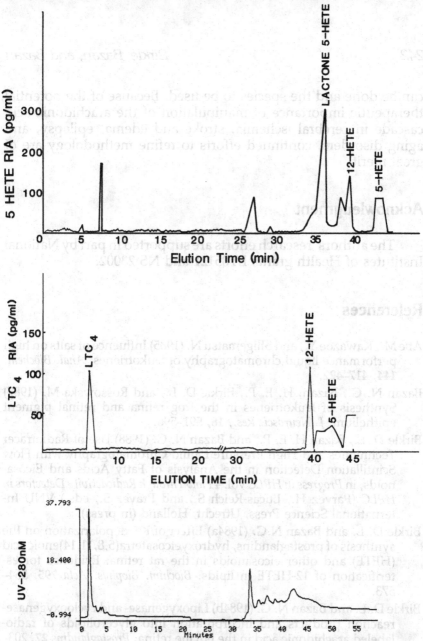

Fig. 5. RIA of fractions collected from reverse phase HPLC analysis of LTC_4 (middle panel) and 5-HETE (upper panel) isolated from frog retina. Ultraviolet absorbance at 280 nm is shown in the lower panel. This analysis demonstrated the phenomenon of cross-reaction of the 5-HETE antibody with 5-HETE lactene, 12-HETE and two unknown compounds, and the cross-reaction of the LTC_4 antibody with 12-HETE and 5-HETE.

can be done and the species to be used. Because of the potential therapeutic importance of manipulation of the arachidonic acid cascade in cerebral ischemia, stroke and edema, epilepsy, and aging disorders, continued efforts to refine methodology are of great merit.

Acknowledgment

The authors' research efforts are supported in part by National Institutes of Health grants EY05121 and NS 23002.

References

Abe M., Kawazoe Y., and Shigematsu N. (1985) Influence of salts on high performance liquid chromatography of leukotrienes. *Anal. Biochem.* **144,** 417–422.

Bazan N. G., Bazan, H. E. P., Birkle D. L., and Rossowska M. (1987) Synthesis of leukotrienes in the frog retina and retinal pigment epithelium. *J. Neurosci. Res.,* **18,** 591–596.

Birkle D. L., Bazan H. E. P., and Bazan N. G. (1988) Use of Radiotracer Techniques and High Pressure Liquid Chromatography with Flow Scintillation Detection in the Analysis of Fatty Acids and Eicosanoids, in *Progress in HPLC* vol. 4 *Flow Through Radioactivity Detectors in HPLC* (Parvez H., Lucas-Reich S., and Pravez S., eds.) VNU International Science Press, Utrecht, Holland (in press).

Birkle D. L. and Bazan N.G. (1984a) Effect of K^+ depolarization on the synthesis of prostaglandins, hydroxyeicosatetra(5,8,11,14)enoic acid (HETE) and other eicosanoids in the rat retina: Evidence for esterification of 12-HETE in lipids. *Biochim. Biophys. Acta* **795,** 564–573.

Birkle D. L. and Bazan N. G. (1984b) Lipoxygenase- and cyclooxygenase-reaction products and incorporation into glycerolipids of radiolabeled arachidonic acid in the bovine retina. *Prostaglandins* **27,** 203–206.

Careaga M. M. and Sprecher H. (1984) Synthesis of two hydroxy fatty acids from 7,10,13,16,19-docosahexaenoic acid by human platelets. *J. Biol. Chem.* **259,** 14413–14417.

Claeys M., Bazan H. E. P., Birkle D. L., and Bazan N. G. (1986) Diacylglycerols interfere in straight phase HPLC analysis of lipoxygenase products of docosahexaenoic or arachidonic acid. *Prostaglandins* **32,** 813–827.

Claeys M., Kivits G. A. A., Christ-Hazelhof E., and Nugteren D. H. (1985) Metabolic profile of linoleic acid in porcine leukocytes through the lipoxygenase pathway. *Biochim. Biophys. Acta* **837,** 35–51.

Eling T., Tainer B., Ally A., and Warnock R. (1986) Separation of arachidonic acid metabolities by high-pressure liquid chromatography. *Meth. Enzymol.* **86,** 511–517.

Green K., Granstrom E., Samuelsson B., and Axen U. (1973) Methods for quantitative analyses of $PGF_{2\alpha}$, PGE_2, $9\alpha,11\alpha$-dihydroxy-15 ketoprost-5-enoic acid, and $9\alpha,11\alpha$, 15-trihydroxyprost-5-enoic acid from body fluids using deuterated carriers and GC-MS. *Anal. Biochem.* **54,** 434–453.

Hamilton J. G. and Karol R. J. (1982) High performance liquid chromatography (HPLC) of arachidonic acid metabolites. *Prog. Lipid Res.* **21,** 155–170.

Hammarstrom S. (1983) Leukotrienes. *Ann. Rev. Biochem.* **52,** 355–377.

Hensby C. (1977) Physical Methods in Prostglandins Research, in *Prostaglandin Research,* (Crabbe, P., ed.) Academic, New York.

Lands W.E.M. (1979) The biosynthesis and metabolism of prostaglandins. *Ann. Rev. Physiol.* **41,** 633–652.

Luderer J. R., Riley D. L., and Demers L. M. (1983) Rapid extraction of arachidonic acid metabolites utilizing octadecyl reversed-phase columns. *J. Chromatogr.* **273,** 402–409.

Metz S. A., Hall M. E., Timothy W. H., and Murphy R. C. (1982) Rapid extraction of leukotrienes from biologic fluids and quantitation by high-performance liquid chromatography. *J. Chromatogr.* **233,** 193–201.

Nicosia S. and Galli G. (1976) Evaluation of prostaglandin biosynthesis in rat cerebral cortex by mass fragmentography. *Adv. Mass Spectrom. Biochem. Med.* **1,** 457–464.

Pediconi M. F., Rodriguez de Turco E. B., and Bazan N. G. (1982) Diffusion of intracerebrally injected $[1\text{-}^{14}C]$ arachidonic acid and $[2\text{-}^{3}H]$glycerol in the mouse brain. Effects of ischemia and electroconvulsive shock. *Neurochem. Res.* **7,** 1435–1456.

Pediconi M. F., Rodriguez de Turco E. B., and Bazan N. G. (1983) Effects of post decapitation ischemia on the metabolism of $[^{14}C]$arachidonic acid and $[^{14}C]$palmitic acid in the mouse brain. *Neurochem. Res.* **8,** 835–845.

Reddy T. S. and Bazan N. G. (1983) Kinetic properties of arachidonoyl coenzyme A synthetase in rat brain microsomes. *Arch. Biochem. Biophys.* **226,** 126–133.

Salmon T. A. and Flower R. J. (1982) Extraction and thin layer chromatography of arachidonic acid metabolites. *Meth. Enzymol.* **86,** 477–493.

Saunders R. and Horrocks L. A. (1984) Simultaneous extraction and preparation for high performance liquid chromatography of prostaglandins and phospholipids. *Anal. Biochem.* **143,** 71–75.

Schulz R. and Seeger W. (1986) Release of leukotrienes into the perfusate of calcium-ionophore stimulated rabbit lungs. *Biochem. Pharmacol.* **35,** 183–193.

Smith B. J., Ross R. M., Ayers C. R., Wills M. R., and Savoy C. R. (1983) Rapid separation of prostaglandins by linear high performance thin layer chromatography. *J. Liquid Chromatogr.* **7,** 1265–1272.

Van Horne K. C. (1985) *Sorbant Extraction Technology* Analytichem, International, California, pg. 82–83.

Varhagen J., Walstra P., Veldink G. A., and Vliegenthart J. F. G. (1984) Separation and quantitation of leukotrienes by reversed-phased high-performance liquid chromatography. *Prostaglandins Leukotrienes Med.* **13,** 15–20.

Wolfe L. S. (1982) Eicosanoids: Prostaglandins, thromboxanes, leukotrienes and other derivatives of carbon-20 unsaturated fatty acids. *J. Neurochem.* **38,** 1–13.

HPLC Analysis of Neutral Glycosphingolipids and Sulfatides

M. David Ullman and Robert H. McCluer

1. Introduction

1.1. General Definition and Biological Importance of Glycosphingolipids and Sulfatides

Glycosphingolipids are important constituents of both neural and non-neural tissues. Because of their complex heterogeneous structures, individual glycosphingolipids have been difficult to name and quantify. All glycosphingolipids are glycosides of N-acylsphingosine (ceramide). The variation in fatty acid, sphingosine, and carbohydrate moieties results in a large number of distinct molecular species. The classification of glycosphingolipids has been based primarily on their carbohydrate structures, rather than on their ceramide residues (Macher and Sweeley, 1978), and several classification schemes are in use. They are classified as neutral glycosphingolipids and acidic glycosphingolipids, which include sulfatides and gangliosides. They are also classified according to the number of carbohydrate residues present (glycosylceramides with 20–50 carbohydrate residues have been detected). The most useful classification depends on the structure of the carbohydrate moieties, i.e., the globo, lacto, muco, ganglio, and gala series. The globo series contains globotriaosylceramide, Gal(α1-4 or 1-3)Gal(β1-4)Glc(β1-1)Cer; the lacto series contains lactotriaosylceramide, GlcNAc(β1-3 or 1-4)Gal(β 1-4)GlcCer; the muco series contains mucotriaosylceramide, Gal(β1-4 or 1-3)Gal(β 1-4)GlcCer; the ganglio series contains gangliotriaosylceramide, GalNAc(β1-4)Gal (β 1-4)GlcCer; and the gala series contains galabiosylceramide, Gal(α1-3 or 1-4)GalCer as the core structures. Details on the structural variations, nomenclature, and distribution of glycosphingolipids have been summarized (Kanfer and Hakomori, 1983).

Galactosylceramides (galactosylcerebrosides) and sulfatides are major components of the myelin sheath, and disturbances in their metabolism lead to severe neurological dysfunction. The leukodystrophies—globoid cell leukodystrophy (Krabbe's disease) and metachromatic leukodystrophy (sulfatidosis)—result from inherited deficiences of lysosomal enzymes responsible for the degradation of galactosylceramide and sulfatide, respectively. Lactosylceramide is also known to accumulate in the brain of individuals with globoid cell leukodystrophy, probably as a characteristic component of the globoid cells. Glucosylceramide accumulates in the extraneural tissues of type 1 Gaucher's disease as a result of the deficiency of glucosylceramide β-glucosidase. In type 2 and 3 Gaucher's disease, neurological abnormalities are present, and glucosylceramide, which in brain is primarily a neuronal component, has been shown to accumulate in neural tissue (Stanbury et al., 1983; Kaye et al., 1986). Neutral fucoglycolipids accumulate in the brain of fucosidosis patients, and, more recently, complex neutral glycosphingolipids have been shown to be components of normal fetal brain (Yamamoto et al., 1985).

This chapter reviews HPLC methodology developed for the quantification of glycosphingolipids and sulfatides. The older chromatographic techniques for the separation and subsequent quantification of glycosphingolipids and sulfatides are tedious and subject to the inconsistencies of thin-layer chromatography. HPLC techniques are sensitive and reproducible. Thus, the use of HPLC in the separation and quantification of glycosphingolipids is proving to be a welcome addition to neuromethods.

1.2. Extraction and Isolation of Lipid Classes

Glycosphingolipids and sulfatides can be extracted along with other lipids and some nonlipid materials by homogenizing wet tissues with 19 vol of chloroform-methanol (2:1) (Folch et al., 1957). Gangliosides and polyglycosylceramides (more than five carbohydrate residues), along with nonlipid contaminants in the extract, can be removed from the bulk of the lipids by the partition formed from chloroform:methanol:water (8:4:3) (Folch et al., 1957). The presence of salts (0.73% NaCl or 0.88% KCl) prevents the partition of acidic phospholipids and sulfatides into the upper phase (Suzuki, 1964). Repeated extraction of the lower phase provides a recovery of about 90% of GM_3 in the upper phase, whereas about 90% of GM_4 remains in the lower phase. The gangliosides and polyglycosylceramides in the upper phase can be separated from the nonlipid contaminants by dialysis (Folch et al., 1957), gel filtration (Christie, 1983), or use of a reverse-phase chromotography (Williams and McCluer, 1980). The gangliosides and neutral polyglycosylceramides can be separated by DEAE-Sephadex chromatography (Ledeen and Yu, 1982).

The lipids in the lower chloroform phase are fractioned by a variety of procedures (Christie, 1983), but one widely used method employs chromatography on silicic acid columns (Vance and Sweeley, 1967). Lower phase lipids are dissolved in chloroform and applied to a silicic acid column, nonpolar neutral lipids are eluted with chloroform, the glycosphingolipids are eluted with acetone:methanol (9:1), and phospholipids are eluted with methanol. Sulfatides in the acetone:methanol fraction are separated from the neutral glycosphingolipids by DEAE-sephadex chromatography (Kanfer and Hakamori, 1983).

The partition of lipids between the solvent phases and lipid recoveries from the chromatographic steps depends to some extent on the lipid composition of the tissue source and on the concentrations of ions in the extract. Thus, the procedures for extraction and fractionation of lipids need to be optimized for both the tissue and the quantity of material to be processed.

1.3. Conventional Analytical Techniques

Neutral glycosphingolipids and sulfatides can be resolved into individual components by TLC. Classical methods for the separation and analysis of the glycosphingolipids and sulfatides have depended primarily on class separation. The degree of heterogeneity in terms of molecular species has generally been established by analysis of the products of hydrolysis such as fatty acids, long chain bases, and sugar residues. A standard method for the analysis of plasma and urine neutral glycosphingolipids involves separation by TLC, elution from the TLC powder, methanolysis, and subsequent analysis of the methylglycosides by GLC of their silanated derivatives (Vance and Sweeley, 1967).

1.4. Background of HPLC Technology and Methods

HPLC technology, which implies the use of injection ports for sample application, reusable columns packed with microparticulate (down to 3 μm) chromatographic adsorbants, pumps for uniform solvent flow, and on-line sample detection, provides the potential for rapid and highly sensitive methods for the separation and analysis of lipids. Numerous books (Snyder and Kirkland, 1979; Johnson and Stevenson, 1978; Walber et al., 1977; Yost et al.,

1980), reviews, and courses are now available that present the theoretical and practical aspects of HPLC and a historical perspective of the development of HPLC technology. HPLC techniques, used with combined chromatographic modes, have the potential to separate and quantitate the hundreds of individual molecular species.

Usually, glycosphingolipid and sulfatide fractions are obtained by solvent partition and chromatographic procedures as outlined above. Individual classes of glycosphingolipids or sulfatides are then separated by normal phase chromatography according to their polarity. Separation of glycosphingolipid molecular species can subsequently be obtained by reverse-phase HPLC, which separates them according to their nonpolar moieties. Argentation chromatography can further resolve components by their degree of unsaturation (Smith et al., 1981).

1.5. General Comments on HPLC Analysis of Glycosphingolipids and Sulfatides

Although the technology exists to perform the separation of native glycosphingolipids by HPLC, the lack of a characteristic chromophore or easily detectable functional group on the compounds makes their detection and subsequent quantification insensitive. Therefore, it is useful to utilize derivatives that allow sensitive detection of the compounds of interest. The perbenzoylated glycosphingolipids have been used because their high extinction coefficients permit detection of picomolar amounts with flow-through UV detectors.

The initial studies with perbenzoyl derivatives utilized 10% benzoyl chloride in pyridine at 60°C for 1 h for the derivatization of cerebrosides (McCluer and Evans, 1973). Interestingly, the derivatives could be separated into two components by HPLC on pellicular silica columns. The two components were shown to be the derivatives of hydroxy fatty acid (HFA) and non-hydroxy fatty acid (NFA) galactosylceramides. Attempts to recover the parent galactosylceramides by treatment of the perbenzoylated derivatives with mild alkali were successful with the HFA-galactosylceramides, but the NFA-galactosylceramides gave rise to both the native NFA-galactosylceramides and benzoyl psychosine (McCluer and Evans, 1973). Subsequently, it was determined that the diacylamines, as N,O-benzoyl derivatives of NFA-galactosylceramides, were formed during the perbenzoylation

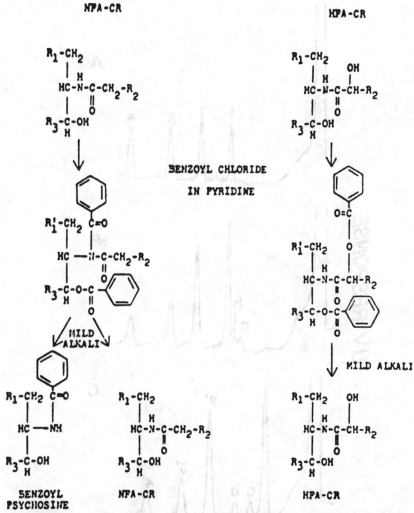

Fig. 1. Formation of perbenzoylated NFA- and HFA-cerebroside derivatives by reaction with benzoyl chloride in pyridine and their degradation with mild alkali.

(McCluer and Evans, 1976). Further, the diacylamine derivatives randomly lost N-acyl groups during treatment with mild alkali (Fig. 1). Thus, the native NFA-galactosylceramides and other sphingolipids that contain NFA or N-acetyl amino sugars were not recovered in high yields.

An alternative method for the exclusive production of per-O-benzoylated glycosphingolipids utilized 10% benzoic anhydride in

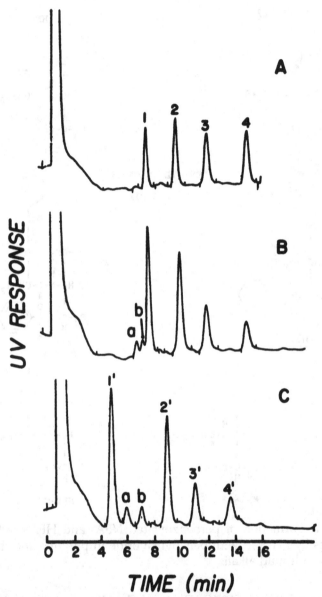

Fig. 2. HPLC of benzoylated standard and plasma glycosphingolipids. The derivatized glycosphingolipids were injected into a Zipax column (2.1 mm × 50 cm) and eluted with a 13-min linear gradient of 2.5–25% dioxane in hexane with detection at 230 nm. (A) Standard glycosphingolipids (GSL) per-*O*-benzoylated with benzoic anhydride and 4-dimethylaminopyridine (DMAP). (B) Plasma GSL per-*O*-benzoylated

5% *N*-dimethylaminopyridine (DMAP) in pyridine at 37°C for 4 h (Gross and McCluer, 1980). With only the O-benzoylated derivatives, native glycolipids were recovered after alkaline hydrolysis. Because sphingolipids that contained only HFA as N-acyl substituents formed the same derivative with both benzoyl chloride and benzoic anhydride, they were easily distinguished from NFA-containing sphingolipids, which formed different derivatives that were distinguishable by HPLC (Fig. 2). The benzoyl chloride reaction is generally used for analytical purposes because resolution of components containing HFA and/or phytosphingosine is superior to that obtained with the O-benzoates formed with benzoic anhydride.

2. Methods

2.1. Quantification of Neutral Glycolipid Mixtures

2.1.1. Derivatization with Benzoyl Chloride

Neutral glycosphingolipids (at least 200 ng of each) and a known amount of internal standard (Ullman and McCluer, 1977), such as *N*-acetylpsychosine, are dried under a stream of nitrogen in a 13 × 100 mm screw-cap culture tube with a Teflon-lined cap. They are per-*O,N*-benzoylated with 0.5 mL of 10% benzoyl chloride in pyridine (v/v), which is added directly to the bottom of the tube. The sample reaction mixture is warmed at 37°C for 16 h. The pyridine is removed with a stream of nitrogen at room temperature until the residue appears as an oil-covered solid. Then, 3 mL of hexane is added to the reaction tube. The hexane is washed three times with 1.8 mL of methanol: water [80:20 (v/v)], which is saturated with sodium carbonate (1.2 g of sodium carbonate in 300 mL of methonal: water 80:20). During the removal of the lower aqueous methanol layer, a slight positive pressure is exerted on the pipet bulb as the pipet passes through the upper (hexane) layer.

← ——————————————————————————

with benzoic anhydride and DMAP. (C) Plasma GSL perbenzoylated with benzoyl chloride. Glycosphingolipid peaks are identified as: (1) glucosylceramide, (2) lactosylceramide, (3) galactosyllactosylceramide, and (4) *N*-acetylgalactosaminylgalactosyl-lactosylceramide. Peak A is unidentified, and peak B is hydroxy fatty acid containing galactosylceramide (reproduced from Gross and McCluer, 1980).

The hexane layer is then washed with 1.8 mL of methanol:water [80:20 (v/v)]. The lower phase is withdrawn and discarded, and the hexane is evaporated with a stream of nitrogen. The derivatives are dissolved in 100–500 μL of carbon tetrachloride, and an aliquot is injected into the HPLC column. The derivatives are quantified by use of the external standards method. Purified neutral glycosphingolipids to be used as standards can be isolated from erythrocytes or other tissues or obtained from commercial sources (Macher and Sweeley, 1978; Ullman and McCluer, 1977, 1978). It is critical that the standard preparations are chromatographically pure and free of closely related isomers and inorganic material such as silicic acid. The hexose or long chain base content of standard preparations can be used to establish and verify the peak area response factor of each glycolipid.

The derivatives can also be isolated from the reagents and reaction byproducts with a C-18 reversed-phase cartridge (C-18 Bond Elut, Analytichem International, Harbor City, CA) by a procedure that has been developed for per-O,N-benzoylated gangliosides (Ullman and McCluer, 1985). The dried reaction mixture is dissolved in 0.8 mL of methanol and transferred to a C18 reverse-phase cartridge that is prewashed with 2 mL of methanol. The reaction vial is rinsed with an additional 0.8 mL of methanol, and the rinse solvent is transferred to the cartridge. The collected and combined eluates are passed through the cartridge again. The cartridge is eluted with 4 mL of methanol to remove reaction byproducts. Perbenzoylated neutral glycosphingolipids are eluted from the cartridge with 3 mL of methanol:benzene (8:2). The derivatives are collected in a 13 × 100 mm screw-cap culture tube, and the solvent is removed at room temperature with a stream of nitrogen. They are then dissolved in 100–500 μL of carbon tetrachloride, and an appropriate aliquot is injected into the HPLC column. N-Acetylpsychosine cannot be used as an internal standard with this method of isolation because the partition characteristics of its per-O,N-benzoylated derivative on the reverse-phase cartridge are substantially different from those for the per-O,N-benzoylated derivatives of neutral glycosphingolipids. The per-O,N-benzoylated products are stable for months providing they are completely free of alkali and stored at −70°C.

It is important that neutral glycosphingolipid samples to be perbenzoylated are relatively free of silica or silicic acid particles.

These particles arise from either dissolved silica after extraction of the glycosphingolipids from TLC or from "fines" that avoid filtration. Moisture (e.g., atmospheric water) reacts instantaneously with benzoyl chloride to form benzoic acid and hydrochloric acid. The reagents and reaction mixture must be protected from the atmosphere and other sources of moisture. Thus, it is important to store benzoyl chloride in dry surroundings (a 500-mL bottle can usually be used for 3–6 mo), and pyridine must be dried by storage over 4 Å molecular sieves. Some batches of pyridine may have a very high moisture content, and the molecular sieves may not trap all of the water. This situation is evident upon mixing the pyridine with the benzoyl chloride because a crystalline precipitate forms almost immediately. There are other indications that moisture has been introduced into the reaction medium. For example, the "solvent front" of the chromatogram may be broader than usual or there may be a broad peak that elutes just after the "solvent front." These anomalies are caused by benzoic acid. The sample(s) must be dried *in vacuo* over phosphorus pentoxide at least 1 h before the perbenzoylation is performed. The benzoyl chloride reaction also benzoylates sulfatides if they are present in the glycolipid fraction. During the solvent partitioning employed to remove reagents, however, the benzoylated sulfatides do not distribute into the hexane phase and thus do not interfere with the analysis of glycosphingolipids.

The derivatives are separated on a pellicular (Zipax, E. I. Dupont de Nemours, Inc., Wilmington, DE) column (2.1 mm id × 500 mm L) (dry packed by the "tap-fill" method (Snyder and Kirkland, 1979; McCluer et al., in press) with a 10-min linear gradient of 2–17% water-saturated ethyl acetate in hexane and a flow rate of 2 mL/min and detected by their UV absorption at 280 nm (Ullman and McCluer, 1977). The minimum level of detection (twice baseline noise) by this procedure is approximately 70 pmol of each neutral glycosphingolipid.

Cerebrosides can be eluted isocratically with 7% ethyl acetate in hexane with detection at 280 nm (McCluer and Evans, 1973). The procedure was utilized to measure cerebrosides by direct benzoylation of chloroform-methanol extracts of adult brain and for the analysis of more purified cerebroside samples. Total lipid extracts of adult brain can be analyzed directly for cerebrosides without interference from other lipids. However, tissue or tissue fractions that contained low concentrations of cerebrosides required the use of larger aliquots of total lipids and, consequently, inadequate chromatographic separations were obtained.

The sensitivity of the procedure can be increased by detection at 230 nm, the absorption maximum of the derivatives that requires the use of a mobile phase that is transparent at this wavelength. A 13-min linear gradient of 1–20% dioxane in hexane and a flow rate of 2 mL/min is used (Fig. 2) (Ullman and McCluer, 1978). Per-O,N-benzoylated ceramides (derivatized with benzoyl chloride) and neutral glycosphingolipids can be separated in a single chromatographic run with a 15-min linear gradient of 0.23–20% dioxane in hexane (Chou and Jungalwala, 1981). A consistent separation of per-O,N-benzoylated glucocerebroside from per-O,N-benzoylated galactocerebroside, at all ratios, is obtained by HPLC of the per-O,N-benzoylated derivatives on the pellicular silica column maintained at 60°C and a 10-min linear gradient of 1–20% dioxane in hexane at a flow rate of 2 mL/min (Fig. 3) (Kaye and Ullman, 1984). This system is particularly useful for the analysis of brain cerebrosides since the ratio of galactocerebroside to glucocerebroside is usually very high.

For these dioxane gradients, the mobile phase is directed through a preinjector flow-through reference cell to cancel the residual absorption of the dioxane. Several UV detectors have flow-through reference cells that are pressure-rated high enough to be utilized in this system. The maximum pressure rating required is about 1000 psi. Poor mixing characteristics of the solvents that compose the mobile phase create baseline instabilities. A dual-chamber dynamic (mechanical) mixer or two single-chamber mixers in series are required for baselines sufficiently stable to allow detection at high sensitivity such as 0.04 absorbance units full scale (AUFS). The selectivity and efficiency of this sytem is such that the derivatives yield only one peak for each neutral glycosphingolipid rather than further partial separation of each derivative on the basis of its fatty acid composition. This provides consistent automatic integration of the peaks.

Inadequate insulation around the tubing that leads from the column outlet to the detector inlet will lead to unstable baselines. This piece of tubing should be well insulated. A piece of rubber tubing slit from end to end and slipped over the tubing provides excellent insulation. Some manufacturers have incorporated heat exchangers into their detectors. An easy method to detect ambient temperature problems is to simply place your finger on the tubing in question for a few seconds. If there is a baseline disturbance, there may be an ambient temperature problem.

Finally, for analytical procedures, it is important to know the

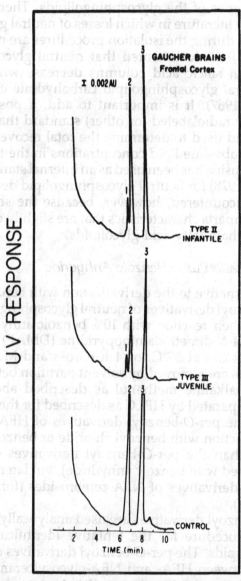

Fig. 3. Separation of perbenzoylated glucocerebroside and galactocerebroside by chromatography on a pellicular (27 μm mean particle diameter) silica column (2.1 mm id × 500 mm L) maintained at 60°C. The derivatives were eluted with a 10-min linear gradient of 1–20% dioxane in hexane at a flow rate of 2 mL/min. Detection of the derivatives was by their UV absorption at 230 nm.

absolute recoveries of the glycosphingolipids. There are several examples in the literature in which losses of neutral glycosphingolipid that occur during the isolation procedures are not taken into consideration. It is documented that neutral glycosphingolipid recoveries from silicic acid columns decrease with increasing length of neutral glycosphingolipid carbohydrate chains (Vance and Sweeley, 1967). It is important to add, if possible, a high specific activity radiolabeled (or other) standard that can be per-benzoylated and used to determine the total recovery, and, consequently, the absolute lipid concentrations in the tissue source. N-Acetylpsychosine has been used as an internal standard (Ullman and McCluer, 1978) for neutral glycosphingolipid determinations. Difficulty is encountered, however, because the short chain N-acetyl group imparts characteristics that are slightly different than those of long chain fatty acid glycolipids.

2.1.2. Derivatization with Benzoic Anhydride

As an alternative to the derivatization with benzoyl chloride, the per-O-benzoyl derivatives of neutral glycosphingolipids can be produced via their reaction with 10% benzoic anhydride in pyridine with 5% 4-N-dimethylaminopyridine (DMAP) as a catalyst. The reaction is run at 37°C for 4 h (Gross and McCluer, 1980). Excess reagents are removed by solvent partition between hexane and aqueous alkaline methanol as described above. The derivatives are separated by HPLC as described for the O,N-benzoyl derivatives. The per-O-benzoyl derivatives of HFA cerebrosides (formed by reaction with benzoyl chloride or benzoic anhydride) elute earlier than the per-O-benzoyl derivatives of NFA cerebrosides (formed with benzoic anhydride), but later than the per-O,N,-benzoyl derivatives of NFA cerebrosides (formed with benzoyl chloride).

The O-benzoyl derivatives are used analytically, especially as an adjunct procedure for the tentative identification of HFA-glycosphingolipids. The per-O-benzoyl derivatives can be used to distinguish between HFA- and NFA-glycosylceramides by comparing their elution times to the per-O,N-benzoylated derivatives. The O-benzoyl derivatives have also been used for preparative purposes since native neutral glycosphingolipids are recovered by mild alkaline methanolysis of the isolated derivatives (McCluer and Ullman, 1980).

2.2. Sulfatides

The analysis of tissue and urinary sulfatides involves the isolation of a glycosphingolipid fraction by silicic acid chromatography and subsequent separation of sulfatides from neutral glycosphingolipids by chromatography on a DEAE-Sephadex column (McCluer and Evans, 1976; Raghavan et al., 1984). The sulfatide fraction is eluted with $0.02M$ ammonium acetate in chloroform: methanol: water (67:17:16), then desulfated in $0.05N$ anhydrous methanolic: HCl at 40°C for 80 min. The resulting glycosphingolipids are recovered by partition in chloroform: methanol: water (Folch et al., 1957) and analyzed by HPLC as their per-O,N-benzoyl derivatives. This procedure has been utilized for the analysis of mono-, di-, and triglycosyl sulfatides in the urine of patients with metachromatic leukodystrophy.

A method for the analysis of perbenzoylated sulfatides has also been developed (Jungalwala et al., 1979). Glycolipid fractions containing sulfatide (1–100 nmol) are perbenzoylated with 100 μL of 0.5% (w/v) benzoic anhydride and 50% saturated 4-dimethyl aminopyridine in tetrahydrofuran for 6 h at 37°C. The sulfatide products are purified on a silicic acid column (0.4 × 30 mm) by elution with chloroform: methanol (19:1). Alternatively, they can be purified on a Bond Elute reverse-phase cartridge. The per-benzoylated sulfatides are analyzed by HPLC on a Micropak-Si-10 column with hexane-2-propanol-propionic acid (100:35:2) at a flow rate of 1 mL/min. HFA-sulfatide is eluted before NFA-sulfatide in the system. Reverse-phase HPLC of HFA- and NFA-sulfatide was performed on a "Fatty Acid Analysis" column (Waters Associates, Milford, MA) (McCluer et al., 1986). The individual molecular species are eluted isocratically with tetrahydro-furan: acetonitrile: water (15:25:40).

The simultaneous analysis of pmole quantities of benzoylated cerebrosides, sulfatides and galactosyl diglycerides has been described (Nonaka and Kishimoto, 1979). This procedure involves benzoylation of total lipid extracts from brain and desulfation with mild acid. In this manner, the cerebrosides derived from sulfatides have one free hydroxyl group. Subsequent chromatography is on a 5-μm porous silica gel column, and a gradient of 2-isopropanol in hexane results in the separation of the totally benzoylated cere-brosides and the cerebrosides from sulfatides that have one free hydroxyl group.

3. Applications

3.1. Overview

Quantification of neutral glycosphingolipids and sulfatides by these HPLC procedures can be used to correlate tissue glycolipid levels with neuropathological changes in the lysosomal storage diseases with neurological involvement (e.g., type 2 or 3 Gaucher's disease). These procedures are also useful as an adjunct in the diagnosis of the glycosphingolipid storage disorders.

3.2. Correlation of Neurochemistry to Neuropathology

The neuropathology in Gaucher's disease types 2 and 3 has been described, but the pathogenesis of the disorders is not understood. Solitary Gaucher cells have been found intermingled with neurons (Banker et al., 1962). Neuronal cell loss and neurophagia in the cerebral cortex, basal ganglia, thalamus, hypothalamus, dentate nucleus, cerebellar Purkinje cells, and brain stem (Bankei et al., 1962; Debre et al., 1951; Norman et al., 1956; Winkleman et al., 1983) have been described. Even though there is an apparent agreement in the literature on the distribution of Gaucher cells, there is not universal agreement on the accumulation of glucocerebroside (the substrate for the deficient enzyme) or its metabolic product, psychosine, or on their respective contributions to the observed neuropathology. Both elevated (Gonzalez-Sastre et al., 1974; Kubota, 1972; Maloney and Cumings, 1960; Nilson and Svennerholm, 1982; Sudo, 1977) and normal (French et al., 1969; Philippart and Menkes, 1976; Rouser et al., 1967) levels of brain glucocerebroside in types 2 and 3 Gaucher's disease have been reported. These studies were all encumbered, however, by the lack of sufficient sensitivity to analyze glucocerebroside concentrations from small, clearly defined brain regions such as the deeper nuclear structures in the basal ganglia, thalamus, brain stem, and cerebellum.

The HPLC procedure described in this chapter provided the necessary sensitivity, resolution, and reproducibility to permit the determination of glucocerebroside levels in specific brain regions and to correlate the findings with the neuropathology. The micro-analytical HPLC procedure for glucocerebroside (Kaye and Ullman, 1984) permits just such analyses on very small (a few mg wet weight) quantities of specific brain regions.

Nine separate brain regions from types 2 and 3 Gaucher's disease were analyzed by HPLC of their perbenzoylated glucocerebrosides (Kaye et al., 1986) (frontal, temporal, occipital, cerebellar corticies, thalamus, corpus striatum, pons, medulla, and dentate nuclei). In both types, the greatest glucocerebroside accumulation, when compared to controls, was found in the occipital cortex, with lesser amounts in the temporal and frontal areas. The cerebellar cortex, corpus striatum, and thalamus of the Gaucher brains had mildly increased levels of glucocerebroside, especially when the values were expressed as a percentage of the total non-hydroxy fatty acid cerebrosides.

The type 2 brainstem structures also showed a slight elevation of glucocerebroside over controls when the data were again expressed as the percentage of the total non-hydroxy fatty acid cerebrosides. The histopathological abnormalities in the type 2 Gaucher brain correlated with elevated tissue glucocerebroside levels. Type 3 Gaucher brain showed no pathological changes, but the glucocerebroside accumulation was similar to that seen in type 2 brain. Indeed, the glucocerebroside accumulation in Gaucher's disease may be a more sensitive measure of the nervous system involvement than the neuropathologic changes.

Thus, the practical application of these HPLC techniques to the study of the neuropathology of the disorders is apparent. The HPLC procedure permitted the accumulation of these data because glucocerebroside could be quantified in the presence of large amounts of galactocerebroside. Although these studies do not confirm that the source of the brain pathology is the accumulated glucocerebroside, they do confirm the correlation of the accumulation to the brain pathology and suggest further avenues of investigation, some of which will utilize this analytical HPLC technique.

3.3. HPLC Analysis as an Adjunct to Diagnosis

HPLC can also be used as an adjunct for the clinical diagnosis of disorders with neurological involvement, even though, for the most part, tissues of non-neural origin must be used. For example, plasma is used to detect the elevated glucocerebroside levels in Gaucher's disease. The values, however, do not distinguish among the three types of the disorder, nor do elevated plasma glucocerebroside levels specifically indicate Gaucher's disease because other disorders (e.g., Niemann-Pick's disease) may have elevated plasma glucocerebroside levels (Dacremont et al., 1974).

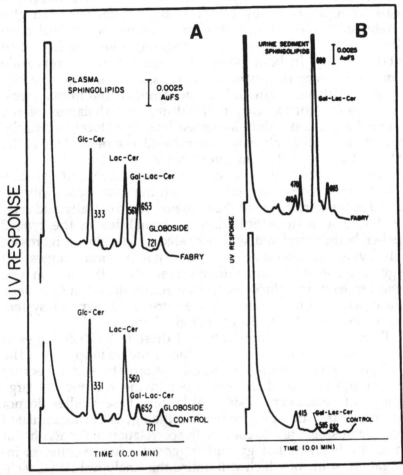

Fig. 4. (A) Perbenzoylated neutral glycolipids from a Fabry patient
and a control that demonstrate the elevation of plasma gal-lac-cer in the
patient. (B) Urinary sediment neutral glycolipids from a Fabry patient and
a control that demonstrate the elevation of urinary sediment gal-lac-cer in
the patient. Numbers adjacent to the peaks are retention times in min ×
100.

The accumulation of globotriaosylceramide in plasma and
urine sediment from Fabry's disease patients can conveniently be
determined by the HPLC methods described above (Fig. 4). The
analysis of urine sediment glycolipids by HPLC can also be used to
detect Fabry's disease heterozygotes (Cable et al., 1982). Although
Fabry hemizygotes can be reliably detected by enzyme analysis

and substantiated by biochemical determinations of plasma and/or urine sediment, Fabry heterozygotes are not reliably established by enzyme analysis. An inverse relationship between leukocyte alpha-galactosidase activity and high renal glycolipid excretion exists in most, but not all, women tested. The discrepancies are best explained by the Lyon hypothesis (Lyon, 1961), which suggests that in each female cell, only one X chromosome is expressed because the other X chromosome is irreversibly inactivated in early embryogenesis, and the pattern persists in all progeny of that embryonic cell. A cellular mosaicism arises, with cells expressing one or the other X chromosome. In X-linked disorders (e.g., Fabry's disease), cells express either normal or absence of enzyme activity. In the past, some (Philippart et al., 1969; Desnick et al., 1971), but not all (Avila et al., 1973; Philippart et al., 1974), heterozygotes were detected by TLC analysis of urine sediment followed by colorimetry or gas-liquid chromatography. In one study (Cable et al., 1982), heterozygotes were reliably detected by the use of HPLC of the perbenzoylated derivatives of urine sediment neutral glycolipids. The procedure is sufficiently sensitive to allow the analysis of the glycolipids from a single urine collection. Because the amount of ceramide trihexoside in the urine varied, a component neutral glycolipid was used as a reference substance for comparison of the relative values of ceramide trihexoside. This reference compound was the major component of normal urine and was determined to be hydroxy fatty acid glucocerebroside. Whether this procedure is more reliable than enzymatic assays or if it is affected by the use of various drugs (e.g., diuretics) remains to be determined.

Certainly, as our understanding of the metabolic dynamics of glycosphingolipids increases, the utilization of analytical HPLC for the delineation of etiologies and clarification of diagnoses will increase.

References

Avila J., Convit J., and Velasquez-Avila G. (1973) Fabry's disease: Normal alpha-galactosidase activity and urinary sediment glycosphingolipid levels in obligate heterozygotes. *Br. J. Dermatol.* **89,** 149–157.

Banker B. Q., Miller J. Q., and Crocker C. A. (1962) The Cerebral Pathology of Infantile Gaucher's Disease, in *Cerebral Sphingolipidosis* (Aronson S. M. and Volk B. W., eds.) Academic, New York.

Cable W. J. L., McCluer R. H., Kolodny E. H., and Ullman M. D. (1982) Fabry disease: Detection of heterozygotes by examination of their urinary sediment glycolipids. *Neurology* **32,** 1139–1145.

Chou K-H. and Jungalwala F. B. (1981) Neutral glycosphingolipids and ceramide composition of ethylnitrosourea induced rat neural tumors. *J. Neurochem.* **36,** 394–401.

Christie W. W. (1983) *Lipid Analysis* Pergamon, New York.

Dacremont G., Kint J. A., Canton D., and Cocquyt G. (1974) Glucosylceramide in plasma of patients with Niemann-Pick disease. *Clin. Chim. Acta* **52,** 365.

Debre R., Bertrand I., Grumback R., and Bargeton E. (1951) Maladie de Gaucher au norrisson. *Arch. Franc Pediat.* **8,** 38–42.

Desnick R. J., Dawson G., Desnick S. J., Sweeley C. C., Krivit W., and Engl N. (1971) Diagnosis of sphingolipidoses by urinary sediment analysis. *J. Med.* **284,** 739–744.

Folch J., Lees M., and Sloan Stanley G. H. (1957) A simple method for the isolation and purification of total lipids from animal tissues. *J. Biol. Chem.* **226,** 497–509.

French J. H., Brotz M., and Poster C. M. (1969) Lipid composition in the brain in infantile Gaucher's disease. *Neurology* **19,** 81–86.

Gonzalez-Sastre F., Pampols T., and Sabater J. (1974) Infantile Gaucher's disease: A biochemical study. *Neurology* **24,** 162–167.

Gross S. K. and McCluer R. H. (1980) High-performance liquid chromatographic analysis of neutral glycosphingolipids as their per-O-benzoyl derivatives. *Anal. Biochem.* **102,** 429–433.

Johnson E. L. and Stevenson R. (1978) *Basic Liquid Chromatography* Varian Associates, California.

Kanfer J. N. and Hakomori S. (1983) *Sphingolipid Biochemistry* Plenum, New York.

Kaye E. M. and Ullman M. D. (1984). Separation and quantitation of perbenzoylated glucocerebroside and galactocerebroside by high-performance liquid chromatography. *Anal. Biochem.* **238,** 380–385.

Kaye E. M., Ullman M. D., Wilson E., and Barranger J. A. (1986) Type 2 and type 1 Gaucher's disease: A correlative study of neuropathologic and biochemical changes. *Ann. Neurol.* **20,** 223–230.

Kubota M. (1972) Studies of neural and visceral glycolipids in a case of infantile form of Gaucher's disease. *Japn. J. Exp. Med.* **42,** 513–526.

Ledeen R. W. and R. K. Yu (1982) Gangliosides: Structure, isolation and analysis. *Meth. Enzymol.* **83,** 139–191.

Lyon M. F. (1961) Gene action in the X chromosome of the mouse (*Mus musculus*). *Nature* **190,** 372–373.

Macher B. A. and Sweeley C. C. (1978) Glycosphingolipids: structure, biological source, and properties. *Methods in Enzymology, Complex Carbohydrates* part C, **50**, 236–251.

Maloney A. F. J. and Cumings J. N. (1960) A case of juvenile Gaucher's disease with intraneuronal lipid storage. *J. Neurol. Neurosurg. Psychiat.* **23**, 207–213.

McCluer R. H. and Evans J. E. (1976) Quantitative analysis of brain galactosylceramides by high performance liquid chromatography of their perbenzoyl derivatives. *J. Lipid Res.* **17**, 412–418.

McCluer R. H. and Evans J. E. (1973) Preparation and analysis of benzoylated cerebrosides. *J. Lipid Res.* **14**, 611–617.

McCluer R. H. and Ullman M. D. (1980) Preparative and analytical high performance liquid chromatography of glycolipids, in *Cell Surface Glycolipids* (Sweeley C. C., ed.) American Chemical Society, Washington, DC.

McCluer R. H., Ullman M. D., and Jungalwala (1986) HPLC of glycosphingolipids and phospholipids. *Adv. Chromatogr.* **25**, 309–353.

McCluer R. H., M. D. Ullman, and F. B. Jungalwala (in press) Analysis of Membrane Lipids with HPLC techniques: Glycosphingolipids and Phospholipids, in *Methods in Enzymology, Biomembranes* part M *Biological Transport*, in press.

Nilson O. and Svennerholm L. (1982) Accumulation of glucosylceramide and glucosylsphingosine (psychosine) in cerebrum and cerebellum in infantile and juvenile Gaucher disease. *J. Neurochem.* **39**, 709–726.

Nonaka G. and Kishimoto Y. (1979) Simultaneous determination of picomole levels of gluco- and galactocerebroside, monogalactosyl diglyceride and sulfatides by high performance liquid chromatography. *Biochim. Biophys. Acta* **572**, 423–431.

Norman R. M., Urich H., and Lloyd O. C. (1956) The neuropathology of infantile Gaucher's disease. *J. Path. Bact.* **72**, 121–131.

Philippart M. and Menkes J. (1976) Isolation and Characterization of the Principal Cerebral Glycolipid in the Infantile and Adult Forms of Gaucher's Disease, in *Inborn Disorders of Sphingolipid Metabolism* Proceedings of the Third International Symposium on the Cerebral Sphingolipidoses (Aronson S. M. and Volk B. W., eds) Pergamon, New York.

Philippart M., Sarlieve L., and Manacorda A. (1969) A urinary glycolipid in Fabry's disease: Their examination in the detection of atypical variants and the presymptomatic state. *Pediatrics* **43**, 201–206.

Philippart M., Kamensky E., Cancilla P., Sparkes R. S., and Cotton M. (1974) Heterozygote detection in Fabry's disease. *Pediatr. Res.* **8**, 393.

Raghavan S. S., Finch E. A., and Kolodny E. H. (1984) HPLC analysis of urinary glycolipids in sphingolipidoses. *Trans. Am. Soc. Neurochem.* **15**, 170.

Rouser G., Galli C., and Kritchevsky G. (1967) Lipid composition of the normal human brain and its variations during various diseases. *Pathol. Biol.* **15**, 195–200.

Smith M., Monchamp P., and Jungalwala F. B. (1981) Separation of molecular species of complex lipids by argentation and reversed-phase HPLC. *J. Lipid Res.* **22**, 697–704.

Snyder L. R. and Kirkland J. J. (1979) *Introduction to Modern Liquid Chromatography* 2nd edn., Wiley Interscience, New York.

Stanbury J. B., Wyngaarden J. B., Fredrickson D. S., Goldstein J. L., and Brown M. S., eds. (1983) *The Metabolic Basis of Inherited Disease* McGraw-Hill, New York.

Sudo M. (1977) Brain glycolipids in infantile Gaucher's disease. *J. Neurochem.* **29**, 379–381.

Suzuki K. (1964) A simple and accurate micromethod for quantitative determination of ganglioside patterns. *Life Sci.* **3**, 1227.

Ullman M. D. and McCluer R. H. (1978) Quantitative microanalysis of perbenzoylated neutral glycosphingolipids by HPLC with detection at 230 nm. *J. Lipid Res.* **19**, 910–913.

Ullman M. D. and McCluer R. H. (1977) Quantitative analysis of plasma neutral glycosphingolipids by high performance liquid chromatography of their perbenzoyl derivatives. *J. Lipid Res.* **18**, 371–377.

Ullman M. D. and McCluer R. H. (1985). Quantitative analysis of brain gangliosides by HPLC of their perbenzoyl derivatives. *J. Lipid Res.* **26**, 501–506.

Vance R. E. and Sweeley C. C. (1967) Quantitative determination of the neutral glycosylceramides in human blood. *J. Lipid Res.* **8**, 631–630.

Walker J. Q., Jackson M. T., Jr., and Maynard J. B. (1977) *Chromatographic Systems: Maintenance and Troubleshooting* Academic, New York.

Williams M. A. and McCluer R. H. (1980) The use of Sep-Pak (TM) C18 cartridges for the isolation of gangliosides. *J. Neurochem.* **35**, 266–269.

Winkelman M. D., Banker B. Q., Victor V., and Moser H. W. (1983) Non-infantile neuronopathic Gaucher's disease: A clinicopathologic study. *Neurology* **33**, 994–1008.

Yamamoto M., Boyer A. M., and Schwarting G. (1985) Fucose containing glycolipids are stage and region-specific antigens in developing embryonic brain of rodents. *Proc. Natl. Acad. Sci. USA* **82**, 3045–3049.

Yost R. W., Ettre L. S., and Conlon R. D. (1980) *Practical Liquid Chromatography: An Introduction* Perkin-Elmer Connecticut.

Methods to Study the Biochemistry of Gangliosides

Allan J. Yates

1. Introduction

The gangliosides consist of a class of glycosphingolipids that by definition contain a sphingosine to which is attached both a fatty acid linked through an amide bond and a sialic acid containing oligosaccharide. The term ganglioside was originally used by Klenk (1942) to indicate that these compounds were specifically located in neurons and that they were glycosidic in nature. Subsequently it was found that ganglosides are widely distributed in both extraneural and neural tissues of echinoderms and chordates, but the highest concentrations are in mammalian cerebral gray matter. The structures, nomenclature, and tissue distributions of gangliosides have been well summarized by Ledeen and Yu (1982).

Investigations into the neurobiology of gangliosides have proceeded along several different lines. Initially, this consisted almost exclusively of chemical analyses to determine the structures and quantities of different gangliosides in neurological tissues, cells, and subcellular fractions. In the past 5–10 years, however, there has been a continually increasing amount of research into the biology of gangliosides (Yates, 1986). From this a considerable amount has been learned about the effects of exogenously administered gangliosides on neurological functions and the effects of altered neurological states on ganglioside composition and metabolism, but the physiological roles of gangliosides are still unknown.

Concurrent with the expanding range of experimental designs has been the development of many new techniques to study gangliosides. Investigators planning experiments involving gangliosides now have the choice of several different methods for isolating, quantitating, and chemically modifying these compounds. Of course, the nature of the individual experiment dictates which technique (or custom-designed combination of steps derived from different techniques) is most appropriate, but the correct choice requires an understanding of the principles upon which these

265

methods are based. It is the purpose of this chapter to review the principles and chemical bases and to summarize the important characteristics of the methods that should find the broadest application in this area. Reference to the original publications will be made where the specific details can be obtained. Specialized techniques for chemical structural analyses are beyond the scope of this review, so the reader interested in these is referred to the review of Ledeen and Yu (1982), as well as to several recent papers (Dabrowski et al., 1980; Koerner et al., 1983; Gasa et al., 1983; Kushi and Handa, 1982; Kushi et al., 1983; Ariga et al., 1982, 1984; Carr and Reinhold, 1984; Arita et al., 1984; Rauvala et al., 1980; Gunnarson et al., 1984; Tanaka et al., 1984).

2. Extraction and Purification of Total Gangliosides

2.1. Initial Tissue Extraction

Several different solvents have been used to extract gangliosides from tissues and body fluids. Chloroform:methanol (C:Me) (2:1) has been commonly used to extract total lipids, including gangliosides, from nervous tissues since it was first described by Folch et al. (1951a). It was initially suggested that at least 20 vol (i.e., 20 mL solvent per g tissue) be used. Suzuki (1965) found that extracting the tissue first with 19 vol of C:Me (2:1) and reextracting the insoluble pellet with a similar amount of C:Me (1:2) containing 5% water removed more ganglioside, especially the more polar ones. This has been one of the most commonly used extraction solvents until Svennerholm and Fredman (1980) found that chloroform:methanol:water (C:Me:W) (4:8:3) gave the highest ganglioside yields from brain tissues. The recently described method of Ledeen and coworkers (Byrne et al., 1985) uses C:Me:W (5:5:1) and also gives very favorable yields of gangliosides, but this was described for quantitative rather than preparative purposes. Extraction of an acetone powder of nervous tissue with C:Me (2:1) has also been reported (Li et al., 1978; Kanfer, 1969). The cation content and pH of tissues affects the extractability of gangliosides from whole brain and subcellular fractions (Spence and Wolfe, 1967; Spence, 1969). Specimens depleted of monovalent cations by dialysis yield less ganglioside, especially those with two or more sialic acids, than undialyzed specimens, and either Na^+ or K^+ added to the extraction mixture increase the ganglioside yield. This is an important point to consider when extracting subcellular fractions washed with cation-depleted solutions. Very low pH of tissues also results in low yields of gangliosides, but this is of less practical significance. Extraction of gangliosides using a solution of tetrahydrofuran (THF) and $0.01M$ potassium phosphate buffer at a pH of 6.8 has also been used extensively (Tettamanti et al., 1973). The total lipid extract in this solution is partitioned against diethyl ether, and gangliosides are recovered from the aqueous (lower) phase. A comparison was made of the amounts of gangliosides extracted from mammalian brains with C:Me:W as described by Suzuki (1965). The THF method extracted 8–15% more ganglioside, much of which consisted of the more polar type (GD_{1a}, GT_{1b}, GQ_{1b}). An additional advantage was that the THF method extracted less glycoprotein. It was contaminated with four times as much phospholipid, however, requiring an additional purification step. Such a step is now usually also employed with C:Me:W extractions. As mentioned above, by increasing the proportion of water, the amounts of gangliosides extracted with C:Me:W are comparable or slightly higher than with THF (Svennerholm and Fredman, 1980).

2.2. Separation of Gangliosides from Other Lipids

Most of the commonly used solvents composed of chloroform and methanol extract virtually all of the tissue lipids as well as some proteins, peptides, and considerable amounts of low molecular weight materials such as hexoses and nucleotide sugars. Therefore, several steps following lipid extraction have been used to separate gangliosides as completely as possible from these substances. Folch et al. (1957) first described the method of removing gangliosides from the bulk of the other lipids by washing the total lipid extract (in C:Me, 2:1) with 0.2 vol of water. This results in most of the brain gangliosides and other water soluble substances partitioning into the upper phase composed of C:Me:W (3:48:47), whereas the bulk of the other lipids remains in the lower phase (C:Me:W, 86:14:1).

The cation content of the initial 0.2 vol aqueous phase affects the partitioning of both gangliosides and other lipids. Partitioning with $0.1M$ KCl reduces the amount of acidic lipids that contaminates the aqueous phase (Folch et al., 1957). It also results in incomplete partitioning of the less polar gangliosides (GM_4, GM_3,

GM$_2$), however, significant amounts of which (especially GM$_4$) remain in the lower phase. Different partitioning schemes have been devised depending upon the importance of recovering these less polar gangliosides. One that we have found useful is to partition initially with 0.1M KCl and then two or three additional times with theoretical upper phase (C:Me:W, 3:48:47) lacking KCl. Although the yields of GM$_2$ and GM$_3$ are quite good, GM$_4$ remains in the lower phase. CaCl$_2$ also has profound effects upon the partitioning of gangliosides (Quarles and Folch-Pi, 1965). Below 0.002M and above 0.16M CaCl$_2$, all major brain gangliosides partition into the aqueous phase. Between these concentrations nearly all gangliosides remain in the lower phase. For samples that contain significant amounts of Ca^{2+}, this effect must be considered. For most neural tissues, however, this has not been a problem, and most brain gangliosides quantitatively partition into the aqueous phase. Svennerholm and Fredman et al. (1980) reported that the best way of separating gangliosides from total lipid extracts is solvent partition using C:Me:W (1.0:1.2:1.4), repeating the partitioning once. This is especially useful for large preparative scale isolation.

In addition to the above procedures there are some steps that can be used in the isolation of large quantities of gangliosides. Many investigators have taken advantage of the insolubility of gangliosides in acetone to reduce the mass of the sample and remove acetone-soluble materials (Svennerholm, 1963a, 1972; Tettamanti et al., 1964; Momoi et al., 1976; Iwamori and Nagai, 1978; Li et al., 1978).

A recent publication by Ladisch and Gillard (1985) addresses the problem of isolating minute quantities of ganglioside from total lipids extracted from sources such as cultured cells and fluids. This consists of partitioning the total lipid extract in diisopropyl ether:1-butanol:50 mM aqueous NaCl (6:4:5). The gangliosides are recovered from the lower aqueous phase and separated from low molecular weight contaminants using a Sephadex G-50 column. An advantage of this procedure is that it completely separates gangliosides from all phospholipids; therefore, base treatment is not necessary and could make this very useful in studying alkali-labile gangliosides. The authors mention three caveats, however: (1) Less polar gangliosides are not quantitatively recovered; (2) there is the potential for contamination with some nucleotide sugars; and (3) complex neutral glycosphingolipids may copurify with the gangliosides.

An alternative to isolating gangliosides by phase partition is the use of anion-exchange chromatography. In this method the total lipid extract is applied to a column of anion-exchange resin from which uncharged (neutral and zwitterionic) lipids are eluted with a nonpolar solvent. Acidic lipids are then eluted with a solvent containing a salt, such as sodium acetate, which in appropriate concentrations can displace acidic lipids. For samples isolated from nervous tissues, the latter are mainly sulfatides, acidic phospholipids, and gangliosides. Isolation of gangliosides from the other acidic lipids usually requires three steps: (1) alkaline methanolysis (Yu and Ledeen, 1972; Dawson, 1972) liberates fatty acyl groups from glycerophospholipids as fatty acid methyl esters; (2) desalting, which has been done by three methods [(a) dialysis against water—this also removes water-soluble glycerophospholipid, (b) gel filtration chromatography, (c) reverse phase chromatography]; (3) silicic acid column chromatography removes fatty acids, fatty acid methyl esters, sulfatide, and some other impurities.

DEAE-Sephadex is the most commonly used anion exchange resin for this purpose. It was first used by Marcus and Cass (1969) and widely applied by Ledeen and Yu (Ledeen et al., 1973; Ledeen and Yu, 1982). DEAE-cellulose (Rouser et al., 1965) and DEAE-silica gel chromatography (Kundu and Roy, 1978; Kundu et al., 1979) have also been employed, however. The latter is claimed to have several advantages over DEAE-Sephadex: faster flow rates, rapid solvent equilibration, easier regeneration, greater economy. It has not been used very extensively for isolating gangliosides, however, possibly because it has to be prepared by conjugating DEAE to porous silica gel.

The major advantage of anion exchange chromatography is that all gangliosides, including less polar ones such as GM_3 and GM_4, can be isolated quantitatively. There are two disadvantages compared with the partitioning method. The first is that there is a considerable amount of salt to remove from the ganglioside fraction. The second is that alkaline methanolysis precludes subsequent analyses of acidic phospholipids. This may be no problem if there is sufficient material to remove an aliquot for phospholipid analyses prior to base treatment. In samples with very small amounts of ganglioside, however, the entire sample may be required.

Two procedures have been described for separating gangliosides from other lipids using silicic acid column chromatography.

The first (Kawamura and Taketomi, 1977) was designed for relatively large amounts of starting brain (5 g). The total lipid was applied to a column of silica gel G, and most of the nonganglioside lipids eluted with C:Me:28% ammonia. Gangliosides were then eluted with C:Me:W (60:35:4), but were contaminated with some other lipids. Removal of the latter required alkaline methanolysis, dialysis, and repeat chromatography. The second procedure (Irwin and Irwin, 1979) was developed to isolate total gangliosides from small (1–15 mg dry weight) brain samples in a simple rapid fashion in sufficient purity to study them by TLC. Total lipid extracts were applied to a small Unisil column, and nonganglioside lipids eluted with either C:Me (2:1) (method A) or C:Me:W (65:25:4) (method B). Gangliosides were then eluted with C:M:W (50:50:15). Ganglioside samples prepared by method A contain contaminants that interfere in both the resorcinol and thiobarbituric acid assays. Method B produces purer preparations, but of slightly lower yields. The amounts of sulfatide and phospholipid contaminating either preparations were not reported, but it is possible that significant amounts of these may be present. A novel method for separating glycolipids from other lipids is the use of a cross-linked polystyrene matrix to which phenylboronic acid is covalently bound (Krohn et al., 1978). Total lipid extracts are loaded onto a column of this material, and nonglycolipids are eluted with C:Me (5:1). With increasing amounts of water, glycolipids are sequentially eluted. Although this method has been neither fully examined nor utilized, it seems that it might be of some value in isolating gangliosides.

2.3. Removal of Low Molecular Weight Contaminants

Removal of water-soluble contaminants from crude ganglioside extracts has been approached in several ways. Dialysis of the upper phase constituents against distilled water has been commonly used since its suggestion by Folch et al. (1957). There are several pitfalls that must be recognized, however. First, if the gangliosides are below their critical micelle concentration (CMC), they will dialyze across the membrane and be lost. The CMC is still not established, but is somewhere between 10^{-4} and $10^{-8}M$ (Corti et al., 1982). When dealing with small amounts of gangliosides, the use of mini-dialysis bags is recommended. If small amounts of radiolabeled gangliosides are being isolated by this method, and chemical quantification is not planned, it may be necessary to add cold carrier ganglioside to increase the amount of total ganglioside above the CMC. Second, it should be noted that some standard dialysis bags contain contaminants that can co-purify with the gangliosides and appear as brown spots on TLC. Pretreatment of the bags with Na_2CO_3 and EDTA can decrease, but not eliminate, these contaminants (Coleman and Yates, 1978). Third, it is necessary to take the upper phase specimens to dryness and reconstitute them in water before dialysis. The presence of methanol in the dialyzing solvent can result in the loss of significant amounts of gangliosides during dialysis. Dialysis of the upper phase directly will result in low ganglioside recoveries from this step. Therefore, the dialysis bag contents should always be dissolved in distilled water.

Although exhaustive dialysis (48 h with eight water changes) removes almost all glucosamine, and probably other hexoses, only 50–90% of nucleotide sugars are removed. Treatment with alkaline phosphatase and phosphodiesterase (Kanfer, 1969) followed by dialysis removes over 99% of nucleotide diphosphate sugars, but only 85% of CMP-N-acetylneuraminic acid (Yates and Warner, 1984). Alkaline methanolysis of the crude ganglioside preparation with 0.3N NaOH in C:Me (1:1) for 1 h at room temperature is more effective than dialysis alone in removing nucleotide diphosphate sugars, but not as good as enzyme treatment and dialysis (Yates and Warner, 1984). Alkaline methanolysis also cleaves the acyl side chains from contaminating phospholipids. The fatty acid methyl esters that are produced are subsequently removed by silicic acid column chromatography, and the glycerol phosphate derivatives are dialyzable. An alternative to dialysis for removing water-soluble contaminants is reverse phase chromatography. Originally described by Williams and McCluer (1980) for Sep-Pak cartridges, it has been found that other small C18 columns such as Bond-elute (Analytichem) and that made by J. T. Baker give more reproducible results. For the successful application of this procedure, the following is required: (1) the column must be conditioned with a methanol-containing solvent to coat the stationary phase with methanol; (2) it may be necessary to pass the loading solution through the column more than once to transfer all of the gangliosides to the column; (3) it is essential to dry the column completely before eluting so that the eluting solvent can reach the adsorbed ganglioside. This appears to be an effective technique for removing both nucleotide diphosphate sugars and CMP-N-acetylneuraminic acid from ganglioside preparations. When possible, however, the recovery of gangliosides should be monitored with trace amounts of radiolabeled gangliosides.

C18 Sep-Pak cartridges have also been used to remove salts from gangliosides isolated using ion exchange chromatography (Kundu and Suzuki, 1981). In this procedure, acidic lipids are obtained by passing the total lipid extract through a DEAE-Sephadex column. This fraction is subjected to an alkaline methanolysis, taken to dryness, reconstituted in water, and neutralized with HCl. The volume is adjusted to $0.1M$, which results in a total volume of 115 mL. This is passed once through a conditioned Sep-Pak cartridge, which is then washed with 25 mL of water to remove water-soluble contaminants. Gangliosides are then eluted with 5 mL of methanol and 25 mL of C:Me (1:2). It is claimed that separation of gangliosides from glucose and N-acetylneuraminic acid is nearly complete, and recovery of gangliosides up to 148 µg sialic acid is over 97%. We have tried similar procedures with Bond-Elut and J. T. Baker C18 columns and confirm the excellent separation of gangliosides and CMP-N-acetylneuraminic acid. Further experience with this method should prove interesting.

Removal of low molecular weight, water-soluble contaminants has also been accomplished by gel filtration. Originally described by Wells and Dittmer (1963), the method has gone through several modifications. A frequently used method is that of Ledeen and Yu (1982), who used Sephadex G-50. A special modification of this procedure is necessary to remove CMP-N-acetylneuraminic acid and involves chromatography in dilute sodium chloride or acetic acid (Yohe et al., 1980). A more recent modification used Sephadex LH-20 (Byrne et al., 1985). This method was specifically developed to isolate gangliosides in a highly purified form from small amounts of tissue. This involves passing the ganglioside preparations through Sephadex LH-20 columns under two different conditions. Initially, the total lipid extract is made slightly acidic to dissociate the gangliosides from basic peptides and proteins that are removed on the first Sephadex LH-20 column. The acidic lipid fraction obtained from a DEAE-Sephadex column is subjected to an alkaline methanolysis and then desalted using Sephadex LH-20 columns under different conditions than the first. The gangliosides are finally purified using silicic acid column chromatography. The final ganglioside preparation is extremely pure, but may contain a small amount of dextran from the columns. We have found that this is easily removed using a C18 reverse phase column. Two points about this method deserve emphasis. First, the Sephadex LH-20 columns always require calibration. Second, it has recently been found that the efficiency of separating gangliosides from inorganic salts with these columns varies among different batches of Sephadex LH-20. Therefore, before purchasing large amounts of this material for this purpose, each batch should be tested with standards.

Precipitation of gangliosides with trichloroacetic and phosphotungstic acids has also been used (Dunn, 1974). The ganglioside patterns on TLC change, however, as a result of internal esters formed during this procedure (Seyfried et al., 1977; Gross et al., 1977; Mestrallet et al., 1976). Therefore, this method of removing low molecular weight contaminants is not recommended if isolation or quantitation of specific gangliosides is planned.

2.4. Final Purification

A final step commonly employed in purifying total ganglioside mixtures is silicic acid column chromatography. This removes less polar contaminants such as fatty acids and sulfatide. Both Unisil (Clarkson Chemical Co.) and Iatrobeads (Iatron Industries, Inc.) have been used. The size of the column can vary considerably, but we have used a 1-g column for amounts of ganglioside up to 6 mg of N-acetylneuraminic acid with quantitative recoveries. Several different elution schemes have been reported (Ledeen and Yu, 1982; Byrne et al., 1985; Hofteig et al., 1981). A final step of LH-20 column chromatography is also sometimes employed to remove final trace impurities (Byrne et al., 1985).

3. Isolation of Classes and Individual Species of Gangliosides

3.1. Column Chromatography

For over two decades improved procedures have been published to extract and separate gangliosides into major classes and individual species that can be isolated on a preparative scale. Initial attempts at separation were made using columns of paper rolls (Svennerholm, 1963b), silicic acid (Svennerholm, 1963b), silica gel G and H (Svennerholm, 1972), and cellulose powder (Kuhn and Wiegandt, 1963; Svennerholm, 1963b). Tailing and overlapping resulted in separations that were rarely clean so that rechromatography was usually required. Large-scale purification of the major and minor gangliosides was accomplished by both column chromatography and steps based on differential solubility (Svennerholm, 1963b, 1972; Tettamanti et al., 1964).

Anion exchange chromatography allows the separation of gangliosides on the basis of charge, i.e., the number of sialic acids per molecule. This was first done using columns of DEAE-cellulose eluting different groups of gangliosides with discontinuous gradients of ammonium acetate in methanol that yielded five ganglioside fractions (Winterbourn, 1971). Subsequent isolation of some individual gangliosides required preparative TLC. A modification of this scheme was used to isolate GM_4, GM_3, and GD_3 from chicken egg yolk (Li et al., 1978). Sequential column chromatography on DEAE-Sephadex and Iatrobeads, large porous silica spheres, allowed the isolation of milligram quantities of the major brain gangliosides starting with 4.6 kg of bovine gray matter (Momoi et al., 1976). A crude ganglioside preparation is loaded onto a DEAE-Sephadex A-25 column and eluted with a linear gradient of ammonium acetate in methanol. Small fractions are collected, monitored using TLC, and pooled to give five major fractions containing mono-, di-, or tri-sialogangliosides. Each of these is then chromatographed on an Iatrobead column to obtain the purified four major gangliosides in large yield.

"Ganglioside mapping" is the term coined by Iwamori and Nagai (1978) and applied by them to the sequential use of DEAE-Sepharose and TLC for the examination and isolation of gangliosides. They found that separation of a mixture of gangliosides into four major classes based on sialic acid content occurred faster and with better resolution using DEAE-Sepharose than with either DEAE-Sephadex or QAE-Sephadex. A comparative study of the abilities of three anion-exchange resins to separate ganglioside mixtures on the basis of sialic acid content with a linear gradient of ammonium acetate in methanol was also made by Kundu et al. (1979). Of the three studied (DEAE-Sephadex, DEAE-controlled porous glass, and DEAE-silica gel), the last provided the best resolution. Even with that resin, however, the tetrasialoganglioside fraction was contaminated with some trisialogangliosides. Apparently only one gradient solvent system was used in their study for all three resins.

A critical study of the abilities of several anion exchange resins (Spherosil-DEAE-Dextran, DEA-Spherosil, DEAE-Sephacel, DE-

AE-Sephadex, DEAE-Sepharose) to separate gangliosides on the basis of charge was made by Fredman et al. (1980). The Spherosil-DEAE-Dextran column was eluted with discontinuous gradients of potassium acetate in methanol, collecting five fractions. These corresponded to mono-, di-, tri-, tetra-, and penta-sialogangliosides. DEA-Spherosil had a low binding capacity for gangliosides, and DEAE-Sephacel did not separate the gangliosides on the basis of sialic acid number. DEAE-Sephadex bound all of the gangliosides, but there was tailing in the elution profile of some classes, and recoveries were a bit lower than with DEAE-Sepharose or Spherosil-DEAE-Dextran. DEAE-Sepharose and Spherosil-DEAE-Dextran both had excellent binding, resolution, and recoveries of gangliosides. The latter was superior to the former in three respects, however: maximal binding capacity was twice that of the latter; better resolution of polysialogangliosides; and easier regeneration. They mentioned two important points for use with these columns. First, for gangliosides to be completely retained on the columns, they must be completely salt-free. Second, internal ester formation can be prevented by adjusting the pH of the sample being loaded onto the column to 8. In addition, these investigators compared silica gel Merck 230 and <230 with Iatrobeads for their abilities to separate gangliosides with uniform oligosaccharides in fractions eluted from the anion exchange resins. Neither had an advantage over the other, and although some gangliosides were well resolved, others were not. Some required preparative TLC for isolation from the silica column fractions.

Separation of gangliosides on the basis of sialic acid number has also been achieved using DEAE-Sephacel with a linear gradient of sodium acetate in methanol (Itoh et al., 1981). Mono-, tri-, and tetra-sialogangliosides are recovered as one peak each; disialogangliosides elute in three peaks. Separations of individual species of gangliosides isolated from anion exchange columns have also been accomplished using Silica gel 60 (200–325 mesh) (Iwamori et al., 1982).

Intermediate pressure column chromatography (FPLC-Pharmacia) using the anion exchange resin Mono Q appears to be promising for the separation of gangliosides into classes on the basis of sialic acid number (Mansson et al., 1985). A stepwise gradient of potassium acetate in methanol gave good class separation with the exception that GQ_{1b} eluted in the monosialoganglioside fraction when sample loading was increased from 2.75 to 7.75 μmol N-acetylneuraminic acid. This is still unexplained, but may be caused by internal ester formation. Recoveries varied between 93 and 99%, with the major loss being caused by adsorption of gangliosides with oxidized ceramide to the column. The column lifetime is not yet determined, but is at least 6 mo with daily use. Its major advantage over other anion exchange resins is its higher binding capacity for gangliosides, which is 5–10 times higher than those discussed above. Although this study was made using the smallest column available, there is reason to believe that this method could be scaled up considerably using larger columns.

3.2. Preparative Thin Layer Chromatography

Gangliosides that can not be resolved using column chromatography have frequently been isolated in sufficient quantities for chemical analyses using preparative TLC (Svennerholm et al., 1973; Ando and Yu, 1979). For this to be successful, resolution of the bands on the plate must be sufficient for the band to be scraped without any contaminating material from adjacent bands, and there must be enough ganglioside in the band to be detected using a nondestructive detection method. Ganglioside has been detected by spraying the plate with dilute aqueous solutions of rhodamine 6G [0.001% (w/v)] and visualizing with UV light (Ledeen and Salsman, 1965; Ledeen et al., 1973). The rhodamine must be separated from the ganglioside; this can be done using a Unisil column. When larger amounts of ganglioside are present, spraying the plate with water can result in visualization, but this is relatively insensitive (Winterbourn, 1971). Iodine vapor has been frequently used to visualize gangliosides for preparative purposes. A permanent record of the chromatogram can be made using diazo paper (Dawson et al., 1972). Prior to elution the iodine is allowed to sublimate in a stream of warm air. A spray of bromthymol blue will result in visualization of gangliosides, but this reagent must be removed from the ganglioside sample. This can be accomplished by washing the scraped gel containing the gangliosides with acetone before eluting the gangliosides (Wherrett et al., 1964), or by eluting the gangliosides with C:Me:W (30:60:20) using small columns with sintered discs (Fredman et al., 1980). Several types of thin layer plates have been used: silica gel G (Ledeen and Salsman, 1965; Ledeen et al., 1973), Kieselgel 60, HPTLC Kieselgel 60 (Fredman et al., 1980), or HPTLC silica gel 60 (Itoh et al., 1981). Wetting the gel using an atomizer can facilitate scraping of plates to which the gel is tightly bound (Yohe et al., 1980). Elution of gangliosides from the gel has also been done using Me:C:P:W (P = pyridine) (56:40:2:12) (Ledeen and Salsman, 1965; Ledeen, 1966), C:Me:W (10:4:0.5) (Dawson et al., 1972), or C:Me:W (10:10:1) (Ledeen et al., 1973). The latter authors found that recoveries of gangliosides from the gel were higher when an amount of Dowex-50(Na$^+$) equal to half the weight of the gel was added to the eluting solvent. It has been found that eluting with solvents containing higher proportions of methanol gives greater recoveries, but the sample is contaminated with larger amounts of silica gel. This can be removed by passing the sample through a reverse phase column (Kubo and Hoski, 1985).

3.3. Radial Thin Layer Chromatography (RTLC)

Although this technique was originally described over two decades ago (Rosmus et al., 1964), it has only recently been applied to the separation of glycosphingolipids (Das et al., 1984). The apparatus consists of a motor-driven glass rotor plate that can be coated with silica gel over which there is a quartz cover-plate. The instrument used by these investigators was a model 7924 chromatron (Harrison Research Co., Palo Alto, CA). The apparatus is placed at a 45° angle, the plate is rotated at approximately 750 rpm, and the sample is applied dropwise at the center of the glass plate in a region scraped clean of silica gel. The eluate drips from the bottom of the plate and is collected in a fraction collector. Aliquots of each fraction are monitored using TLC. This technique was used to separate 25 mg of a mixture of GM_1, GM_2, and GD_{1a} on silica gel 660 PF 254. Although it appears that some tailing occurred, pooling of appropriate fractions can give samples of reasonable purity with a recovery of 80%. The silica gel can be regenerated and used at least twice. The advantage of this method compared with preparative TLC is its larger sample capacity. Compared with most column chromatographic procedures, it is fast and uses small volumes of solvents.

4. Quantitative Assays

4.1. Spectrophotometric Assays

In the past four decades since the discovery of gangliosides, several methods for quantitating these compounds have been described. Most of these methods are based on the detection of sialic acids because these are the characteristic components of gangliosides. Some of the earlier methods, although of great value at the time, were less specific and are now of historical interest only (reviewed in Svennerholm, 1964). The "browning reaction" to quantitate strandin (later determined to be identical with ganglioside) was described in detail by Folch et al. (1951b). This consisted of heating the ganglioside sample in concentrated HCl, adding acetone to the cooled solution, and reading the absorbance at 475 nm. More specific was the adaptation of Bial's reaction using orcinol-hydrochloric acid for the quantitation of sialic acids (Svennerholm, 1957a). Although this method is not specific for sialic acids, correction for interfering carbohydrates can be made by reading the absorbance at two wavelengths.

The orcinol-HCl method was soon superceded when the resorcinol-HCl method was reported by Svennerholm (1957b). Adaptations of this have been the most commonly used methods for quantitating gangliosides since that time. Not only is it 50% more sensitive than the orcinol-HCl method, but it is more specific for sialic acids. The molar absorbance for N-acetylneuraminic acid is 8000 when assayed for 15 min in a boiling waterbath and read at 580 nm in amyl alcohol. Thus 1 µg of N-acetylneuraminic acid yields an absorbance of approximately 0.030. It is important to note that the molar absorbance of N-glycolylneuraminic acid is 30% greater than for N-acetylneuraminic acid. Therefore, if a ganglioside contains only the former sialic acid, the final value must be multipled by 0.77 if the standard used is N-acetylneuraminic acid (Svennerholm, 1963c). Another major advantage is that this method equally estimates free and lipid-bound sialic acids, making it unnecessary to cleave all sialic acids to obtain accurate quantitative estimates of gangliosides. The main modification of the original method commonly employed is the use of butyl acetate : butanol (85:15) instead of amyl alcohol to extract the chromophore (Miettinen and Takki-Luukkainen, 1959). Although the molar absorbancies for sialic acid chromophores extracted in this solvent are slightly higher than in amyl alcohol, the major advantage is that the absorbancies for the chromophores of other potentially contaminating monosaccharides such as glucose are much lower, especially at their absorbance peak, which is shifted to a slightly longer wavelength (480 nm). If a ganglioside preparation contains significant amounts of such contaminants, a correction of the spectrophotometric reading at 580 nm can be made using the two-wavelength method. Ledeen and Yu (1982) suggest the following formula:

$$A_{580} = \frac{[(A_{580})(R_H - A_{450})]R_s}{(R_s)\, R_H - 1}$$

where A = absorbance, R_s = ratio of absorbance for sialic acid at 580 nm and 450 nm, respectively, R_H = absorbance of interfering substance at 450 nm and 580 nm, respectively.

Based on the same principles, Wherrett (Wherrett and Brown, 1969) has developed the following formula (J. R. Wherrett, personal communication) to correct for interference by galactose:

Corrected A_{580} = observed $A_{580} \times 0.265\,[1.073 - (A_{470}/A_{580})]$

For most ganglioside preparations of reasonable purity, we have found such correction factors to be close to 1.0. It was of considerable value, however, when individual ganglioside bands on TLC were scraped and assayed by the resorcinol method (MacMillan and Wherrett, 1969; Yates and Thompson, 1977). More recently, Svennerholm and Fredman (1980) have quantitated gangliosides with the resorcinol reaction reading at 620 nm. Although this is slightly less sensitive for sialic acids, interference from other sugars is much lower. Therefore, reliable quantitative estimates of gangliosides can be made without a two-wavelength correction, even in the presence of considerable amounts of other lipids. It should be noted that at room temperature the absorbance of the chromophore decreases by 5% in the first hour and 1%/h subsequently, so the absorbance should be determined as soon as possible after color development.

The sensitivity of the resorcinol assay can be increased by a factor of 3–6 by mild periodate oxidation of the sialic acid prior to the resorcinol reaction (Jourdian et al., 1971). This makes it half as sensitive as the periodate-thiobarbituric acid techniques discussed below. The molar extinction coefficients for N-acetylneuraminic acid and N-glycolylneuraminic acid, respectively, for each of these are: resorcinol-HCl, 8000 (Svennerholm, 1957b) and 10,400 (Svennerholm, 1963c); periodate-resorcinol, 27,900 and 27,300 (Jourdian et al., 1971); thiobarbituric acid, 57,700 and 46,000 (Warren, 1959), or 70,700 and 44,400 (Aminoff, 1961). The chromogen reacting with resorcinol is not known, but is possibly a furfural or pyrrole derivative of sialic acid. The chromophore has an absorption maximum at 630 nm whether derived from free or glycosidically linked sialic acids. The chromogen derived from free sialic acid is much less stable than that from the glycosidically linked form, however, especially at 37°C. This property forms the basis for differential quantitation between free and bound forms. Total sialic acid is determined by conducting the periodate oxidation at 0°C for 35 min; for quantitating glycosidically linked sialic acids, oxidation is performed at 37°C for 60 min. There is virtually no interference from sugars, amino acids, or lipids commonly found in neurological tissues. As pointed out by Ledeen and Yu (1982), however, this method is unsuitable for quantitating total sialic acid in samples of gangliosides with sialosyl (2→8)sialosyl linkages because these are resistant to periodate oxidation. This is of special importance in studies on nervous system gangliosides, which contain considerable amounts of such compounds. Sialic acids substituted at carbon-8 are also resistant to periodate. Therefore, considerable caution must be exercised before employing this method to quantitate gangliosides, especially if their structures are unknown.

A two-step procedure involving periodate oxidation of sialic acid, and a coupling reaction of the oxidation product (β-formylpyruvic acid) with thiobarbituric acid (TBA) was developed by Warren (1959) and Aminoff (1961). It yields a chromophore that is eluted into an organic solvent in which the absorption maximum is 549 nm. Although it is relatively sensitive for quantitating free sialic acid, it does not estimate the bound form, so for it to yield reliable estimates of gangliosides, a hydrolysis reaction must first occur. Svennerholm (1963c) compared the results of the TBA assay (following hydrolysis in 0.05M H_2SO_4 at 80°C for 2 h) with the resorcinol reaction for gangliosides. The yields with TBA were much lower (50–70%) because of both incomplete hydrolysis of the internal sialic acid from hexosamine-containing gangliosides and partial destruction of the liberated sialic acid (Hess and Rolde, 1964). Therefore, if this method is used to quantitate gangliosides, a standard as close as possible to the chemical structure of the sample should be employed to correct for these factors. Some biological constituents can interfere with this method: 2-deoxyribose, malonaldehyde, L-fucose, and unsaturated lipids. Corrections for the former can be made using the two-wavelength method (532 and 549 nm). A method for reducing interference caused by unsaturated lipids has been described by Diringer (1972) in which bromine addition to the double bonds of contaminating lipids is performed prior to hydrolysis. Bromine treatment does not affect the assay itself, but can give yields of GM_3 near 100% even in the presence of large amounts of phospholipids.

Colorimetric quantitation of gangliosides on the basis of sphingosine content using trinitrobenzene sulfonic acid (TNBS) has been described (Siakotos et al., 1970). This involves a BF_3-catalyzed methanolysis reaction to remove the fatty acid from sphingosine that is extracted into benzene using a buffered hyamine solution. The detergent effect of hyamine is used to dissolve the sphingosine in the organic solution, and the buffer is to neutralize BF_3. Aqueous TNBS is added and reacts with the free amino group of sphingosine to form the chromophore, which has an absorption maximum at 340 nm. HCl (1.2M) quenches the unreacted TNBS. The method can be directly applied to pure ganglioside samples or gangliosides on silica gel scraped from TLC plates. Because the reaction with TNBS involves free amino groups, any substance bearing these, such as amino-containing lipids, will interfere with the quantitation of gangliosides. The limit of sensitivity is approximately 10 nmol, and the reaction is linear to at least 200 nmol.

4.2. Fluorometric Assays

Highly fluorescent quinaldines are produced from the reaction of 2-deoxy sugars with 3,5-diaminobenzoic acid. When heated in the presence of mineral acids, N-acetylneuraminic acid will undergo decarboxylation and deacylation to form a compound of this general type (2-deoxy-4-amino-octose). Based on these reactions, Hess and Rolde (1964) developed a highly sensitive microfluorometric assay for gangliosides that measures both free and bound sialic acid in the range of 1 nmol. By carefully controlling the reaction conditions, interference from hexoses can be decreased to less than 1.6% on a molar basis. The fluorophore exhibits intense green fluorescence with an excitation maximum at 405 nm and an emmission maximum at 510 nm. The authors used a mercury lamp, and primary and secondary barrier filters that transmitted at 436 and 525 nm, respectively. This method was successfully applied by Harris and Klingman (1972) who quantitated gangliosides in very small samples of autonomic ganglia of rat and cat. They emphasize the strict controls required to correct for fluorescent contaminents from silica gel, vessels, and connectors such as Tygon tubing.

Gangliosides bind to a wide range of biological and nonbiological substances, one of which is serotonin. This has been used in a simple, rapid, nondestructive, but nonspecific assay to quantitate total gangliosides (Price and Frame, 1981). The sample to be quantitated is taken to dryness in the presence of a known amount of serotonin, transfered to a dialysis bag in water, and dialyzed for 30 min against 40 mL of water. Fluorescence of the diffusate is measured with excitation and emission wavelengths of 285 and 340 nm, respectively. The amount of serotonin in the diffusate is inversely proportional to the amount of ganglioside in the dialysis bag. Ganglioside values are determined from standard curves prepared with known amounts of ganglioside. Neither pH nor temperature affect the assay, but any contaminants that bind serotonin will cause interference. To be reproducible, the ratio of serotonin to ganglioside must be within the range of 0.1–0.5. The limit of sensitivity is reported to be 20 nmol. One potential problem not addressed by the authors, however, is the loss of gangliosides from the dialysis bag at concentrations below the CMC, which could introduce large errors. Furthermore, standard curves must be prepared from gangliosides similar to those being analyzed or serious errors will result. Although this method is simple, sensitive, and nondestructive, it will probably find very restricted application.

4.3. Gas-Liquid Chromatography

Several methods have been described to quantitate and determine the carbohydrate compositions of gangliosides using gas-liquid chromatography (GLC). The procedure originally described by Vance and Sweeley (1967) involves an initial methanolysis in methanolic HCl. Fatty acid methyl esters are extracted with hexane, and HCl is neutralized either by passing the solution through Amberlite CG-4B resin or exposure to silver carbonate. D-Mannitol has been frequently used as an internal standard. It was found, however, that it can bind to silver carbonate, making treatment with Amberlite IR-45 a preferable way to neutralize the methanolic HCl (Rickert and Sweeley, 1978). O-Trimethylsilyl (TMS) methyl glycosides are prepared and separated using temperature programming on either 2% SE-30 or 3% OV-1 columns and detected using a flame ionization detector. This method was extensively studied and improved upon by Clamp et al. (1967) who introduced a re-N-acetylation step. This allowed accurate quantitation of N-acetylhexosamines as well as neutral sugars in small amounts of sample (80 nmol). Although the derivatives of mannitol and *N*-acetylneuraminic acid each elute as one peak, those of glucose appear as two peaks, and galactose and *N*-acetylgalactosamine, as three peaks. Two of the peaks are caused by the α and β anomers of the methyl pyranosides; the third is probably the preferred anomer of the methyl furanoside. With temperature programming, separation of these multiple peaks can be accomplished (Sweeley and Tao, 1972). In our experience, however, this does not occur in every run, and peaks for the hexosamines are frequently incompletely resolved. *N*-acetylneuraminic acid and *N*-glycolylneuraminic acid can not be differentiated with this method because they become N-deacetylated during the methanolysis procedure. A method similar to this, but with slightly different methanolysis conditions, was described by Iwamori and Nagai (1978).

Quantitation of *N*-acetylneuraminic acid and *N*-glycolylneuraminic acid can be accomplished with a modification of this method developed by Yu and Ledeen (1970). A short, mild, acid methanolysis releases most of the ganglioside sialic acid, but does not N-deacetylate it. Fatty acids are extracted with hexane, and the internal standard (phenyl *N*-acetyl- α -D-glucosaminide) added. TMS derivatives are prepared and separated isothermally on either 3% OV-1 or 3% OV-225 columns. Better separations of *N*-acetylneuraminic acid and *N*-glycolylneuraminic acid were obtained with the former. Although both α- and β-ketosidic anomers are formed, for both *N*-acetylneuraminic acid and *N*-glycolylneuraminic acid the latter is by far the dominant one, and quantification is based entirely on its peak area. The N-deacylated derivative is also produced, but in only minor quantities and is not included in quantitative analyses. The methanolysis conditions are too harsh to preserve O-acetylated sialic acids, which can not be identified with this method. The relative detector response is constant when the ratio of sialic acid to internal standard is between 10 and 0.6. The sensitivity of the procedure is 0.05 μg sialic acid per injection, below which there is loss caused by adsorption onto the column. Reliable estimates can be made on as little as 0.3 μg of sialic acid for which six replicate injections can be performed. Therefore, this method supercedes the level of sensitivity of the resorcinol reaction, with the added advantage of being able to differentiate *N*-glycolylneuraminic acid from *N*-acetylneuraminic acid directly.

Ando and Yamakawa (1971) developed a method for determining monosaccharide ratios in gangliosides using acid methanolysis and *N,O*-trifluoroacetylation. The trifluoroacetyl (TFA) derivatives have the advantages of being more stable to heat and acid and having higher volatility than the TMS derivatives. Samples are separated on a stationary phase consisting of either XE-60/SE-30 (3:200) or 6% SP-2401 and 0.5% OV-225 on Gas chrom Q (Itoh et al., 1981), with a linearly programmed temperature gradient. Derivatives of all monosaccharides except sialic acid emerge as multiple peaks. Although glucose and galactose peaks are resolved, there is overlap of the major and minor peaks for the two hexosamines; N-acetylneuraminic acid and *N*-glycolylneuraminic acid can not be differentiated. Sphingosines elute as numerous peaks, the areas of which must be totaled for quantitative purposes. The lower limits of sensitivity and linear range were not described, but the former is at the most 5 µg of sialic acid (Itoh et al., 1981).

Analysis of hexoses and hexosamines of gangliosides as their alditol acetate derivatives has also been used. Initially, the ganglioside samples are subjected to an acid hydrolysis and the sugars reduced to their alditol derivatives, which are then acetylated. These are separated on a 1% OV-225 column using temperature programming, and quantitated, using an internal standard, with a flame ionization detector. Following a critical study of the variables involved in this method, an optimized procedure was suggested that can quantitate as little as 1 µg of sugar with an average error of 11% (Torello et al., 1980). An advantage of this method is that each hexose and hexosamine elutes as only one well-resolved peak. This makes peak identification easier and the method slightly more sensitive than the TMS method. Sialic acids have not been analyzed, however, so that this method is better used for determining sugar ratios than for quantitative purposes.

In all of the GLC methods, the sugar moieties must be cleaved from the ceramide portion of the molecule. Different hexoses, hexosamines, and sialic acids are not all removed at the same rate, and once they are liberated, they begin to form secondary products that are not quantitated. Therefore, the recovery of these moieties is different for different gangliosides. Correction for breakdown of free sugars can be estimated by the use of external standards for these sugars run with each assay, but this does not correct for the different rates of liberation of the sugars among the different gangliosides. Corrections for this can only be made by using ganglioside external standards of the same or similar structure as those being quantitated. This poses a problem for samples composed of mixtures of gangliosides whose structures are not known. Yu and Ledeen (1970) have developed a series of correction factors for their method that can be used in situations in which estimates of the ganglioside composition can be made based on preliminary evidence from TLC. Without such estimates, the maximum error using their method would be approximately 15%.

4.4. Chemical Ionization Mass Fragmentography

A procedure for quantitating sialic acid requiring more elaborate instrumentation was described by Ashraf et al. (1980). They developed the method to quantitate sialic acids cleaved from whole red blood cell ghosts, but it could almost certainly be used to quantitate ganglioside sialic acids. Following a mild acid methanolysis, the liberated monosaccharides are converted to their TMS-derivatives. These are separated on a 3% OV-1 column that is run isothermally, and the GLC-MS data are collected. The sum of the intensities of the m/e 610 signal across the GLC peak for the TMS-methylglycoside of N-acetylneuraminic acid and of the m/e 498 signal across the peak of the internal standard (TMS-phenyl-N-acetyl-α-D-glucosaminide) is used to quantitate the amount of N-acetylneuraminic acid present in the sample. As little as 1 pmol can be quantitated, but the useful range is 10 ng–1 µg. It seems that the same precautions required for GLC analyses of gangliosides discussed above will be necessary for this method.

4.5. Enzymatic Assay

Quantitation of free sialic acids can be done in two different ways using the enzyme NANA-aldolase (Araki and Yamada, 1984). This enzyme splits N-acetylneuraminic acid into pyruvate and N-acetylmannosamine. Both methods base their quantitations on determining the amount of pyruvate released through this reaction. In one, the pyruvate reacts with NADH and H$^+$ through a reaction catalyzed by lactate dehydrogenase, which yields L-lactate and NAD$^+$. The latter is estimated spectrophotometrically by measuring the absorbance at 339 nm, which is proportional to the amount of free N-acetylneuraminic acid in the sample. In the other method, pyruvate oxidase converts pyruvate, Pi, and oxygen to acetylphosphate and H_2O_2. The latter is quantitated colorimetrically through an enzymatically catalyzed reaction involving a toluidine derivative. The absorption at 550 nm is proportional to the amount of free sialic acid present in the original aliquot. This method has been adapted to estimate the amount of protein-bound sialic acid by treating the sample first with neuraminidase. It has

been less useful for estimating gangliosides because of the resistance of the internal sialic acid to enzymatic hydrolysis. It may be
possible, however, to obtain more accurate estimates of ganglioside sialic acid with this method by performing the enzymatic
hydrolysis in the presence of a detergent. A commercial kit to
perform these assays is available from Boehringer Mannheim
Biochemicals.

4.6. Radioassay

A radioassay for GM_1 concentration in human CSF, which
exploits the high affinity interaction between GM_1 and cholera
enterotoxin, was reported by Ginns and French (1980). The assay is
conducted by preincubating ^{125}I-labeled cholera toxin with 20–50
μL of CSF, followed by a second incubation with a liver membrane
preparation. The sample is filtered through a Millipore membrane
that is placed in a gamma counter and radioactivity determined.
Results are expressed as percent inhibition of cholera toxin binding
to the liver membranes and compared with a standard curve cali-
brated against different amounts of GM_1. The method is very
sensitive, with a lower limit of detection of 2.5 ng/mL CSF. Glycoproteins can interfere with this reaction, but it is reliable up to a
total CSF protein concentration of 1000 mg/dL. Although this
method probably can not be applied to many biological samples, its
sensitivity, allowing determinations on very small volumes of CSF,
could make it of considerable potential value for some clinical
studies.

5. Thin Layer Chromatography

5.1. Separation

The value of TLC in separating gangliosides was recognized
early in the 1960s. Since that time numerous modifications and
innovations have been reported. Initially, the stationary phase
plates were prepared by the experimenters in their own laboratories. These plates generally were of silica gel and required heat
activation prior to use. Approximately 15 years ago, commercially
prepared silica gel plates became available. Although some of these
gave separations similar to the home-made type, other brands of
factory-coated plates with binder systems that did not include

CaSO$_4$ gave unsatisfactory separations of gangliosides in most of the solvent systems then used. Van Den Eijnden (1971) found that the addition of cations to the developing solution greatly improved the separation of gangliosides on these commercially prepared plates. More recently, high performance TLC (HPTLC) plates have provided better resolution, and much smaller samples can be visualized. Quantitative results using scanning densitometry are also superior with these plates.

Although separations using one solvent in one dimension have been most common, multiple developments with more than one solvent and two-dimensional TLC have also been employed to advantage. Table 1 lists some of the solvent systems that have been used to separate gangliosides. The choice of a solvent depends mainly on which gangliosides are the most important to resolve. Although the separations achieved are largely empirical, there are some general principles that can be applied. In general, gangliosides are separated on the basis of polarity: more polar gangliosides move more slowly than less polar ones. Furthermore, less polar solvents tend to separate less polar gangliosides better, whereas more polar gangliosides are more poorly resolved. Thus, solvents containing a larger proportion of chloroform, but less methanol or water, will separate GM$_4$, GM$_3$, and GM$_2$ well. Conversely, more polar solvents will separate tri-, tetra-, and penta-sialogangliosides better than less polar solvents. Separation can also occur on the basis of the position of sialic acids on the oligosaccharide, e.g., GD$_{1a}$ and Gd$_{1b}$. Doublets of a single ganglioside type may occur because of differences in either the sialic acid or ceramide moieties: N-acetylneuraminic acid vs. N-glycolylneuraminic acid; hydroxy vs nonhydroxy fatty acids; sphingosine vs dihydrosphingosine.

Allowing the solvent to stand in the tank 1–2 h before developing the plate is generally employed, and lining the tanks with filter paper to facilitate saturating the atmosphere with solvent has been frequently used (Ledeen and Yu, 1978; Zanetta et al., 1977; Brownson and Irwin, 1982; Pennick et al., 1966). Its value has never been clearly established, however. Similarly, activation of self-made and standard commercial TLC plates was essential for most ganglioside separations (Ledeen and Yu, 1978), but its value with HPTLC, and for some separations even with standard commercial plates, is less definite (Ledeen et al., 1981). Nevertheless, many experienced investigators use heat activation of plates for some separations (Ando et al., 1978; Chigorno et al., 1982; Mullin et al., 1983; Hunter et al., 1981). For some quantitative studies, prewash-

Table 1
Solvents Used for TLC Separation of Gangliosides

Single development	
Solvents	Reference
B:P:0.55% KCl (6:3:2)	Iwamori and Nagai (1978)
B:P:W (3:2:1)	Klenk and Gielen (1961)
B:P:0.1% KCl (9:6:4)	Ando and Yu (1979); Itoh et al. (1981)
Pr:W (7:3)	Kuhn et al. (1961)
Pr:A:W (6:2:1)	Weicker et al. (1960)
Pr:12.5% A (4:1)	Jatzkewitz (1961)
Pr:P:W (7:1:2)	Itoh et al. (1981)
Pr:0.25% KCl (3:1)	Svennerholm and Fredman (1980)
Pr:0.25% KCl (7:3)	Svennerholm and Fredman (1980)
C:Me:0.25% KCl (60:35:8)	Harth et al. (1978)
C:Me:W: (50:50:8)	Svennerholm (1963b)
C:Me:W (14:16:1)	Jatzkewitz (1961)
C:Me:W (60:35:8)	Wagner et al. (1961)
C:Me:W (70:30:4)	Harth et al. (1978)
C:Me:W:CaCl$_2$·2H$_2$O (55:45:10:0.0.02%)	Ando and Yu (1979)
C:Me:W:CaCl$_2$·2H$_2$O (60:35:8:0.02%)	Van den Eijnden (1971)
C:Me:W:0.2% CaCl$_2$ (50:45:10:0.02%)	Saito et al. (1985)
C:Me:2.5 M-A (60:35:8)	Wherrett and Cumings (1963)
C:Me:3.5 M-A (55:40:10)	Wherrett et al. (1964)
C:Me:5 N-A (55:45:10)	Iwamori et al. (1982)
C:Me:0.25% KCl (50:40:10)	Higashi et al. (1984)
C:Me:2.5 N-A (60:32:7)	Svennerholm and Fredman (1980)
C:Me:2.5 N-A (60:40:9)	Svennerholm and Fredman (1980); Ledeen and Salsman (1965)
THF:W (5:1)	Eberlein and Gercken (1975)
Methyl acetate:2-propanol:2.5% KCl (45:30:20)	Zanetta et al. (1977)

Multiple development	
Solvents[a]	Reference
a) Chloroform	
b) C:Me:W (70:30:4) or C:Me:0.25% KCl (60:35:8)	Harth et al. (1978)
c) C:Me:0.25% KCl (60:35:8)	

a) C:Me:W:12 mM MgCl$_2$, 15M   Rosner (1980, 1981)
   A (60:35:7.5:3)
b) C:Me:12 mM MgCl$_2$ (58:40:9)
   at 38°C

---

[a]a, b, and c refer to first, second, and third solvent developments.

Two-dimensional TLC

Solvents[a]	Reference
a) C:Me:0.2% CaCl$_2$ (60:35:8)	Brownson and Irwin (1982)
b) C:Me:2.5 M A (50:50:15) with 0.2% KCl	
a) C:Me:0.2% CaCl$_2$ (50:40:10)	Chigorno et al. (1982)
b) P:17M A:water (6:2:1)	
a) C:Me:0.2% CaCl$_2$ (55:45:10)	Hunter et al. (1981)
b) C:Me:2.5 M A with 0.2% KCl (50:40:10)	
a) C:Me:0.2% CaCl$_2$ (50:40:10)	Hunter et al. (1981)
b) C:Me:2.5 M A with 0.2% KCl (50:40:10)	
a) C:Me:0.2% CaCl$_2$ (50:40:10)	Ohashi et al. (1979)
b) C:Me:0.25% KCl (50:40:10)	
b) C:Me:28% A/W (60:40:3:6)	
a) C:Me:0.2% CaCl$_2$ (50:40:10)	Ledeen et al. (1981)
a) C:Me:0.25% KCl (50:40:10)	
b) C:Me:2.5 M A (50:40:10)	
b) C:Me:2.5 M A with 0.25% KCl (50:40:10)	

---

[a]a) refers to the solvent used in the first dimension and b) refers to the solvent used in the second dimension. In some cases, two solvents are used sequentially in the same dimension.

Prewashing TLC plates

Solvents	Reference
C:Me (2:1)	Mullin et al. (1983)
Corresponding solvent system	Hunter et al. (1981)

ing of plates may be necessary (Table 1), but again this is not universally considered necessary (Chigorno et al., 1982). For each of these three steps, it is best for the investigator to determine its value for the type of sample being separated. In our laboratory, we prewash the HPTLC plates in the solvent to be used prior to spotting the sample, which will be quantitated using scanning densitometry. This is done simply by allowing the solvent to move up the plate in the same way as when developing it during sample separation. It is not necessary to remove the plate as soon as the solvent front reaches the top of the plate, but the plate should not stay in the solvent any more than 24 h. The plate is then removed, air dried, and activated immediately prior to use by heating at 100°C for 30 min. This is of greatest importance on days of high humidity. The sample is dissolved in approximately 25 μL of C/Me (2:1) and spotted using a 25 μL Hamilton syringe, rinsing the sample tube a few times with very small quantities of solvent, which are also spotted. We pour the solvent into the developing tank at least 1 h before the plate is developed, but we do not line the tanks with paper. For small plates (10 × 10 cm), we use tanks 6 × 11 inches or 6 × 7 inches. A maximum of two plates is run at one time in a tank.

Two interesting systems deserve attention. Harth et al. (1978) developed a TLC method for quantitating gangliosides that did not require purification of the gangliosides from other lipids prior to chromatography. Total lipid containing 5–8 μg of ganglioside N-acetylneuraminic acid is spotted on a DC-Fertigplatten Kieselgel 60 plate (Merck), which is developed sequentially in three different solvents: (1) Chloroform—moves nonpolar lipids to the top of the plate; (2) C:Me:W (70:30:4)—moves phospholipids ahead of the gangliosides; (3) C:Me:0.25% KCl (60:35:8)—separates the gangliosides. The plate is then sprayed with resorcinol reagent, and gangliosides are quantitated by scanning densitometry (*see below*). This approach has not been widely employed, but deserves further attention.

Some minor gangliosides in nervous tissues change their chromatographic mobility after exposure to alkaline conditions. Those that are susceptible to such treatment contain either internal esters (lactones) or O-acetylated sialic acids. Sonnino et al. (1982) devised a two-dimensional TLC method to identify these. A mixture of gangliosides containing the alkali-labile gangliosides is spotted in the corner of an HPTLC plate, which is developed first in one dimension, air-dried, and exposed to ammonia vapor for 30 min at room temperature. It is then developed in the second dimension with the same solvent and the gangliosides located with p-aminobenzaldehyde reagent. Quantitation of the spots is done with scanning densitometry as described below. Gangliosides that are not susceptible to the base treatment lie on a straight diagonal line. Those that are alkali-labile, however, migrate off of this straight line. Spots are tentatively identified in relation to known standards that are run in in one dimension only with each development. Obviously, this method is of value only if the sample being studied has not been exposed to base.

Overpressured TLC (OPTLC) is a recent modification of TLC (Pick et al., 1984a) that has been applied to ganglioside separation and analysis (Pick et al., 1984b). Mixtures of gangliosides or gangliosides and sulfatide are applied to Merck Si 60 HPTLC plates, which are prewashed in methanol. Specimens are applied in 200 nL, methyl red is spotted as an indicator at both ends of the plate, and solvents for development are pumped under pressure using a Chrompes 10 overpressured layer chromatograph. Gangliosides can be separated from sulfatides with an isocratic run of C:Me (7:3) in 7 min. Individual ganglioside species are separated using a step gradient of C:Me (7:3) and C:Me:0.25% KCl in water (55:36:9). The separated gangliosides are visualized with resorcinol reagent and quantitated with scanning densitometry. The amounts of material that can be separated by this method were not reported, making it difficult to assess its relative merit in comparison with other available methods. Its speed and the reproducibility of separations make it of interest for future studies, however.

5.2. Detection

Several methods of detecting the separated ganglioside bands have been advocated over the past two decades. The most specific and commonly used way is to spray the developed plate with resorcinol reagent (Svennerholm, 1957b), cover the plate with a clean glass plate, and heat them until the gangliosides appear as purple bands (Wherrett and Cumings, 1963). Unfortunately this procedure destroys the gangliosides so that subsequent structural studies can not be performed. Other destructive methods also used include orcinol-HCl reagent (Zanetta et al., 1977), orcinol-sulfuric acid (Eberlein and Gercken, 1971), benzidine reagent (Bischel and Austin, 1963), and p-dimethylaminobenzaldehyde (Chigorno et al., 1982), all of which are sprayed onto the plates. Charring will also detect separated gangliosides, but this is the least specific of all methods. It can be done by spraying the plate with concentrated sulfuric acid or 2% sulfuric acid in glacial acetic acid (Sandhoff et al., 1968) and heating at 120°C until the spots appear. Nondestructive but less specific methods include spraying with bromthymol blue (BTB) (Wherrett et al., 1964), or exposure of the plate to I_2 vapors. Both of the latter two are less sensitive techniques, but the silica gel containing the bands can be scraped from the plate, and the gangliosides extracted from the gel without being chemically modified (Ledeen, 1966).

5.3. Quantitation

Quantitation of individual separated bands has been done in two ways. Either the identified bands are scraped into test tubes and the amount of ganglioside determined colorimetrically, or the plate with visualized bands is subjected to scanning densitometry. Gel containing gangliosides detected with BTB is first eluted with acetone in which BTB is soluble, but gangliosides are not. The gangliosides are then extracted from the gel with C:Me (2:1) and methanol. Quantitation can then be done using one of the colorimetric methods. Although it was claimed that 95% of the ganglioside is recovered using this method, this procedure has not been carefully studied using small amounts of ganglioside. Suzuki (1964) separated samples containing about 40 μg of ganglioside sialic acid on TLC and detected the bands with I_2 vapors. Bands are marked off with a needle in the gel, and the I_2 removed by sublimation, which can be facilitated using a warm air blower such as a hair dryer. Circumscribed portions of gel are then individually scraped into test tubes. Water is added to the gel to elute the ganglioside, and ganglioside sialic acid quantitated directly using the resorcinol reaction (Svennerholm, 1957b). With this method, 2 μg of sialic acid can be accurately determined in each band.

MacMillan and Wherrett (1969) improved upon this method in two ways. First, they detected the gangliosides by spraying the developed plate lightly with resorcinol reagent and heating at 150°C until the purple spots just became visible. Each spot was immediately scraped into separate test tubes and the resorcinol reaction continued directly without prior elution of the gangliosides. In addition they used a scaled down resorcinol assay that allowed detection of as little as 0.75 μg of sialic acid. The conditions affecting this procedure were studied by Yates and Thompson

(1977) who found that recoveries could reach 100% when the plates were heated at 135°C for 6 min. This was the method of choice for quantitating gangliosides separated by TLC until the development of scanning densitometry was widely employed.

The value and limitations of scanning densitometry have been documented using different methods of visualizing and scanning the ganglioside bands. Table 2 summarizes some of the important characteristics of different methods reported. Sandhoff et al. (1968) first employed scanning densitometry to determine the proportions of individual gangliosides within a mixture separated by TLC. Gangliosides are visualized by charring with 2% sulfuric acid in glacial acetic acid and heating the plate for 30 min at 180°C. The spots are quantitated by scanning densitometry using the transmission mode. The method is linear between 1 and 100 µg of ganglioside, and there is an error of only 2% with replicate runs between 5 and 25 µg of ganglioside. Although this method is simple and fast, the ganglioside sample chromatographed must be very pure or the amounts of contaminating substances that char will be erroneously included in the final values as gangliosides. Brady et al. (1969) used a Zeiss M4 QIII chromatogram spectrophotometer coupled to an electronic integration circuit to determine the absorption at 580 nm of ganglioside spots visualized by the resorcinol reagent. Over the range of 0.5–5 nmol, the error was ±2%. Ando et al. (1978) conducted a detailed study of the characteristics of scanning gangliosides on TLC visualized with resorcinol. Using the transmittance mode, they found a linear relationship between the detector response and amount of sialic acid between 0.5 and 10 nmol (0.15–3.0 µg). The lower limit of sensitivity was 0.09 nmol. Mullin et al. (1983) found that the range of linearity depended upon the thickness and density of staining of the separated band. Thus the linear range for different gangliosides differed both from each other in the same solvent system and for the same ganglioside run in different systems. They determined that this was caused by differences in band geometry, and from this they suggested that ganglioside standards should be run on the same plate as the unknown mixtures. Using their system, the limit of sensitivity was 10 pmol; the linear range varied between 10 and 900 pmol. Chigorno et al. (1982) developed a scanning densitometric method for quantitating gangliosides separated by two-dimensional TLC and visualized using *p*-dimethylaminobenzaldehyde. The plates were scanned using a Camag TLC densitometer interfaced with a 3390 Hewlett-Packard integrator.

Table 2
Scanning Densitometry

Visualization	Specificity	Scanning mode	Sensitivity[a]	Linear range[a]	Reference
Charring	Nonspecific	Transmittance	5 μg Ganglioside	5–100 μg Ganglioside	Sandhoff et al. (1968)
Resorcinol	Sialic acid	Not stated	Not stated	0.5–5.0 nmol	Brady et al. (1969)
Resorcinol	Sialic acid	Reflectance	10 nmol	0–40 nmol	Smid and Reinisova (1973)
Orcinol-HCl	Carbohydrate	Not stated	At least 0.5 μg	Not stated	Zanetta et al. (1977)
Resorcinol	Sialic acid	Reflectance	Not stated	Not stated	Harth et al. (1978)
Resorcinol	Sialic acid	Transmittance	0.09 nmol	0.5–10 nmol	Ando et al. (1978)
p-Dimethylamino Benzaldehyde	Sialic acid	Not stated	0.1 nmol	0.1–6 nmol	Chigorno et al. (1982)
Resorcinol	Sialic acid	Transmittance	10 pmol (maximum)	10–900 pmol	Mullin et al. (1983)

[a]Amount per spot.

294

The entire plate was scanned in 1-mm strips. The limit of sensitivity was 0.1 nmol, and the linear range was 0.1–6 nmol.

5.4. Identification

The most common way that ganglioside spots separated by TLC have been "identified" is by chromatographic mobility in relation to known standards run on the same plate. Although this is to a degree reasonable when working with tissues in which the gangliosides have been well characterized structurally, it should be remembered that even then such identification is tentative because structurally different compounds can co-chromatograph. When working with tissues, especially pathological ones, in which the structural nature of the ganglioside is not known, the positive identification of a ganglioside on the basis of thin layer chromatographic behavior alone is not justified. Unfortunately, it is under just such circumstances that the amount of material may be so limited that it precludes complete structural characterization. In such cases separating the sample in two different solvent systems will increase the likelihood of correct identification. Such comparisons are often made using both a neutral and a basic solvent system.

Recently, immunological methods have become available that can directly yield information about the oligosaccharide portion of glycolipids separated by TLC. The technique was initially described by Magnani et al. (1981), who developed it to demonstrate binding of a specific anti-ganglioside monoclonal antibody to its ganglioside antigen on a thin layer chromatogram. This was done by soaking the air-dried chromatogram in 1% polyvinylpyrrolidine with 0.1% sodium azide in phosphate buffer, which was then exposed to the antibody. ^{125}I-labeled F(ab')2 of rabbit immunoglobulin G antibodies to mouse immunoglobuins were layered on top and visualized with autoradiography. A modification of this consists of using a horseradish-conjugated secondary antibody and 4-chloro-1-naphthol as a chromogenic substrate for the peroxidase reaction (Higashi et al., 1984; Harpin et al., 1985). This yields a very low level of background, and the colored product can be quantitated by scanning densitometry at 578 nm. Using an affinity-purified chicken anti-N-glycolylneuraminic acid-lactosylceramide antibody, color development was linear to 10 pmol and semilinear to 50 pmol with a sensitivity of 0.5 pmol. The sensitivity varies among different antigens, however, and probably depends to a great deal upon the affinity of the antibody. A modification of this general approach was made by Towbin et al. (1984) who transferred the separated glycolipids from TLC to nitro-cellulose sheets by diffusion. The plates were then incubated with anti-glycolipid antibodies, which were detected using a second antibody coupled to peroxidase and a color reaction. The advantages of this and the previous method are their simplicity, speed, use of nonradioactive reagents, and applicability for screening large numbers of samples for antibodies.

The value of this general approach was extended by Saito et al. (1985). They developed a method for digesting the separated gangliosides on TLC with glycosidases prior to exposing them to antibodies directed toward the carbohydrate portions of the molecules. The oligosaccharide products are then detected using ^{125}I-staphylococcal protein A (SPA) and autoradiography. The procedure was studied using *Arthrobacter ureafaciens* neuraminidase, which cleaves all of the sialic acid moieties, including the internal one, from gangliosides on the thin layer plates. Following such neuraminidase treatment on TLC of several gangliosides (GM_1, GD_{1a}, GD_{1b}, GT_{1a}, GT_{1b}, GQ_{1b}), anti-Ggose$_4$ antibody reacted with all of the products and was detected using ^{125}I-SPA and auto-radiography. It should be noted that similar treatment with *Clostridium perfringens* neuraminidase failed to convert any of these gangliosides to Ggose$_4$. The authors mention that, with this method, jack bean β-galactosidase converts GD_{1b} on TLC into GD_2, which can then be detected with a monoclonal antibody directed against GD_2. Through the combined use of several glycosidases and antibodies, this method has a great deal of potential for gaining considerable amounts of information about the structural characteristics of gangliosides using extremely small amounts of material. Therefore, it should find wide application in many areas.

Another recent development is the use of matrix-assisted secondary ion mass spectrometry to analyze lipids separated by TLC while they are still adsorbed to the silica gel (Kushi and Handa, 1985). In this procedure, lipids are separated on aluminum- or plastic-backed thin layer silica gel plates. The lipid of interest is visualized by a nondestructive method (e.g., iodine vapors); the area of the plate containing the lipid is excised and placed into contact with the secondary ion mass spectrometer probe tip. A lipid solvent and matrix liquid are applied to the probe tip, which is inserted into the mass spectrometer. The authors claim that mass spectra comparable to those obtained using conventional methods can be obtained using as little as 1 μg of GM_2. This method could be of considerable value in determining the structure of unknown gangliosides separated by TLC.

6. High Performance Liquid Chromatography

High performance liquid chromatography (HPLC) has been used to great advantage in the analysis and preparation of a wide range of compounds. The potential advantages of sensitivity, speed, resolution, reproducibility, and adaptability for analytical and preparative purposes have prompted several investigators to explore ways to use this method to study gangliosides. Table 3 summarizes several important characteristics of the published methods to date, which are discussed below.

Tjaden et al. (1977) separated underivatized mono-, di-, and tri-sialogangliosides on silica columns with acidified C:Me:W solutions. Acidification was essential to neutralize the negative charge that interacted with the stationary phase and distorted the peak shapes. Compounds were detected using a moving wire detection system with a flame ionization detector. Although mono-sialogangliosides were well separated, the level of purity of iso-lated di- and tri-sialogangliosides was lower. With GM_1, peak areas were linear between 2 and 200 μg, but it was essential to keep wire speed and solvent flow rates constant. For this method to be quantitative, it is necessary to know the carbon content of each compound. Therefore, it is of less value in quantitating mixtures of unknown gangliosides, and standard curves of known ganglio-sides must be prepared. It has also been used for micropreparative purposes, but ethanol had to be substituted for methanol because the latter solubilizes silica. With an ethanol system, separation of di- and tri-sialogangliosides is less satisfactory.

Tettamanti and coworkers have developed both analytical and preparative HPLC procedures for underivatized gangliosides that are detected at 195 nm using a flow-through cell. Up to 5 mg of either GM_1 or GD_{1a} can be separated on a semi-preparative reverse phase Spherisorb-S5 column with acetonitrile-sodium phosphate buffer (Sonnino et al., 1984). Following HPLC, the samples re-quired a few simple purification steps, but this allows isolation of samples containing a single type of long-chain base. The procedure was scaled down for analytical purposes, and it was found to be reliable with starting materials between 1 and 10 μg. This method

Table 3
Characteristics of Several HPLC Systems[a]

Derivative	Columns	Solvents	Detector	Scale	Sensitivity	Time	Compounds[a] studied	Reference
None	Silica Si 60	Acidified C:Me:W	Moving wire and flame ionization detector	Analytical; micropreparative	2 μg ganglioside	40 min	Mono, di, tri	Tjaden et al. (1977)
None	Spherisorb-S5	Acetonitrile:phosphate buffer	UV (195 nm)	Semipreparative (5 mg)	—	40–60 min	GM_1; GD_{1a}	Sonnino et al. (1984)
None	Spherisorb-S5	Acetonitrile:phosphate buffer	UV (195 nm)	Analytical (1–10 μg)	—	40 min	GM_1;GD_{1a}	Sonnino et al. (1984)
None	Lichrosorb RP-8	Acetonitrile:phosphate buffer	UV (195 nm)	Analytical	0.1 nmol	6–16 min	Mono, di, tri, tetra	Gazzotti et al. (1984)
None	μBondapak RP-18	Acetonitrile:phosphate buffer:tetrahydrofuran	UV (195 nm)	Preparative (50 mg)	0.1 nmol	—	GM_1; GD_{1a}	Gazzotti et al. (1984)
None	Ultrasphere ODS	Methanol:water	UV (205 nm)	—	5 μg	—	Mono	Kadowaki et al. (1984)
Perbenzoyl	Silica (3 μm)	2-Propanol:hexane	UV (230 nm)	Analytical (8 nmol)	5 pmol	15 min	Mono, di, tri, tetra	Ullman and McCluer (1985)
Perbenzoyl	Zipax	Dioxane:hexane	UV (230 nm)	Qualitative	—	30 min	Mono, di	Lee et al. (1982)

Derivatization	Column	Solvent system	Detection	Scale	Sensitivity / time	Gangliosides analyzed	Reference
p-Bromophenacyl	Zorbax Sil; Zorbax C8	2-Propanol:hexane:water; Acetonitrile:methanol	UV (261 nm)	Quantitative	10 ng N-acetylneuraminic acid	Mono	Nakabayashi et al. (1984)
Dinitrophenyl hydrazides	Silica	Chloroform:methanol:water:acetic acid	UV (342 nm)	Quantitative (0.02–1.6 nmol)	0.02 nmol	Mono, di, tri, tetra	Miyazaki et al. (1986)
p-Nitrobenzyloxyamine	µBondapak C18	Methanol:water	UV (254 nm)	Analytical	—	Major brain gangliosides	Traylor et al. (1983)
None	Zorbax Sil	2-propanol:hexane:water	Monitor by TLC	Semipreparative (50–500 µg)	1–2 h	Mono, di, tri	Kundu and Scott (1982)
None	DEAE-CPG; CPG	Chloroform:methanol; Chloroform:methanol:water; chloroform:methanol:Li acetate	Monitor by TLC	—	140 min	Mono, di, tri, tetra; neutral glycolipids	Watanabe and Tomono (1984)
None	Lichrosorb-NH2	Acetonitrile:water K2PO4	UV (195 nm)	25 nmol	0.6 nmol; 80 min	Sialooligosaccharides	Bergh et al. (1981)

"Mono, di, tri, tetra = monosialo-, disialo-, trisialo-, and tetrasialo-gangliosides. See text for other abbreviations.

was also found to be useful in purifying radiolabeled gangliosides of any one base type.

Another system developed by the same group can separate a mixture of gangliosides varying from one to four sialic acids per molecule (Gazzotti et al., 1984). For analytical purposes, 20–100 nmol of underivatized mixed brain gangliosides are separated on a reverse phase Lichrosorb RP-8 column using acetonitrile-phosphate buffer. The limit of detection using a flow-through cell at 195 nm was 0.1 nmol. Each ganglioside standard separated into four peaks on the basis of the long-chain base (C18 and C20 sphingosines; C18 and C20 sphinganines). Although each of these four peaks separated well, there were some that overlapped when complex mixtures of gangliosides were analyzed. The peak areas were linear in the range 0.1–50 nmol, but the slope of this line varied among different gangliosides. Therefore, standards of all gangliosides in the mixture must be run with each analysis for quantitative purposes.

Reverse phase chromatography of underivatized samples has also been used to separate individual species of monosialogangliosides (Kadowaki et al., 1984). This method employs an Ultrasphere ODS column and different proportions of Me:W solutions to separate purified monosialogangliosides on the basis of both fatty acid and long-chain base contents. The UV detector is set at a wavelength of 205 nm. This method can be used directly for qualitative analyses, or samples from individual peaks can be collected, perbenzoylated, and quantitated using normal phase HPLC (Bremer et al., 1979). In establishing this method, the authors conducted extensive structural characterization, including analyses of long chain bases (LCBs) by HPLC. In this method the LCBs are hydrolyzed from gangliosides in aqueous acetonitrile-HCl and derivatized with biphenylcarbonylchloride. Derivatized LCBs are separated on the reverse phase column (Ultrasphere-ODS, 5 μm) and detected at 280 nm (Kadowaki et al., 1983).

Separation and quantitation of complex ganglioside mixtures can also be accomplished using perbenzoyl derivatives (Ullman and McCluer, 1985). It is essential to heat the 3-μm mean particle size silica column to 90°C and employ a linear gradient of 2-propanol in hexane. The UV detector is set at 230 nm, and quantitation is by the external standard method, with peak areas being linear between 5 pmol and 8 nmol. All major and several minor brain gangliosides can be analyzed by this method with standard deviations of less than 3%. Lee et al. (1982) were able to obtain

information concerning the structures of several neutral glycolipids and gangliosides by separating perbenzoyl derivatives of
original and glycosidase-treated compounds on a Zipax column
with dioxane in hexane.

UV-absorbing p-bromophenacyl derivatives of gangliosides
have been studied by Nakabayashi et al. (1984). Good separations
of major brain monosialogangliosides were achieved on a silica gel
column using 2-propanol:n-hexane:water. GD_{1a} was not well resolved, however, and results of separations of polysialogangliosides were not reported. Separations of some species on the basis
of long-chain bases were obtained with reverse phase
chromatography. There appear to be good prospects for quantitating separated gangliosides with this method because of the high
sensitivity and linearity of the UV absorbing constituent. One
major advantage is the simplicity and speed of derivatization.

A rapid and convenient method to extract and analyze gangliosides using HPLC has recently been described (Miyazaki et al.,
1986). Initially, the tissue is extracted with water to remove interfering amino acids, then the pellet is extracted with C:Me (1:1)
at 70°C. The mixture of lipids is reacted with 2,4-dinitrophenylhy-
drazine HCl under conditions that attach the 2,4-dinitrophenylhy-
drazine to the carboxyl groups of sialic acids. The reactants are
passed through a small Unisil column to remove excess reagents,
and the resulting mixture is directly injected into the HPLC. Separation is on a Resolve 5-μm Spherical silica column with solvents
composed of C:Me:W:acetic acid. The solvent proportions are
different for separating gangliosides with lower or higher carbohydrate contents. Detection is by UV absorbance (342 nm), and the
detector response is linear between 0.020 and 1.6 nmol.

Traylor et al. (1983) developed a method for quantitating gangliosides on the basis of their olefinic content. This involves a
preliminary ozonolysis step before reacting them with triphenyl-
phosphine to cleave the ozonide and form aldehydes. The latter are
reacted with p-nitrobenzyloxyamine to form oxime derivatives.
These ganglioside derivatives are first isolated from other interfering substances in the reaction mixture on a DEAE-Sephadex A-25
column, and then separated by means of HPLC on μBondapak C18
reverse phase column using methanol:water. Detection is by UV
absorbance (254 nm). Using this method, quantitative analyses of
gangliosides isolated from human brain gave similar results to
analyses using TLC and scanning densitometry. Although the
limit of sensitivity was not investigated, it appears to be less than 1

μg of sialic acid in the aliquot injected into the HPLC. It was suggested that the fluorescence characteristic of the compound might be exploited to increase the sensitivity by two orders of magnitude. The limitation of this method is that the olefinic content of gangliosides studied must be known.

Silica gel HPLC of underivatized gangliosides has been used to isolate individual species of gangliosides (Kundu and Scott, 1982). First, a mixture of purified gangliosides is separated into mono-, di-, and tri-sialo compounds. Each of these groups is then chromatographed on a Zorbax Sil column using a linear gradient of 2-propanol:hexane:water. One-milliliter fractions are collected and monitored by TLC. Separations of major monosialogangliosides are very good on the initial run, but purification of minor ones and some of the polysialogangliosides requires appropriate pooling and rechromatography. Each run lasts 1–2 h. Although it was reported that yields were 50–500 μg, the authors were optimistic that this method could be considerably scaled up.

A one-step separation of neutral glycolipids and gangliosides using HPLC has been described (Watanabe and Tomono, 1984). This involves three serially connected columns. The first consists of DEAE conjugated to controlled pore glass (CPG); the second and third are unconjugated CPG. Mixtures of prepared neutral glycolipid and ganglioside standards are injected into the HPLC, and the two classes separated with C:Me (95:5). The columns are then eluted with progressively more polar solvents that elute neutral glycolipids first and then gangliosides of increasing polarity. The effluent is collected in fractions that are monitored by TLC. Although several different types of neutral glycolipids and gangliosides were studied, results of samples directly extracted from tissues were not reported. Therefore, its application to specific biological problems has yet to be demonstrated.

Bergh et al. (1981) described a method to separate and quantitate underivatized mixtures of oligosaccharides isolated from glycoproteins. Approximately 25 nmol of each oligosaccharide is injected onto a Lichrosorb-NH_2 column and eluted with acetonitrile:water:potassium phosphate. The eluate is monitored by UV absorption. It was emphasized that the procedure is nondestructive so that structural characterizations can be performed on the purified samples. Although this was designed to study glycoproteins, it seems that it could also be used to investigate oligosaccharides isolated from gangliosides.

7. Labeled Gangliosides

7.1. Radiolabeled Gangliosides

7.1.1. Preparation

The obvious value of radiolabeled gangliosides in studies on the metabolism and transport of these compounds has led to several different methods of producing them. Burton et al. (1963) administered ^{14}C-labeled precursors of gangliosides intraperitoneally to young rats and studied subsequent labeling patterns of brain gangliosides. They found that maximum synthesis of gangliosides occurred between 10 and 12 d of age, and that maximum levels of radiolabeled gangliosides were obtained 24 h after injection. Radioactivity from glucose and galactose was incorporated into all the carbohydrate moieties of the gangliosides, but glucosamine labeled primarily galactosamine and sialic acid. D,L-[3-^{14}C]serine was a very nonspecific precursor, labeling sialic acid sphingosine and all of the carbohydrates. Radioactivity of D-[1-^{14}C]glucose and D-[^{14}C]galactose labeled all of the carbohydrate residues of gangliosides. Based on their results, we have produced ^{14}C-labeled gangliosides by injecting 17 μCi of N-[^{14}C]acetylglucosamine intracerebrally into 12-d-old rats that were sacrificed 24 h later. Gangliosides isolated from the brains of these animals had a specific activity of 415 cpm per μg of N-acetylneuraminic acid. It should be noted that extensive purification of these is required to remove radiolabeled contaminants (see above). Kolodny et al. (1970) devised a variant of this technique to produce GM_2 labeled with tritium in the sialic acid moiety. They injected N-acetyl-D-[^{3}H]mannosamine intracerebrally into 8-d-old rats that they sacrificed 2 d later. Total brain gangliosides were extracted and converted to GM_1 with neuraminidase. The GM_1 was then converted to GM_2 with galactosidase. The resulting GM_2 was radiolabeled only in the N-acetylneuraminic acid portion of the molecule, but the specific activity was somewhat low (745–1200 cpm per nmol).

In vitro biosynthesis of gangliosides has also been achieved (Tallman et al., 1977). GM_3 labelled in the N-acetylneuraminic acid residue was synthesized from CMP-N-[^{14}C]-acetylneuraminic acid and lactosylceramide using a sialyltransferase preparation isolated from 12-d-old rat brains. GD_{1a} labeled in the terminal sialic acid was synthesized from GM_1 and CMP-N-[^{14}C]-acetylneuraminic acid using a sialyltransferase preparation from cultured human skin fibroblasts. Following incubation, the reaction mixtures were extracted with C:M (2:1), and the products isolated using Sephadex G-25 column chromatography and preparative TLC. The authors obtained a total of 9×10^6 cpm [^{14}C]GM$_3$ and 6×10^5 cpm [^{14}C]GD$_{1a}$. These products have been found suitable for use in assays for sialidase activities.

In preparing radiolabeled gangliosides from biological sources, one of the major difficulties is removing radiolabeled contaminants such as nucleotide sugars and peptides. This has been addressed above, but the reader is also referred to other publications in which this is discussed in detail (Byrne et al., 1983, 1985; Yates and Warner, 1984).

Labeling of glycosphingolipids with ^{14}C can also be accomplished using the method of Higashi and Basu (1982). First the amino sugars are N-deacetylated or N-deglycolylated by hydrazinolysis. Although this is very effective in removing the appropriate acetyl groups, it causes very little N-deacylation of ceramide, and the fatty acid and sugar compositions of the glycosphingolipids are unchanged. The N-deacetylated products are then re-N-acetylated using [1-^{14}C]acetic anhydride, forming [acetyl-^{14}C]glycosphingolipids, which are then purified using a Bio-Sil column. The latter two steps are quantitative (over 96%), and the resulting specific radioactivities achieved were $9–15.6 \times 10^6$ cpm/μmol. The major advantage of this method is that most of the radiolabeled products are not chemically modified in the process. The exceptions are those gangliosides with N-glycolylneuraminic acid, which are converted to N-acetylneuraminic acid.

Tritium-labeled glycosphingolipids can be prepared by the catalytic reduction of the unsaturated bonds using tritium gas. Di Cesare and Rapport (1974) accomplished this using platinum as a catalyst to prepare radiolabeled mixtures of several glycolipids, which were subsequently individually purified from the reaction mixture. They arranged to have the reduction step performed commercially. The final specific activities were relatively high: GM$_1$ = 23 μCi/mg; GM$_3$ = 850 μCi/mg. Although this method is fairly simple, they caution that the physical and chemical properties of these compounds may be different from the natural products. They also indicate that there is probably some destruction of the gangliosides during tritium addition that may be greater than for neutral glycosphingolipids.

Catalytic addition of tritium to the double bonds of ceramide has also been accomplished using palladium as a catalyst and potassium boro[^{3}H]hydride as a reducing agent (Schwarzmann, 1978). All gangliosides studied yielded single spots on TLC and autoradiography. The method is quite simple and fast and yields radiolabeled gangliosides of quite high specific activity (54–1910 Ci/mol). The recoveries varied between 90 and 97%, and no loss of sialic acid was detected. The authors caution that excess amounts of boro[^{3}H]hydride are necessary to ensure optimum labeling. Although others have reported that tritium-labeled neutral glycosphingolipids are unstable over the period of 1–2 mo (Hungund, 1981; Seyama et al., 1968), Schwarzmann found that GM$_1$ labeled using his method was stable over at least a 9-mo period. The location of the radiolabel (i.e., ceramide vs oligosaccharide) may be important in this regard, but the stability of radiolabeled gangliosides has not been extensively studied.

A method to tritiate gangliosides at the 3-position of sphingosine was developed by Ghidoni et al. (1981). This involves two steps: (1) oxidation of the 3-position of sphingosine using 2,3-dichloro-5,6-dicyanobenzoquinone (DDQ); and (2) reduction of the 3-keto-ganglioside with [^{3}H]NaBH$_4$. The oxidation step requires the presence of Triton X-100 and is performed in toluene, which creates inverted mixed micelles. In this condition the N-acetylneuraminic acid moiety of the ganglioside is unexposed to the DDQ and is thus not susceptible to the oxidation process. DDQ specifically oxidizes allylic hydroxyl groups, forming the 3-keto derivatives of gangliosides. A major advantage of this method is that C18 and C20 sphingosine containing gangliosides are equally labeled. It is uncertain, however, to what degree the reduction step creates the threo form of gangliosides, which may have different physical properties (such as hydrogen bonding capacity) than the erythro forms. Following such treatment, GM$_1$ and GD$_{1a}$ migrate as single spots on TLC and have a radiochemical purity of over 95%. The specific radioactivity of these compounds was 1.25 Ci/mmol, and less than 1% of the radiolabel was in the saccharide portion of the molecule. Therefore this method should have wide application to studies on the biochemical and biophysical properties of gangliosides.

Ganglioside analogs (gangliosidoides) with tritium in the aliphatic portion of the molecule were chemically synthesized by Wiegandt and collaborators. The first one synthesized (1-deoxy-1([9- or 10-^{3}H]stearoylamido)-II^{3}monosialogangliotetraitol) is an analog of GM$_1$ and contains only one hydrocarbon chain linked to the oligosaccharide through an amide bond (Wiegandt and Ziegler, 1974). They later reported the synthesis of another analog of GM$_1$, 1-deoxy-1-2-([9- or 10-^{3}H]stearoylamido)stearoylamido II^3-monogangliotetraitol, which contains two hydrocarbon chains, but only one of them is labeled with tritium (Mraz et al., 1980).

Veh et al. (1977) developed a method to tritiate the sialic acid moiety of ganglioside. This involves periodate oxidation and subsequent boro[^{3}H]hydride reduction. They thoroughly investigated the conditions for oxidation using GM$_1$ and GD$_{1a}$ and determined the conditions that produced mainly C8-N-acetylneuraminic acid or C7-N-acetylneuraminic acid. This is of importance because these two derivatives are cleaved at different rates by several neuraminidases and may impart different physical properties to the labeled gangliosides. Low periodate concentrations (1 mM) will oxidize the side chain of N-acetylneuraminic acid, but cause minimal destruction of other monosaccharides in the ganglioside molecule. Using their optimized technique, they obtained specific activities for GM$_1$ and GD$_{1a}$ of 3.5 Ci/mol and 9.1 Ci/mol, respectively, and this could be increased by two orders of magnitude using boro[^{3}H]hydride of higher specific activity.

Galactose oxidase will oxidize the terminal carbon of galactose and N-acetylgalactosamine of gangliosides that are not substituted on the terminal galactose moiety. Boro[^{3}H]hydride will reduce the aldehyde of the product back to a carbinol, thus introducing a tritium atom at that position. Suzuki and Suzuki (1972) successfully applied this method for producing radiolabeled GM$_2$, asialo-GM$_2$, and globoside. Ghidoni et al. (1974) improved this technique by performing the oxidation step in the presence of Triton X-100. The detergent markedly changed the reaction kinetics, yielding a radiolabeled product with a 10-fold higher specific activity (1.31 × 10^5 cpm/nmol) than the original method produced. The authors thought that the detergent formed mixed micelles with the ganglioside, resulting in a more efficient interaction of the enzyme with the substrate. Using GM$_1$ it was found that, of the total radioactivity in the molecule, 90% was in the terminal galactose, 6% in the subterminal N-acetylgalactosamine, 8% in the internal galactose, and 0.4% in the glucose moieties. Although the oxidation and reduction steps are relatively easy to perform, the products require several purification steps, and the method is practical only for those compounds having an unsubstituted terminal galactose or galactosamine residue.

A variation of the previous method was devised by Klemm et al. (1982) to radioiodinate GM$_1$. Following oxidation with galactose oxidase, the carbonyl is reacted with tyramine to form an imine that in turn is reduced with cyanoborohydride to a secondary amine. The tyramine linked to the ganglioside is then iodinated using $Na^{125}I$. The radiolabeled GM_1 is purified either by dialysis or TLC. The advantages are that the final product has a high specific radioactivity (1.3×10^7 cpm/nmol) and is labeled with a gamma-emitting atom. The disadvantage is that the compound is chemically altered by the addition of the tyramine residue that will change many of its chemical and physical properties.

7.1.2. Detection of Radiolabeled Gangliosides

7.1.2.1. RADIOACTIVE COUNTING. The amphipathic nature of gangliosides permits them to dissolve in a wide range of solvents both organic and aqueous. Therefore, radioactive counting of gangliosides has been done in scintillation cocktails of different types: water and Ready-Solv HP, Beckman (Miller-Podraza et al., 1982); water and Aquasol, New England Nuclear (Yohe et al., 1980); water and Hydrofluor, National Diagnostics (Byrne et al., 1985); Instagel, Packard (Ghidoni et al., 1983); Unisolve scintillation fluid (Radsak and Wiegandt, 1984); dioxane with SDS, diphenyloxazole, and naphthalene (Cortassa et al., 1984). Combustion of the ganglioside prior to liquid scintillation counting has also been done (Forman and Ledeen, 1972). The physical state of gangliosides in many of these solutions is uncertain, expecially those that contain detergents. It is likely, however, that in the water-based ones that they exist as micelles, probably mixed micelles, with detergent molecules. This provides two possible sources for quenching: (1) solute self quenching of molecules within the micelle; and (2) phase quenching caused by the nonhomogeneity of the micelle and solvent. The latter could be variable depending upon both where the radiolabel is in the ganglioside molecule and upon the nature of the micelle. For gangliosides with the label in the part of the molecule facing the outside of the micelle, phase quenching may be negligible, but the reverse situation could lead to considerable quenching. To our knowledge, this problem has never been adequately addressed. Until it is more completely understood, it seems most prudent to use as a standard for determining quenching a compound radiolabeled in a position and counted at a concentration as close as possible to the test ganglioside.

Not only the type of scintillation cocktail, but the way in which gangliosides are solubilized, can affect counting efficiency. We compared the relative counting efficiencies of [^{3}H]GM$_1$ labeled in the terminal galactose using PPO-POP in toluene and Aquasol-2 (New England Nuclear) counted in a Beckman LS-7000 correcting for quenching with an external standard (Warner and Yates, unpublished results). The highest counts were obtained by dissolving the gangliosides in 1 mL of water and then adding 10 mL of Aquasol-2. Compared with this, PPO-POP in toluene gave only half as many counts (Table 4). Aquasol-2 alone yielded 76%, and increasing amounts of water up to 1 mL gave progressively higher counts. Adding the Aquasol-1 before the water gave only 94% as many counts as the reverse order, however. A similar finding was obtained when radiolabeled gangliosides spotted on silica gel TLC plates were scraped and radioactivity counted. When scrapings were placed in the scintillation vials, 1 mL of water added, and 10 mL of Aquasol-2 added immediately, the counting efficiency was only 54% of that of standards counted directly without being applied to silica plates. When water was added to the scrapings in the vial first and Aquasol-2 added 24 h later, the efficiency calculated in the same way was 89–100%. When water was in contact with the gel for only 1 h, efficiency was 6% lower (Guzman-Harty and Yates, unpublished results).

Relative distributions of radioactivity in ganglioside bands separated by TLC have been determined using TLC radioscanning instruments (Meldolesi et al., 1976; Mullin et al., 1976; Moss et al., 1977; Ghidoni et al., 1983). This method can be very useful in localizing radiolabeled gangliosides on TLC plates and for determining the relative amounts of radioactivity in different bands. There are some limitations, however. Most instruments require fairly large amounts of tritium for detection. For example, acceptable tracings were obtained by Mullin et al. (1976), but the amount of radioactive ganglioside applied to the plate was 176,000 cpm. Advances in design and computer software have improved the quality of tracings (Ghidoni et al., 1983), and further developments in technology allowing signal averaging of multiple scans of the same lane should considerably increase the sensitivity level (Tettamanti, personal communication). The cost and availability of such instruments is a current limiting factor to their general application, however.

7.1.2.2. AUTORADIOGRAPHY AND AUTOFLUOROGRAPHY. These methods have been commonly used to locate gangliosides separated by TLC. Good visualization of the spots can be obtained even with relatively small amounts of radioactivity when the ganglio-

Table 4
Counts of [^{3}H]GM$_1$ in Various Scintillation Cocktails[a]

Scintillation cocktail	Avg. dpm	Recovery, %
11 mL Aquasol-2	9788	76.0
11 mL PPO-POP in Toluene	6490	50.4
0.1 mL H$_2$0, 10.9 mL Aquasol-2	11,414	88.6
0.3 mL H$_2$0, 10.7 mL Aquasol-2	12,401	96.3
0.5 mL H$_2$0, 10.5 mL Aquasol-2	12,586	97.7
1.0 mL H$_2$0, 10 mL Aquasol-2[b]	12,878	100
1.0 mL H$_2$0, 10 mL Aquasol-2[c]	12,051	93.6

[a]All samples were prepared in triplicate. Counts per minute (cpm) converted to dpm using H number conversions.
[b]H$_2$0 added first, then Aquasol-2.
[c]Aquasol-2 added first, then H$_2$0.

sides are [^{14}C]-labeled. Kodak XR-2 X-ray film has been used, but, because it lacks sensitivity, the exposure time at room temperature is several weeks for spotted samples containing 1–2 × 10^3 cpm radioactivity (Yates et al., 1980). Kodak No Screen NS54T film provides visualization of separated bands in less time; Yogeeswaren et al. (1981) found that with 2–6 × 10^3 cpm, only 2 wk development time was required. Kodak X-Omat S, Kodak X-Omat AR (Miller-Podraza et al., 1982; Yao and Rastetter, 1985), Kodak X-Omat SO-282 (Ghidoni et al., 1983), and Kodak SB-5 (Ledeen et al., 1981) films have also been used. The sensitivity of these techniques can be increased by three additional steps. Preflashing the film by exposing it in the dark at arms length to a small photographic flash increases the sensitivity, but in our experience also increases the background density of the film, necessitating a very bright light to view the spots of interest. After TLC the dry plate can be sprayed with an autoradiography enhancer, such as PPO-DMSO or EN3HANCE (New England Nuclear) and wrapped in an airtight material such as Glad Wrap or Saran Wrap. Exposure at −70°C can also increase the sensitivity of these films (Miller-Podraza et al., 1982; Yao and Rastetter, 1985).

Because of the low-energy beta radiation, tritium-labeled gangliosides separated by TLC are less easily detected using autoradiographic methods. Autofluorography, preflashing of films, and long exposure times at low temperatures have all been employed to increase the sensitivity of the X-ray film to tritium. We have found that a mixture of gangliosides containing 5000 dpm tritium, and which resolves on HPTLC plates into about eight bands, can be visualized with the following method (Guzman-Harty and Yates, unpublished results). HPTLC plates are removed from the chromatography tank, air-dried, and sprayed three times with EN3HANCE. Between each spraying the plate is allowed to air-dry to the point of dampness and is rotated 90°. The plate is placed in contact with a Kodak SB-5 X-ray film, wrapped tightly in Saran Wrap, placed in a Kronex Lightning Plus screened cassette, and kept at –70°C for 2 wk. The film is then developed manually. Although the bands are not extremely dark with this amount of tritium, the fairly clear background allows easy visual detection of the radioactive areas. Momoi and Wiegandt (1980) used auto-fluorography to detect tritium-labeled mono- and oligo-saccharides separated on HPTLC cellulose plates. They dipped the plate twice into a 12% (w/v) solution of PPO in ether, allowed it to air-dry, then exposed it to Kodak XO X-ray film for 7–10 d at –80°C. By this method they were able to visualize single spots containing 1000 dpm tritium.

Most of the commonly used X-ray films have a layer of gelatin, which prevents scratching of the film. It has been found that when such films are used for autoradiographic detection of tritium that 90% of the tritium radiation is absorbed by this layer. Therefore an X-ray film without this anti-scratch layer of gelatin has been designed specifically for the detection of tritium (Ehn and Larsson, 1979). This is commercially available as Ultrofilm 3H (LKB Produkter AB, Bromma, Sweden). No fluorescent enhancer is used, and exposure is at room temperature. A ganglioside mixture (5000 dpm) separating as eight bands on HPTLC can be visualized after an exposure time of 4 wk. With much longer exposure times, the clarity of the bands deteriorates (Guzman-Harty and Yates, unpublished results). Because the background is clear, contrast between the developed bands and background is good. The film should be developed by hand because the absence of an antiscratch layer makes it very susceptible to artifacts. For the same reason considerable care should also be exercised when applying and removing the film from contact with the TLC plate.

7.2. Fluorescent-Labeled Gangliosides

Gangliosides have also been labeled with fluorescent compounds. Wilchek et al. (1980) devised a two-step method that can be used to label isolated gangliosides and glycoconjugates at the cell surface. First the substrates are oxidized using either sodium metaperiodate or galactose oxidase to form aldehyde groups on sialic acid, or terminal galactose or galactosamine moieties, respectively. The oxidized substrates are then reacted with one of five fluorescent hydrazides: fluoresceine-β-alanine hydrazide; rhodamine-β-alanine hydrazide; lissamine rhodamine-β-alanine hydrazide; lissamine rhodamine-β-α -aminobutyric hydrazide (Wilchek et al., 1980); Lucifer yellow CH (Spiegel et al., 1983). When added exogenously to medium containing cells, fluorescent gangliosides synthesized by this method become attached to their plasmalemma, but the nature of this attachment is still not completely characterized. Following oxidation of thymocytes and nematodes by either method, sufficient binding of the fluorescent reagents to the surface membrane occurred to visualize it with a fluorescent microscope and conduct flow cytometry on the cells. Although no fluorescence was seen inside the cells with the second two reagents, a small amount was seen with the first two listed above. These methods could be very useful in studying the distribution, uptake and movement of glycoconjugates within cells. The method for labeling cells *in situ* does not specifically label glycolipids, however. As the authors show, glycoproteins also become fluorescently labeled. Furthermore, the fluorescent group is relatively large and undoubtedly will change some chemical and physical properties of the gangliosides; periodate oxidation of gangliosides by itself changes the rates of reactions with some sialidases. Some of the fluorescent reagents used in these methods are commercially available from Miles-Yeda Company, Kiryat Weizmann, Rehovot, Israel.

7.3. Spin-Labeled Gangliosides

The first spin-labeled glycolipid to be synthesized was a derivative of galactosylcerebroside (Sharom and Grant, 1975). It was prepared from stearic acid, which was spin-labeled at the C-16 position (16-nitroxyl stearic acid), and galactosylpsychosine. Spin-labeling of gangliosides has been done in several ways. Through the formation of phosphate esters, spin labels can be incorporated into the oligosaccharide portion of the molecule (Sharom and Grant, 1977). The disadvantage of this method is that it adds both a nitroxide-containing ring and a negative charge to the oligosaccharide, which could change the physical and chemical properties of the molecules to such a degree that they do not behave the same as underivatized gangliosides in either natural or artificial membranes. Sharom and Grant (1978) also developed a method for introducing a spin label into gangliosides by esterifying primary alcohols of carbohydrates to nitroxide-containing rings through an attached carboxyl group. This has two advantages over the former method: the spin label is separated from the sugar ring by only two bonds about which free rotation can occur; the spin-labeled region is uncharged. Both of the previous two methods randomly label the carbohydrates, however. This problem has been overcome with the method of Lee et al. (1980) for adding a nitroxide-containing ring to the sialic acid moieties of gangliosides. This involves an initial periodate oxidation followed by reacting the product with 4-amino-2,2,4,4-tetramethylpiperidine-1-oxyl in Me_2SO_2. The advantages and disadvantages of these derivatives have been discussed by the investigators who adapted this technique for gangliosides.

Chemically synthesized spin-labeled analogs of GM_1 have also been studied by Wiegandt and colleagues (Kanda et al., 1982a,b). The methods to prepare these are essentially the same as those to prepare the tritiated analogs (see above), except that spin-labeled instead of tritiated stearic acid is used. By beginning with stearic acids containing the spin label on either carbon-5 [5-(N-oxyl-4,4'-dimethyloxazolidine)-stearic acid] or carbon-12 [12-(N-oxyl-4,4'-dimethyloxazolidine)-stearic acid] it is possible to prepare gangliosidoides with the label at different positions of the aliphatic chain.

Spin-labeling of sialic acids attached to glycoconjugates within biological membranes has also been accomplished (Feix and Butterfield, 1980). This involves sequential periodate oxidation and reductive amination with $NaBH_3CN$. The conditions of the former step are such that only the C-7, C-8, and C-9 vicinal hydroxyls of sialic acid are oxidized to a terminal aldehyde. This method applied to red blood cell ghost membranes labels both glycoproteins and gangliosides. Of the total electron spin resonance signal intensity from these membranes, gangliosides were responsible for about 30%. This will, of course, vary among different types of membranes.

8. Attachment to Solid Supports

For many types of studies it is necessary to have gangliosides attached to solid supports, and several methods for accomplishing this have been described. Cuatrecasas et al. (1973) coupled ganglio-sides through their sialic acid carboxyl groups to the amino groups of derivatized agarose. This was done in three ways: (1) Coupling with carbodiimides (using either carbodiimide reagent or di-cyclohexylcarbodiimide); (2) by means of an active N-hydroxysuc-cinimide ester of ganglioside; (c) through an activated mixed an-hydride of the ganglioside. Gangliosides have also been coupled to water-soluble copolymers of lysine and alanine by means of carbo-diimides. The disadvantage of these methods is that the carboxyl groups of at least some of the sialic acids are involved in the coupling process and not available for interactions with other materials such as antigens. Specific interactions of biologically active molecules with the oligosaccharide portion of ganglioside bound in this way can occur, however, as has been demonstrated with interleukin 2 (Parker et al., 1984).

This problem does not exist with another method that pro-duces carboxyl-containing glycolipids through oxidative ozonoly-sis of the olefinic bond in sphingosines, which are then bound to amino groups on solid supports (Laine et al., 1974). A modification of this approach that is simpler was subsequently reported by this same group of investigators (Young et al., 1978, 1979). This in-volves an initial step to protect sialic acids by forming methyl esters. The products are acetylated and then subjected to crown ether-solubilized permanganate oxidation in benzene of the olefin-ic bond. Using N-hydroxysuccinimide and dicyclohexyl carbo-diimide, an activated glycolipid ester is formed that is then attached to alkylamine glass beads. The portion of attached glyco-lipid is then treated with sodium methoxide in methanol to remove the acyl and methyl ester groups. The entire oligosaccharide of the bound ganglioside is free for interactions, but it should be noted that dihydrosphingosine containing gangliosides can not be attached using this method.

Gangliosides can be easily attached to polystyrene (Holmgren et al., 1980a,b). They are dissolved in phosphate-buffered saline and then incubated with the polystyrene material for 18–20 h at room temperature in a humid atmosphere. Unattached ganglio-sides are removed by washing the polystyrene three times with 0.15M NaCl, and any remaining exposed binding sites blocked by incubating in a solution of bovine serum albumin (Holmgren et al., 1980a) or egg albumin (Holmgren et al., 1980b) for 30 min at 37°C. Gangliosides have also been dissolved in a solution consisting of 0.1M Tris-hydrochloride buffer, pH 7.5, containing 0.2% (w/v) gelatin (Holmgren, 1973). The tubes and plates used in the original descriptions were from Nunc, Roskilde, Denmark and AB Heger Plastic, Stallarholmen, Sweden, respectively, but several other brands have also been used (Ristaino et al., 1983; Mohapatra et al., 1984). They should be rinsed in 95% ethanol and dried before use.

References

Aminoff D. (1961) Methods for the quantitative estimation of N-acetylneuraminic acid and their application to hydrolysates of sialomucoids. *Biochem. J.* **81,** 384–392.

Ando S. and Yamakawa T. (1971) Application of trifluoroacetyl derivatives to sugar and lipid chemistry. I. Gas chromatographic analysis of common constituents of glycolipids. *J. Biochem.* **70,** 335–340.

Ando S. and Yu R. K. (1979) Isolation and characterization of two isomers of brain tetrasialogangliosides. *J. Biol. Chem.* **254,** 12224–12229.

Ando S., Chang N.-C., and Yu R. K. (1978) High-performance thin-layer chromatography and densitometric determination of brain ganglioside composition of several species. *Anal. Biochem.* **89,** 437–450.

Araki H. and Yamada M. (1984) Sialic Acid, in *Methods of Enzymatic Analysis* (Bergmeyer H. U., ed.) 3rd Edn., vol. 6, Verlag Chemie, Florida.

Ariga T., Yu R. K., Suzuki M, Ando S., and Miyatake T. (1982) Characterization of GM$_1$ ganglioside by direct inlet chemical ionization mass spectrometry. *J. Lipid Res.* **23,** 437–442.

Ariga T., Yu R. K., and Miyatake T. (1984) Characterization of gangliosides by direct inlet chemical ionization mass spectrometry. *J. Lipid Res.* **25,** 1096–1101.

Arita M., Iwamori M., Higuchi T., and Nagai Y. (1984) Positive and negative ion fast atom bombardment mass spectrometry of glycosphingolipids. Discrimination of the positional isomers of gangliosides with sialic acids. *J. Biochem.* **95,** 971–981.

Ashraf J., Butterfield D. A., Jarnefelt J., and Laine R. A. (1980) Enhancement of the Yu and Ledeen gas-liquid chromatographic method for sialic acid estimation: Use of methane chemical ionization mass fragmentography. *J. Lipid Res.* **21,** 1137–1141.

Bergh M. L., Koppen P., and Van Den Eijnde D. H., (1981) High-pressure liquid chromatography of sialic acid-containing oligosaccharides. *Carbohydr. Res.* **94,** 225–229.

Bischel M. D. and Austin J. H., (1963) A modified benzidine method for the chromatographic detection of sphingolipids and acid polysaccharides. *Biochim. Biophys. Acta* **70,** 598–600.

Brady R. O., Borek C., and Bradley R. M. (1969) Composition and synthesis of gangliosides in rat hepatocyte and hepatoma cell lines. *J. Biol. Chem.* **244,** 6552–6554.

Bremer E. G., Gross S. K., and McCluer R. H. (1979) Quantitative analysis of monosialogangliosides by high-performance liquid chromatography of their perbenzoyl derivatives. *J. Lipid Res.* **20,** 1028–1035.

Brownson E. and Irwin L. (1982) A highly reproducible method for two-dimensional thin-layer chromatography of multiple ganglioside samples. *J. Neurosci. Meth.* **5,** 305–307.

Burton R. M., Garcia-Bunuel L., Golden M., and Balfour Y. M. (1963) Incorporation of radioactivity of D-Glucosamine-1-C^{14}, D-Glucose-1-C^{14}, D-Galactose-1-C^{14}, and DL-Serine-3-C^{14} into rat brain glycolipids. *Biochemistry* **2,** 580–585.

Byrne M. C., Ledeen R. W., Roisen F. J., Yorke G., and Sclafani J. R. (1983) Ganglioside-induced neuritogenesis: Verification that gangliosides are the active agents, and comparison of molecular species. *J. Neurochem.* **41,** 1214–1222.

Byrne M. C., Sbaschnig-Agler M., Aquino D. A., Sclafani J. R., and Ledeen R. W. (1985) Procedure for isolation of gangliosides in high yield and purity: Simultaneous isolation of neutral glycosphingolipids. *Anal. Biochem.* **148,** 163–173.

Carr S. A. and Reinhold V. N. (1984) Structural characterization of glycosphingolipids by direct chemical ionization mass spectrometry. *Biomed. Mass Spectrom.* **11,** 633–642.

Chigorno V., Sonnino S., Ghidoni R., and Tettamanti G. (1982) Densitometric quantification of brain gangliosides separated by two-dimensional thin layer chromatography. *Neurochem. Int.* **4,** 397–403.

Clamp J. R., Dawson G., and Hough L. (1967) The simultaneous estimation of 6-deoxy-1-galactose (L-fucose), D-mannose, D-galactose, 2-acetamido-2-deoxy-D-glucose (N-acetyl-D-glucosamine) and N-acetylneuraminic acid (sialic acid) in glycopeptides and glycoproteins. *Biochim. Biophys. Acta* **148,** 343–349.

Coleman M. T. and Yates A. J. (1978) Interference in analyses of glycolipids due to contaminants in dialysis bags. *J. Chromatogr.* **166,** 611–614.

Cortassa S., Panzetta P., and Maccioni H. J. F. (1984) Biosynthesis of gangliosides in the developing chick embryo retina. *J. Neurosci. Res.* **12,** 257–267.

Corti M., Degiorgire V., Ghidoni R., and Sonnino S. (1982) Micellar Properties of Gangliosides in *Solution Behavior of Surfactants* (Mittal K. L. and Fendler E. J., eds.) vol. 1, 573–594, Plenum, New York.

Cuatrecasas P., Parikh I., and Hollenberg M. D. (1973) Affinity chromatography and structural analysis of *Vibrio cholerae* enterotoxin-ganglioside agarose and the biological effects of ganglioside-containing soluble polymer. *Biochemistry* **12**, 4253–4264.

Dabrowski J., Hanfland P., and Egge H. (1980) Structural analysis of glycosphingolipids by high-resolution nuclear magnetic resonance spectroscopy. *Biochemistry* **19**, 5652–5658.

Das K. K., Basu M., and Basu S. (1984) A rapid preparative method for isolation of neutral and acidic glycospingolipids by radial thin-layer chromatography. *Anal. Biochem.* **143**, 125–134.

Dawson G. (1972) Glycosphingolipid levels in an unusual neurovisceral storage disease characterized by lactosylceramide galactosyl hydrolase deficiency: Lactosylceramidosis. *J. Lipid Res.* **13**, 207–219.

Dawson G., Matalon R., and Dorfman A. (1972) Glycosphingolipids in cultured human skin fibroblasts. *J. Biol. Chem.* **247**, 5944–5950.

Di Cesare J. L. and Rapport M. M. (1974) Preparation of some labeled glycosphingolipids by catalytic addition of tritium. *Chem. Phys. Lipids* **13**, 447–452.

Diringer H. (1972) The thiobarbituric acid assay of sialic acids in the presence of large amounts of lipid. *Hoppe Seylers Z. Physiol. Chem.* **353**, 39–42.

Dunn A. (1974) Brain ganglioside preparation: Phosphotungstic acid precipitation as an alternative to dialysis. *J. Neurochem.* **23**, 293–295.

Eberlein K. and Gercken G. (1975) Thin-layer chromatographic separation of gangliosides. *J. Chromatogr.* **106**, 425–427.

Ehn E. and Larsson B. (1979) Properties of an antiscratch-layer-free x-ray film for the autoradiographic registration of tritium. *Sci. Tools* **26**, 24–29.

Feix J. B. and Butterfield D. A. (1980) Selective spin labeling of sialic acid residues of glycoproteins and glycolipids in erythrocyte membranes. *FEBS Lett.* **115**, 185–188.

Folch J., Ascoli I., Lees M., Meath J. A., and LeBaron F. N. (1951a). Preparation of lipide extracts from brain tissue. *J. Biol. Chem.* **191**, 833–841.

Folch J., Arsove S., and Meath J. A. (1951b) Isolation of brain strandin, a new type of large molecule tissue component. *J. Biol. Chem.* **191**, 819–831.

Folch J., Lees M., and Sloane Stanley G. H. (1957) A simple method for the isolation and purification of total lipides from animal tissues. *J. Biol. Chem.* **226**, 497–509.

Forman D. S. and Ledeen R. W. (1972) Axonal transport of gangliosides in the goldfish optic nerve. *Science* **177**, 630–633.

Fredman P., Nilsson O., Tayot J-L., and Svennerholm L. (1980) Separation of gangliosides on a new type of anion-exchange resin. *Biochim. Biophys. Acta* **618**, 42–52.

Gasa S., Mitsuyama T., and Makita A. (1983) Proton nuclear magnetic resonance of neutral and acidic glycosphingolipids. *J. Lipid Res.* **24**, 174–182.

Gazzotti G., Sonnino S., Ghidoni R., Kirschner G., and Tettamanti G. (1984) Analytical and preparative high-performance liquid chromatography of gangliosides. *J. Neurosci. Res.* **12**, 179–192.

Ghidoni R., Tettamanti G., and Zambotti V. (1974) An improved procedure for the in vitro labeling of ganglioside. *Ital. J. Biochem.* **23**, 320–328.

Ghidoni R., Sonnino S., Masserini M., Orlando P., and Tettamanti G. (1981) Specific tritium labeling of gangliosides at the 3-position of sphingosines. *J. Lipid Res.* **22**, 1286–1295.

Ghidoni R., Sonnino S., Chigorno V., and Tettamanti G. (1983) Occurrence of glycosylation and deglycosylation of exogenously administered ganglioside G_{M1} in mouse liver. *Biochem J.* **213**, 321–329.

Ginns E. and French J. (1980) A radioassay for G_{M1} ganglioside concentration in cerebrospinal fluid. *J. Neurochem.* **35**, 977–982.

Gross S., Summer A., and McCluer R. H. (1977) Formation of ganglioside internal esters by treatment with trichloracetic acid-phosphotungstic acid reagent. *J. Neurochem.* **28**, 1133–1136.

Gunnarsson A., Lundsten J., and Svensson S. (1984) Specific cleavage of the glycosidic bond between the carbohydrate and ceramide portions in glycosphingolipids using trifluoroacetolysis. *Acta Chem. Scand. B* **38**, 603–609.

Harpin M. L., Coulon-Morelec M. J., Yeni P., Danon F., and Baumann N. (1985) Direct sensitive immunocharacterization of gangliosides on plastic thin-layer plates using peroxidase staining. *J. Immunol. Meth.* **78**, 135–141.

Harris J. U. and Klingman J. D. (1972) Detection, determination, and metabolism *in vitro* of gangliosides in mammalian sympathetic ganglia. *J. Neurochem.* **19**, 1267–1278.

Harth S., Dreyfus H., Urban P. F., and Mandel P. (1978) Direct thin-layer chromatography of gangliosides of a total lipid extract. *Anal. Biochem.* **86**, 543–551.

Hess H. H. and Rolde E. (1964) Fluorometric assay of sialic acid in brain gangliosides. *J. Biol. Chem.* **239**, 3215–3220.

Higashi H. and Basu S. (1982) Specific ^{14}C labeling of sialic acid and N-acetylhexosamine residues of glycosphingolipids after hydrazinolysis. *Anal. Biochem.* **120**, 159–164.

Higashi H., Fukui Y., Ueda S., Kato S., Hirabayashi Y., Matsumoto M., and Naiki M. (1984) Sensitive enzyme-immunostaining and densitometric determination on thin-layer chromatography of N-glycolylneuraminic acid-containing glycosphingolipids, Hanganutziu-Deicher antigens. *J. Biochem.* **95**, 1517–1520.

Hofteig J. H., Mendell J. R., and Yates A. J. (1981) Chemical and morphological studies on garfish peripheral nerves. *J. Comp. Neurol.* **198,** 265–274.

Holmgren J. (1973) Comparison of the tissue receptors for *Vibrio cholerae* and *Escherichia coli* enterotoxins by means of gangliosides and natural cholera toxoid. *Infect. Immunol.* **8,** 851–859.

Holmgren J., Elwing H., Fredman R., and Svennerholm L. (1980a) Polystyrene-adsorbed gangliosides for investigation of the structure of the tetanus-toxin receptor. *Eur. J. Biochem.* **106,** 371–379.

Holmgren J., Svennerholm L., Elwing H., Fredman P., and Strannegard O. (1980b) Sendai virus receptor: proposed recognition structure based on binding to plastic-adsorbed gangliosides. *Biochemistry* **77,** 1947–1950.

Hungund B. L. (1981) Stability of tritium-labelled glycosphingolipids. *Int. J. Appl. Radiat. Isot.* **32,** 355–356.

Hunter G. D., Wiegant V. M., and Dunn A. J. (1981) Interspecies comparison of brain ganglioside patterns studies by two-dimensional thin-layer chromatography. *J. Neurochem.* **37,** 1025–1031.

Irwin C. C. and Irwin L. N. (1979) A simple rapid method for ganglioside isolation from small amounts of tissue. *Anal. Biochem.* **94,** 335–339.

Itoh T., Li Y-T., Li S-C., Yu R. K. (1981) Isolation and characterization of a novel monosialosylpentahexosyl ceramide from Tay-Sachs brain. *J. Biol. Chem.* **256,** 165–169.

Iwamori M. and Nagai Y. (1978) A new chromatographic approach to the resolution of individual gangliosides. Ganglioside mapping. *Biochim. Biophys. Acta* **528,** 257–267.

Iwamori M., Sawada K., Hara Y., Nishio M., Fujisawa T., Imura H., and Nagai Y. (1982) Neutral glycospingolipids and gangliosides of bovine thyroid. *J. Biochem.* **91,** 1875–1887.

Jatzkewitz H. (1961) Eine neue Methode zur quantitativen Ultramikrobestimmung der Sphingolipoide aus Gehirn. *Hoppe Seylers Z. Physiol. Chem.* **326,** 61–64.

Jourdian G. W., Dean L., and Roseman S. (1971) The sialic acids. XI. A periodate-resorcinol method for the quantitative estimation of free sialic acids and their glycosides. *J. Biol. Chem.* **246,** 430–435.

Kadowaki H., Bremer E. G., Evans J. E., Jungalwala F. B., and McCluer R. H. (1983) Acetonitrile-hydrochloric acid hydrolysis of gangliosides for high performance liquid chromatographic analysis of their long chain bases. *J. Lipid Res.* **24,** 1389–1397.

Kadowaki H., Evans J. E., and McCluer R. H. (1984) Separation of brain monosialoganglioside molecular species by high-performance liquid chromatography. *J. Lipid Res.* **25,** 1132–1139.

Kanda S., Inoue K., Nojima S., Utsumi H., and Weigandt H. (1982a) Incorporation of spin-labeled ganglioside analogues into cell and liposomal membranes. *J. Biochem.* **91**, 1707–1718.

Kanda S., Inque K., Nojima S., Utsumi H., and Wiegandt H. (1982b) Incorporation of a ganglioside and a spin-labeled ganglioside analogue into cell and liposomal membranes. *J. Biochem.* **91**, 2095–2098.

Kanfer J. N. (1969) Preparation of gangliosides. *Meth. Enzymol.* **14**, 660–664.

Kawamura N. and Taketomi T. (1977) A new procedure for the isolation of brain gangliosides, and determination of their long chain base compositons. *J. Biochem.* **81**, 1217–1225.

Klemm N., Su S.-N., Harnacher B., and Jeng I. (1982) A procedure for the radioiodination of a ganglioside derivative. *J. Lab. Comp. Radiopharm.* **19**, 937–944.

Klenk E. (1942) Uber die Ganglioside eine neue Gruppe von zuckerhaltigen Gehirnlipoiden. *Hoppe Seylers Z. Physiol. Chem.* **273**, 76–86.

Klenk E. and Gielen W. (1961) Uber ein chromatographisch einheitliches Hexosaminhaltiges Gangliosid aus Menschengehirn H. S. Z. *Physiol. Chem.* **323**, 126.

Koerner Jr. T. A. W., Prestegard J. H., Demou P. C., and Yu R. K. (1983) High-resolution proton nmr studies of gangliosides. 1. Use cf homonuclear two-dimensional spin-echo j-correlated spectroscopy for determination of residue composition and anomeric configurations. *Biochemistry* **22**, 2676–2687.

Kolodny E. H., Brady R. O., Quirk J. M., and Kanfer J. N. (1970) Preparation of radioactive Tay-Sachs ganglioside labeled in the sialic acid moiety. *J. Lipid Res.* **11**, 144–149.

Krohn K., Eberlein K., and Gercken G. (1978) Separation of glycolipids from neutral lipids and phospholipids. *J. Chromatogr.* **153**, 550–552.

Kubo H. and Hoshi M. (1985) Elimination of silica gel from gangliosides by using a reversed-phase column after preparative thin-layer chromatography. *J. Lipid Res.* **26**, 638–641.

Kuhn R. and Wiegandt H. (1963) Die Konstitution der Ganglio-n-tetraose und des Gangliosids g_1. *Chem. Ber.* **96**, 866–880

Kuhn R., Wiegandt H., and Egge H. (1961) Zum Bauplan der Ganglioside *Angew. Chem.* **73**, 580–581

Kundu S. K. and Roy S. K. (1978) A rapid and quantitative method for the isolation of gangliosides and neutral glycosphingolipids by DEAE-silica gel chromatography. *J. Lipid Res.* **19**, 390–395.

Kundu S. K. and Scott D. D. (1982) Rapid separation of gangliosides by high-performance liquid chromatography. *J. Chromatogr.* **232**, 19–27.

Kundu S. K. and Suzuki A (1981) Simple micro-method for the isolation of gangliosides by reversed-phase chromatography. *J. Chromatogr.* **224,** 249–256.

Kundu S. K., Chakravarty S. K., Roy S. K., and Roy A. K. (1979) DEAE-silica gel and DEAE-controlled porous glass as ion exchangers for isolation of glycolipids. *J. Chromatogr.* **170,** 65–72.

Kushi Y. and Handa S. (1982) Application of field desorption mass spectrometry for the analysis of sphingoglycolipids. *J. Biochem.* **91,** 923–931.

Kushi Y. and Handa S. (1985) Direct analysis of lipids on thin layer plates by matrix-assisted secondary ion mass spectrometry. *J. Biochem.* **98,** 265–268.

Kushi Y., Handa S., Kambara H., and Shizukuishi K. (1983) Comparative study of acidic glycosphingolipids by field desorption and secondary ion mass spectrometry. *J. Biochem.* **94,** 1841–1850.

Ladisch S. and Gillard B. (1985) A solvent partition method for microscale ganglioside purification. *Anal. Biochem.* **146,** 220–231.

Laine R. A., Yogeeswaran G., and Hakomori S. (1974) Glycosphingolipids covalently linked to agarose gel or glass beads. *J. Biol. Chem* **249,** 4460–4466.

Ledeen R. (1966) The chemistry of gangliosides: A review. *J. Am. Oil Chem. Soc.* **43,** 57–66.

Ledeen R. and Salsman K. (1965) Structure of the Tay-Sachs' ganglioside. *Biochemistry* **4,** 2225–2233.

Ledeen R. W. and Yu R. K. (1978) Methods for Isolation and Analysis of Gangliosides, in *Research Methods In Neurochemistry* vol. 4 (Marks N. and Rodnight R., eds.) Plenum, New York.

Ledeen R. W. and Yu R. K. (1982) Gangliosides: Structure, isolation, and analysis. *Meth. Enzymol.* **83,** 139–191.

Ledeen R. W., Haley J. E., and Skrivanek J. A. (1981) Study of ganglioside patterns with two-dimensional thin-layer chromatography and radioautography; detection of new fucogangliosides and other minor species in rabbit brain. *Anal. Biochem.* **112,** 135–142.

Ledeen R. W., Yu R. K., and Eng L. F. (1973) Gangliosides of human myelin: sialosylgalactosylceramide (G_7) as a major component. *J. Neurochem.* **21,** 829–839.

Lee P. M., Ketis N. V., Barber K. R., and Grant C. W. M. (1980) Ganglioside headgroup dynamics. *Biochim. Biophys. Acta* **601,** 302–314.

Lee W. M. F., Estrick M. A., and Macher B. A. (1982) High-performance liquid chromatography of long-chain neutral glycospingolipids and gangliosides. *Biochim. Biophys. Acta* **712,** 498–504.

Li, S-C., Chien J-L., Wan, C. C., and Li Y-T. (1978) Occurrence of glycosphingolipids in chicken egg yolk. *Biochem. J.* **173,** 697–699.

MacMillan V. H. and Wherrett J. R. (1969) A modified procedure for the analysis of mixtures of tissue gangliosides. *J. Neurochem.* **16**, 1621–1624.

Magnani J. L., Brockhaus M., Smith D. F., Ginsburg V., Blaszczyk M., Mitchell K. F., Steplewski Z., and Koprowski H. (1981) A monosialoganglioside is a monoclonal antibody-defined antigen of colon carcinoma. *Science* **212**, 55–56.

Mansson J-E., Rosengren B., and Svennerholm L. (1985) Separation of gangliosides by anion-exchange chromatography on mono Q. *J. Chromatogr.* **322**, 465–472.

Marcus D. M. and Cass L. E. (1969) Glycospingolipids with Lewis blood group activity: Uptake by human erythrocytes. *Science* **164**, 553–555.

Meldolesi M. F., Fishman P. H., Aloj S. M., Kohn L. D., and Brady R. O. (1976) Relationship of gangliosides to the structure and function of thyrotropin receptors: Their absence on plasma membranes of a thyroid tumor defective in thyrotropin receptor activity. *Proc. Natl. Acad. Sci. USA* **73**, 4060–4064.

Mestrallet M. G., Cumar F. A., and Caputto R. (1976) Split chromatographic spot produced by supposedly single gangliosides treated with trichloroacetic acid-phosphotungstic acid reagent. *J. Neurochem.* **26**, 227–228.

Miettinen T. and Takki-Luukkainen I.-T. (1959) Use of butyl acetate in determination of sialic acid. *Acta Chem. Scand.* **13**, 856–858.

Miller-Podraza H., Bradley R. M., and Fishman P. H. (1982) Biosynthesis and localization of gangliosides in cultured cells. *Biochemistry* **21**, 3260–3265.

Miyazaki K., Okamura N., Kishimoto Y., and Lee Y. C. (1986) Determination of gangliosides as 2,4-dinitrophenylhydrazides by high performance liquid chromatography. *Biochem. J.* **235**, 755–761

Mohapatra L. N., Samantaray J. C., Deb M., Stintzing G., Mollby R., Holme T. (1984) Comparative evaluation of GM1-ELISA and tissue culture assays for detection of thermolabile enterotoxin (LT) of *Escherichia coli. Indian J. Med. Res.* **79**, 733–740.

Momoi T. and Wiegandt H. (1980) Separation and micro-detection of oligosaccharides of glycosphingolipids by high performance cellulose thin-layer chromatography-autoradiofluorography. *Hoppe Seylers Z. Physiol. Chem.* **361**, 1201–1210.

Momoi T., Ando S., and Nagai Y. (1976) High resolution preparative column chromatographic system for gangliosides using DEAE-Sephadex and a new porous silica, Iatrobeads. *Biochim. Biophys. Acta* **441**, 488–497.

Moss J., Manganiello V. C., and Fishman P. H. (1977) Enzymatic and chemical oxidation of gangliosides in cultured cells: Effects of choleragen. *Biochemistry* **16**, 1876–1881.

Mraz W., Schwarzmann G., Sattler J., Momoi T., Seemann B., and Wiegandt H. (1980) Aggregate formation of gangliosides at low concentrations in aqueous media. *Hoppe Seylers Z. Physiol. Chem.* **361,** 177–185.

Mullin B. R., Aloj S. M., Fishman P. H., Lee G., Kohn L. D., and Brady R. O. (1976) Cholera toxin interactions with thyrotropin receptors on thyroid plasma membranes. *Proc. Natl. Acad. Sci. USA* **73,** 1679–1683.

Mullin B. R., Poore C. M. B., and Rupp B. H. (1983) Quantitation of gangliosides by scanning densitometry of thin-layer chromatography plates. *J. Chromatogr.* **278,** 160–166.

Nakabayashi H., Iwamori M., and Nagai Y. (1984) Analysis and quantitation of gangliosides as p-bromophenacyl derivatives by high-performance liquid chromatography. *J. Biochem.* **96,** 977–984.

Ohashi M. (1979) A comparison of the ganglioside distributions of fat tissues in various animals by two-dimensional thin layer chromatography. *Lipids* **14,** 52–57.

Parker J., Caldini G., Krishnamurti C., Ahrens P. B., and Ankel H. (1984) Binding of interleukin 2 to gangliosides. *FEBS Lett.* **170,** 391–395.

Penick R. J., Meisler M. H., and McCluer R. H. (1966) Thin-layer chromatographic studies of human brain gangliosides. *Biochim. Biophys. Acta* **116,** 279–287.

Pick J., Vajda J., Anh-Tuan N., Leisztner L., and Hollan S. R. (1984a) Class fractionation of acidic glycolipids and further separation of gangliosides by OPTLC. *J. Liq. Chromatogr.* **7,** 2777–2791.

Pick J., Vajda J., and Leisztner L. (1984b). Neutral lipid class fractionation and further separation of simple neutral glycolipids by OPTLC. *J. Liq. Chromatogr.* **7,** 2759–2776.

Price H. C. and Frame J. H. (1981) A rapid non-destructive fluorometric assay for gangliosides. *Life Sci.* **28,** 2903–2907.

Quarles R. and Folch-Pi J. (1965) Some effects of physiological cations on the behaviour of gangliosides in a chloroform-methanol-water biphasic system. *J. Neurochem.* **12,** 543–553.

Radsak K. and Weigandt H. (1984) Glycosphingolipid synthesis in human fibroblasts infected by cytomegalovirus. *Virology* **138,** 300–309.

Rauvala H., Finne J., Krusius T., Karkkainen J., and Jarnefelt J. (1980) Methylation techniques in the structural analysis of glycoproteins and glycolipids. *Adv. Carbohydr. Chem. Biochem.* **37,** 157–223.

Rickert S. J. and Sweeley C. C. (1978) Quantitative analysis of carbohydrate residues of glycoproteins and glycolipids by gas-liquid chromatography. An appraisal of experimental details. *J. Chromatogr.* **147,** 317–326.

Ristaino P. A., Levine M. M., and Young C. R. (1983) Improved GM1-enzyme-linked immunosorbent assay for detection of *Escherichia coli* heat-labile enterotoxin. *J. Clin. Microbiol.* **18,** 808–815.

Rosmus J., Pavalicek M., and Deyl Z. (1964) Thin layer chromatographic Prsc. Symp. Rome, p. 119.

Rosner H. (1980) A new thin-layer approach for separation of multi-sialogangliosides. *Anal. Biochem.* **109,** 437–442.

Rosner H. (1981) Isolation and preliminary characterization of novel polysialogangliosides from embryonic chick brain. *J. Neurochem.* **37,** 993–997.

Rouser G., Kritchevsky G., Galli C., and Heller D. (1965) Determination of polar lipids: Quantitative column and thin-layer chromatography. *J. Am. Oil Chem. Soc.* **42,** 215–227.

Saito M., Kasai N., and Yu R. K. (1985) *In situ* immunological determination of basic carbohydrate structures of gangliosides on thin-layer plates. *Anal. Biochem.* **148,** 54–58.

Sandhoff K., Harzer K., and Jatzkewitz H. (1968) Densitometrische Mikrobestimmung von Gangliosiden aus dem Gesamtlipidextrakt nach Dunnschichtchromatographie. *Hoppe Seylers Z. Physiol. Chem.* **349,** 283–287.

Schwarzmann G. (1978) A simple and novel method for tritium labeling of gangliosides and other sphingolipids. *Biochim. Biophys. Acta* **529,** 106–114.

Seyama Y., Yamakawa T., and Komai T. (1968) Application of the isotope dilution method to microanalytical determination of five classes of sphingoglycolipids in tissues. *J. Biochem.* **64,** 487–493.

Seyfried T. N., Weber E. J., and Yu R. K. (1977) Influence of trichloroacetic acid-phosphotungstic acid on the thin layer chromatographic mobility of gangliosides. *Lipids* **12,** 979–980.

Sharom F. J. and Grant C. W. M. (1975) A glycosphingolipid spin label: Ca^{2+} effects on sphingolipid distribution in bilayers containing phsophatidyl serine. *Biochem. Biophys. Res. Commun.* **67,** 1501–1506.

Sharom F. J. and Grant C. W. M. (1977) A ganglioside spin label: Ganglioside head group interactions. *Biochem. Biophys. Res. Commun.* **74,** 1039–1045

Sharom F. J. and Grant C. W. M. (1978) A model for ganglioside behaviour in cell membranes. *Biochim. Biophys. Acta* **507,** 280–293.

Siakotos A. N., Kulkarni S., and Passo S. (1970) The quantitative analysis of sphingolipids by determination of long chain base as the trinitrobenzene sulfonic acid derivative. *Lipids* **6,** 254–259.

Smid F. and Reinisova J. (1973) A densitometric method for the determination of gangliosides after their separation by thin-layer chromatography and detection with resorcinol reagent. *J. Chromatogr.* **86,** 200–204.

Sonnino S., Ghidoni R., Chigorno V., and Tettamanti G. (1982) Chemistry of Gangliosides Carrying O-Acetylated Sialic Acid, in *New Vistas*

In Glycolipid Research (Makita A., Handa S., Taketomi T., and Nagai Y., eds) Plenum, New York.

Sonnino S., Ghidoni R., Gazzotti G., Kirschner G., Galli G., and Tettamanti G. (1984) High performance liquid chromatography preparation of the molecular species of GM1 and GD1a gangliosides with homogenous long chain base composition. *J. Lipid Res.* **25**, 620–629.

Spence M. W. (1969) Studies on the extractability of brain gangliosides. *Can. J. Biochem.* **47**, 735–742.

Spence M. W. and Wolfe L. S. (1967) The effect of cations on the extractability of gangliosides from brain. *J. Neurochem.* **14**, 585–590.

Spiegel S., Wilchek M., and Fishman P. H. (1983) Fluorescence labeling of cell surface glycoconjugates with Lucifer Yellow CH. *Biochem. Biophys. Res. Commun.* **112**, 872–877.

Suzuki K. (1964) A simple and accurate micromethod for quantitative determination of ganglioside patterns. *Life Sci.* **3**, 1227–1233.

Suzuki K. (1965) The pattern of mammalian brain gangliosides-ii: Evaluation of the extraction procedures, postmortem changes and the effect of formalin preservation. *J. Neurochem.* **12**, 629–638.

Suzuki Y. and Suzuki K. (1972) Specific radioactive labeling of terminal N-acetylgalactosamine of glycosphingolipids by the galactose oxidase-sodium borohydride method. *J. Lipid Res.* **13**, 687–690.

Svennerholm L. (1957a) Quantitative estimation of sialic acids. I. A colorimetric method with orcinol-hydrochloric acid (Bial's) reagent. *Arkiv. Kemi.* **10**, 577–596.

Svennerholm L. (1957b) Quantitative estimation of sialic acids. II. A colorimetric resorcinol-hydrochloric acid method. *Biochim. Biphys. Acta* **24**, 604–611.

Svennerholm L. (1963a) Isolation of gangliosides. *Acta Chem. Scand.* **17**, 239–250.

Svennerholm L. (1963b) Chromatographic separation of human brain gangliosides. *J. Neurochem.* **10**, 613–623.

Svennerholm L. (1963c) Sialic acids and derivatives: Estimation by the ion-exchange method. *Meth. Enzymol.* **6**, 459–462.

Svennerholm L. (1964) The gangliosides. *J. Lipid Res.* **5**, 145–155.

Svennerholm L. (1972) Gangliosides isolation. *Meth. Carbohydr. Chem.* **6**, 464–474.

Svennerholm L. and Fredman P. (1980) A procedure for the quantitative isolation of brain gangliosides. *Biochim. Biophys. Acta* **617**, 97–109.

Svennerholm L., Mansson J-E., and Li Y-T. (1973) Isolation and structural determination of a novel ganglioside, a disialosylpentahexosylceramide from human brain. *J. Biol Chem.* **248**, 740–742.

Sweeley C. C. and Tao R. V. P. (1972) Gas Chromatographic Estimation of Carbohydrates in Glycosphingolipids, in *Methods in Carbohydrate*

Chemistry vol. 6 (Whistler R. L. and BeMiller J. N., eds.) Academic Press, New York.

Tallman J. F., Fishman P. H., and Henneberry R. C. (1977) Determination of sialidase activities in HeLa cells using gangliosides specifically labeled in N-acetylneuraminic acid. *Arch. Biochem. Biophys.* **182,** 556–562.

Tanaka Y., Yu R. K., Ando S., Ariga T., and Itoh T. (1984) Chemical-ionization mass spectra of the permethylated sialo-oligosaccharides liberated from gangliosides. *Carbohydr. Res.* **126,** 1–14.

Tettamanti G., Bertona L., Berra B., and Zambotti V. (1964) On the evidence of glycolylneuraminic acid in beef brain gangliosides. *Ital. J. Biochem.* **13,** 315–321.

Tettamanti G., Bonali F., Marchesini S., and Zambotti V. (1973) A new procedure for the extraction, purification and fractionation of brain gangliosides. *Biochim. Biophys. Acta* **296,** 160–170.

Tjaden U. R., Krol J. H., Van Hoeven R. P., Oomen-Muelemans E. P. M., and Emmelot P. (1977) High-pressure liquid chromatography of glycosphingolipids (with special reference to gangliosides). *J. Chromatogr.* **136,** 233–243.

Torello L. A., Yates A. J., and Thompson D. K. (1980) Critical study of the alditol acetate method for quantitating small quantities of hexoses and hexosamines in gangliosides. *J. Chromatogr.* **202,** 195–209.

Towbin H., Schoenenberger C., Ball R., Braun D. G., and Rosenfelder G. (1984) Glycosphingolipid-blotting: An immunological detection procedure after separation by thin layer chromatography. *J. Immunol. Meth.* **72,** 471–479.

Traylor T. D. (1983) High-performance liquid chromatographic resolution of p-nitrobenzyloxyamine derivatives of brain gangliosides. *J. Chromatogr.* **272,** 9–20.

Ullman M. D. and McCluer R. H. (1985) Quantitative analysis of brain gangliosides by high performance liquid chromatography of their perbenzoyl derivatives. *J. Lipid Res.* **26,** 501–506.

Vance D. E. and Sweeley C. C. (1967) Quantitative determination of the neutral glycosylceramides in human blood. *J. Lipid Res.* **8,** 621–630.

Van Den Eijnden D. H. (1971) Chromatographic separation of gangliosides on precoated silicagel thin-layer plates. *Hoppe Seylers Z. Physiol. Chem.* **353,** 1601–1602.

Veh R. W., Corfield A. P., Sander M., and Schauer R. (1977) Neuraminic acid-specific modification and tritium labelling of gangliosides. *Biochim. Biophys. Acta* **486,** 145–160.

Wagner H., Horhammer L., and Wolff P. (1961) Dunnschichtchromatographie von Phosphatiden und Glycolipiden. *Biochem Z.* **334,** 175–184.

Warren L. (1959) The thiobarbituric acid assay of sialic acids. *J. Biol. Chem.* **234,** 1971–1975.

Watanabe K. and Tomono Y. (1984) One-step fractionation of neutral and acidic glycosphingolipids by high-performance liquid chromatography. *Anal. Biochem.* **139,** 367–372.

Weicker H., Dain J. A., Schmidt G., and Tannhauser, S. J. (1960) Separation of different components of beef brain ganglioside. *Fed. Proc.* **19,** A219.

Wells M. A. and Dittmer J. C. (1963) The use of Sephadex for the removal of nonlipid contaminants from lipid extracts. *Biochemistry* **2,** 1259–1263.

Wherrett J. R. and Brown B. L. (1969) Erythrocyte glycolipids in Huntington's chorea. *Neurology* **19,** 489–493.

Wherrett J. R. and Cumings J. N. (1963) Detection and resolution of gangliosides in lipid extracts by thin-layer chromatography. *Biochem. J.* **86,** 378–381.

Wherrett J. R., Lowden J. A., and Wolfe L. S. (1964) Studies on brain gangliosides. II. Analysis of human brain ganglioside fractions obtained by preparative thin-layer chromatography. *Can. J. Biochem.* **42,** 1057–1063.

Wiegandt H. and Ziegler W. (1974) Synthetic glycolipids containing glycosphingolipid-derived oligosaccharides. *Hoppe Seylers Z. Physiol. Chem.* **355,** 11–18.

Wilchek M., Spiegel S., and Spiegel Y. (1980) Fluorescent reagents for the labeling of glycoconjugates in solution and on cell surfaces. *Biochem. Biophys. Res. Commun.* **92,** 1215–1222.

William M. A. and McCluer R. H. (1980) The use of Sep-Pak C_{18} cartridges during the isolation of gangliosides. *J. Neurochem.* **35,** 266–269.

Winterbourn C. C. (1971) Separation of brain gangliosides by column chromatography on DEAE-cellulose. *J. Neurochem.* **18,** 1153–1155.

Yao J. K. and Rastetter G. M. (1985) Microanalysis of complex tissue lipids by high-performance thin-layer chromatography. *Anal. Biochem.* **150,** 111–116.

Yates A. J. (1986) Gangliosides in the nervous system during development and regeneration. *Neurochem. Pathol.* **5,** 309–329.

Yates A. J. and Thompson D. (1977) An improved assay of gangliosides separated by thin-layer chromatography. *J. Lipid Res.* **18,** 660–663.

Yates A. J. and Warner J. K. (1984) Behavior of sugar derivatives in procedures for ganglioside isolation. *Lipids* **19,** 562–569.

Yates A. J., Mattison S. L., and Whisler R. L. (1980) Effect of Concanavalin A on ganglioside metabolism of human lymphocytes. *Biochem. Biophys. Res. Commun.* **96,** 211–218.

Yogeeswaren G., Gronberg A., Hansson M., Dalianis T., Kiessling R., and Welsh R. M. (1981) Correlation of glycosphingolipids and sialic acid in Yac-1 lymphoma variants with their sensitivity to natural killer-cell-mediated lysis. *Int. J. Cancer* **28,** 517–526.

Yohe H. C., Ueno K., Chang N.-C., Glaser G. J., and Yu R. K. (1980) Incorporation of N-acetylmannosamine into rat brain subcellular gangliosides: Effect of pentylenetetrazol-induced convulsions on brain gangliosides. *J. Neurochem.* **34,** 560–568.

Young Jr. W. W., Laine R. A., and Hakomori S-I. (1978) Covalent attachment of glycolipids to solid supports and macromolecules. *Meth. Enzymol.* part C, **Vol. L,** 137–140.

Young Jr. W. W., Laine R. A., and Hakomori S. (1979) An improved method for the covalent attachment of glycolipids to solid supports and macromolecules. *J. Lipid Res.* **20,** 275–278.

Yu R. K. and Ledeen R. W. (1970) Gas-liquid chromatographic assay of lipid-bound sialic acids: Measurement of gangliosides in brain of several species. *J. Lipid Res.* **11,** 506–516.

Yu R. K. and Ledeen R. W. (1972) Gangliosides of human, bovine, and rabbit plasma. *J. Lipid Res.* **13,** 680–686.

Zanetta J. P., Vitiello F., and Robert J. (1977) Thin-layer chromatography of gangliosides. *J. Chromatogr.* **137,** 481–484.

Index